Handbook of Alternative Energy

Handbook of Alternative Energy

Editor: Rob Koslowski

Callisto Reference
www.callistoreference.com

Callisto Reference,
118-35 Queens Blvd., Suite 400,
Forest Hills, NY 11375, USA

Visit us on the World Wide Web at:
www.callistoreference.com

ISBN: 978-1-64116-829-8 (Hardback)

Cataloging-in-Publication Data

Handbook of alternative energy / edited by Rob Koslowski.
p. cm.
Includes bibliographical references and index.
ISBN 978-1-64116-829-8
1. Renewable energy sources. 2. Power resources. 3. Renewable natural resources.
4. Agriculture and energy. I. Koslowski, Rob.
TJ808 .A48 2023
333.794--dc23

Table of Contents

Preface

Alternative energy refers to a type of energy that is derived from sources other than fossil fuels. This energy emits little to no greenhouse gases such as carbon dioxide (CO_2) and methane. The energy generated from alternative sources does not lead to the greenhouse effect that is responsible for climate change. There are various types of alternative energy sources including wind power, hydroelectricity, tidal power, solar power, biomass, nuclear power and geothermal power. The carbon footprint of alternative energy sources is far less as compared to coal, natural gas and other fossil fuels. Alternative energy sources can be adopted at a broad level if renewable technologies become more effective and necessary reforms are made in the electric utility sector. This book includes some of the vital pieces of works being conducted across the world, on various topics related to alternative energy. It elucidates new techniques and their applications in a multidisciplinary manner. This book is a vital tool for all researching or studying alternative energy sources as it gives incredible insights into emerging trends and concepts.

This book has been the outcome of endless efforts put in by authors and researchers on various issues and topics within the field. The book is a comprehensive collection of significant researches that are addressed in a variety of chapters. It will surely enhance the knowledge of the field among readers across the globe.

It gives us an immense pleasure to thank our researchers and authors for their efforts to submit their piece of writing before the deadlines. Finally in the end, I would like to thank my family and colleagues who have been a great source of inspiration and support.

Editor

1

A Review on Deep Learning Models for Forecasting Time Series Data of Solar Irradiance and Photovoltaic Power

Rial A. Rajagukguk, Raden A. A. Ramadhan and Hyun-Jin Lee *

Department of Mechanical Engineering, Kookmin University, 77 Jeongneung-ro, Seongbuk-gu, Seoul 02727, Korea; arifinrial@gmail.com (R.A.R.); adhityaraar@kookmin.ac.kr (R.A.A.R.)

* Correspondence: hyunjinlee@kookmin.ac.kr

Abstract: Presently, deep learning models are an alternative solution for predicting solar energy because of their accuracy. The present study reviews deep learning models for handling time-series data to predict solar irradiance and photovoltaic (PV) power. We selected three standalone models and one hybrid model for the discussion, namely, recurrent neural network (RNN), long short-term memory (LSTM), gated recurrent unit (GRU), and convolutional neural network-LSTM (CNN–LSTM). The selected models were compared based on the accuracy, input data, forecasting horizon, type of season and weather, and training time. The performance analysis shows that these models have their strengths and limitations in different conditions. Generally, for standalone models, LSTM shows the best performance regarding the root-mean-square error evaluation metric (RMSE). On the other hand, the hybrid model (CNN–LSTM) outperforms the three standalone models, although it requires longer training data time. The most significant finding is that the deep learning models of interest are more suitable for predicting solar irradiance and PV power than other conventional machine learning models. Additionally, we recommend using the relative RMSE as the representative evaluation metric to facilitate accuracy comparison between studies.

Keywords: deep learning; time series data; solar irradiance; PV power; evaluation metric

1. Introduction

Solar energy is a popular renewable energy source because it is abundant and environment-friendly. The amount of solar energy incident on the earth's surface is approximately 1.5 × 1018 kW h/year, which is approximately 10,000 times the current annual energy consumption of the entire world [1]. Therefore, in recent years, solar photovoltaic (PV) has a significant role in electricity generation. A challenging issue associated with the solar PV is that its power output strongly depends on uncertain and uncontrollable meteorological factors, such as atmospheric temperature, wind, pressure, and humidity [2]. As the solar PV capacity increases, risks caused by the uncontrollable nature of PV power increase. Energy storage could mitigate such risks, but the drawbacks are the installation and management costs. However, solar irradiance forecasting is an inexpensive and immediate solution and effective in microgrid operation optimization such as peak shaving, uncertainty impact reduction, and economic dispatch problem in the power system [3].

Generally, solar irradiance can be forecast from very short terms (several minutes ahead) to long terms (several days ahead)—the requirement for the time horizon changes with applications. For the very short-term and short-term forecast horizons, sudden variations of solar irradiance, namely the ramp events, are of interest. Abrupt and severe variations of solar irradiance have the potential to degrade the reliability and quality of PV power. Hence, forecasting results for the short-term horizon are useful for estimating the largest PV power ramp rates [4]. Meanwhile, forecasting for the medium and long-term horizons helps operation optimization and market participation. For example, Husein et al. [3] demonstrated that day-ahead solar irradiance forecasting could increase annual energy savings for the operation of a commercial building microgrid. Therefore, solar energy forecasting should be targeted for a specific application, and accordingly, an appropriate forecasting method must be selected.

The solar irradiance and PV power forecasting methods are divided into physical and statistical models. The physical model mathematically or numerically manages the interaction of solar radiation in the atmosphere based on the laws of physics. It comprises numerical weather prediction, sky imagery, and satellite image models [5]. The statistical model finds a relationship between input and output variables and consists of conventional statistical models and machine learning models. Conventional statistical models include the fuzzy theory, Markov chain, autoregressive, and regression models. The machine learning model, also known as an artificial intelligence model, can efficiently extract high-dimensional complex nonlinear features and directly map input and output variables. In the past, the well-known machine learning models for predicting solar energy were the support vector machine (SVM), k-nearest neighbors, artificial neural network (ANN), naive Bayes, and random forest. These statistical models rely primarily on historical data to predict future time series. Therefore, the quantity and quality of historical data are essential for an accurate forecast.

Nowadays, the deep learning model becomes more popular in solar irradiance forecasting. The deep learning model, which is the subpart of the machine learning model, was developed to solve a complex problem with a large amount of data. The multiple layers in the deep learning structure automatically learn the abstract features directly from the raw data to discover useful representations

Deep learning models are distinctive from other machine learning models because they outperform as input data scale increases. Ng et al. [6] compared machine learning and deep learning models' performance while changing the amount of input data. The result showed that deep learning models tend to increase their accuracy as the number of training data increases, whereas the traditional machine learning models stop improving at a certain amount of data.

The deep learning models specialized for handling sequential or time-series data such as text, speech, and image have been developed and have been successful. Recurrent neural network (RNN), long short-term memory (LSTM), gated recurrent unit (GRU), and convolutional neural network-LSTM (CNN–LSTM) models are typical. Because solar forecasting is intrinsically based on sequential data, such deep learning models were also applied for solar forecasting. For instance, Zang et al. [7] demonstrated that the accuracies of CNN–LSTM, LSTM, and CNN models are better than those of ANN and SVM in the short-term forecasting of global horizontal irradiance (GHI). Due to popularity and excellency, more applications of the deep learning model for solar forecasting are expected.

Outstanding review papers dedicated to machine learning can be found in solar forecasting literature [2,5,8,9]. However, they encompassed various machine learning models or paid little attention to deep learning models. Because the deep learning models suitable for sequential data and applied for solar forecasting have not been reviewed, the present study discusses basic principles of the RNN, LSTM, GRU, and CNN–LSTM selected as recently developed deep learning models and their comparative analyses. This paper is structured into six sections. Section 2 describes the solar irradiance variability for different time scales. Section 3 summarizes the theoretical background of the models, and Section 4 explains the evaluation metrics used to compare the performance of deep learning models. Then, Section 5 analyzes relevant studies in terms of features and accuracy. Finally, a summary of major findings in this study is presented in Section 6.

2. Solar Irradiance Variability

A single location on earth will experience a high degree of solar irradiance variability. The variability is highly dependent on local weather and atmospheric conditions as well as diurnal and seasonal cycles. The diurnal and seasonal variabilities due to the sun's motion and the earth's distance from the sun are fully predictable. However, the variability due to the moving clouds affected by local weather and atmospheric conditions has stochastic features and consequently is very difficult to predict.

The variability of solar irradiance depends on the time scale. A single datum represents a mean value of more frequent measurement records for a given time period. For instance, an hourly value of GHI may be acquired from 60 pyranometer records at the one-minute interval. Therefore, a long time interval results in the smoothing effect, and thus the variability lessens. Figure 1 shows the smoothing effect on the solar irradiance variability. The data were recorded every minute at Kookmin University, South Korea and presented by different time scales. The finest temporal resolution of one minute reveals a high degree of variability, but the variability smooths out with the time scale. It should be noticed that the smoothing effect also appears when a spatial averaging over a large area is applied.

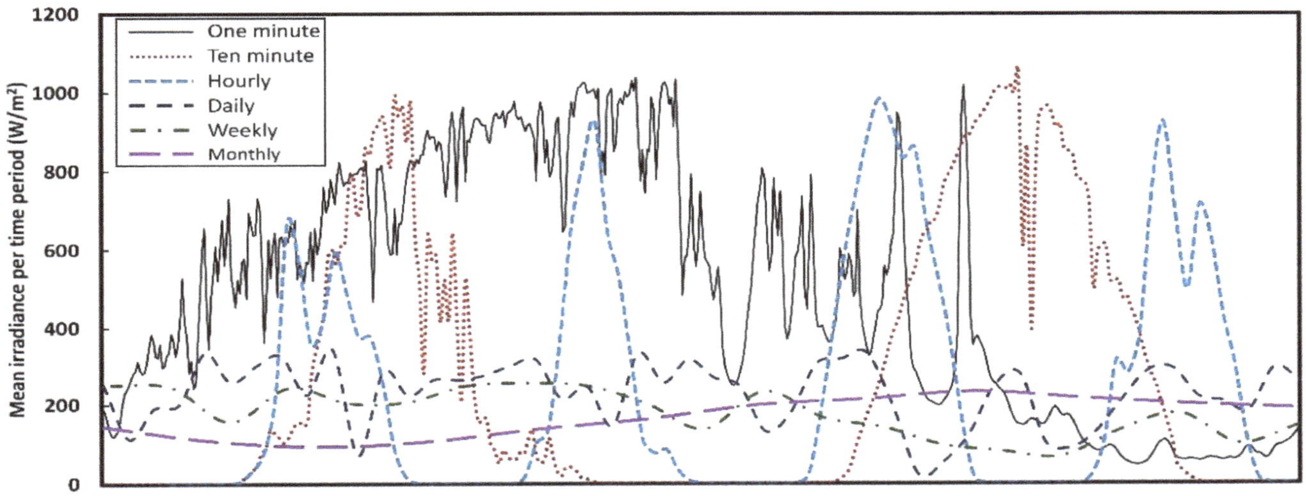

Figure 1. The variability of global horizontal irradiance (GHI) at Kookmin University, South Korea, as a function of time scale. The figure includes 1 day of 1 min data, 2 days of 10 min data, 4 days of 1 h data, 48 days of day data, and 12 months of monthly data.

The variability of solar irradiance directly affects solar PV power systems' performance and more seriously disturbs grid stability. The grid typically absorbs power fluctuations at short time scales as fluctuations in frequency and voltage. Therefore, grid operators can control the power ramp rate by imposing a limit on PV plants; for instance, 10% of the nameplate capacity per minute. In addition, grid operators need to balance supply and demand while complying with power system regulations. If the power supply exceeds demand, curtailment can be applied to disconnect power delivery from PV plants with the grid. Solar irradiance or PV power forecasting can optimize the ramp rate control and schedule the load-following operation more effectively with an energy storage system.

Various solar forecasting techniques have their own applicable regimes in terms of time scale. Thus, depending on solar irradiance variability of interest, an appropriate technique must be selected. Figure 2 illustrates a guideline to select forecasting techniques. For time horizons less than one hour, sky-image-based techniques offer very good forecasting capability [10]. Satellite-image-based techniques are recommended for several hours ahead of forecasting with a spatial resolution around 1–5 km [11]. Numerical weather predictions (NWP) allows long-term forecasting over 1 day and up to 15 days ahead, but its time step and update interval are large. On the other hand, time series forecasting using statistical models covers many applications from short to long terms with fine time steps [12–14]. As an advanced version of statistical models, deep learning models for sequential data are anticipated to expand the regime.

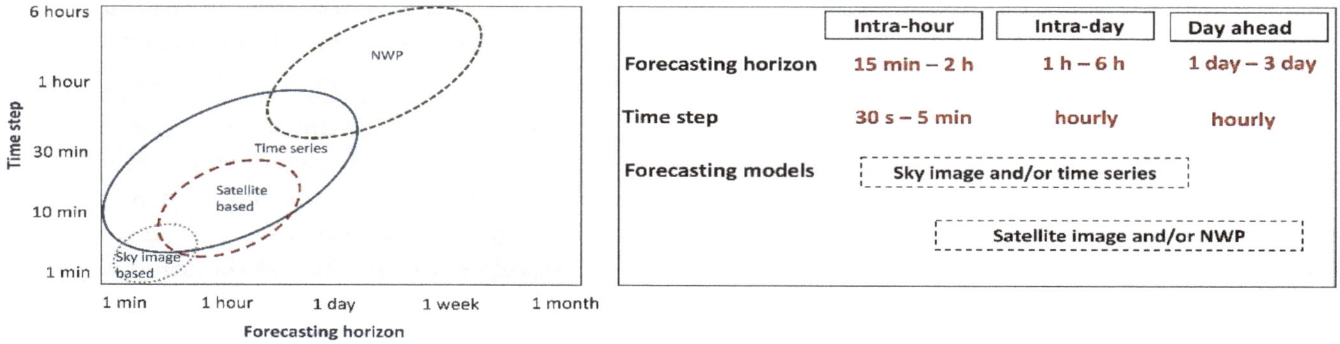

Figure 2. Solar irradiance forecasting techniques in different time horizons and steps.

3. Deep Learning Models

This section provides an overview of the theoretical background of the selected deep learning models. We describe a neural network and activation function because they are fundamental for all types of neural network algorithms, although ANN is not in the scope of this review.

3.1. Neural Network

The neural network is inspired by the structure of the human brain and significantly contributed to machine learning technology development. It is a simplified mathematical model to solve various nonlinear problems. In the past, researchers reviewed ANN models in predicting solar energy, such as Yadav et al. [15] for solar radiation prediction, Mellit et al. [16] for PV applications, and Cheon et al. [17] for solar energy forecasting. One conclusion from these studies is that the ANN model predicts solar radiation more accurately than other conventional models, such as the Angstrom, conventional, linear, nonlinear, and fuzzy logic models.

The neural network model comprises the input, hidden, and output layers with auxiliary components, such as neurons, weight, bias, and activation functions. Figure 3 shows the basic neural network architecture with a multilayer perceptron. The input layer receives input values, and the hidden layer analyzes the input values. The output layer collects the data from the hidden layer and decides the output. In the learning process, the neural network modifies its structure to get the same reference or set point as the supervisor. The training process will be repeated until the difference between the neural network output and the supervisor lies within an acceptable range [18].

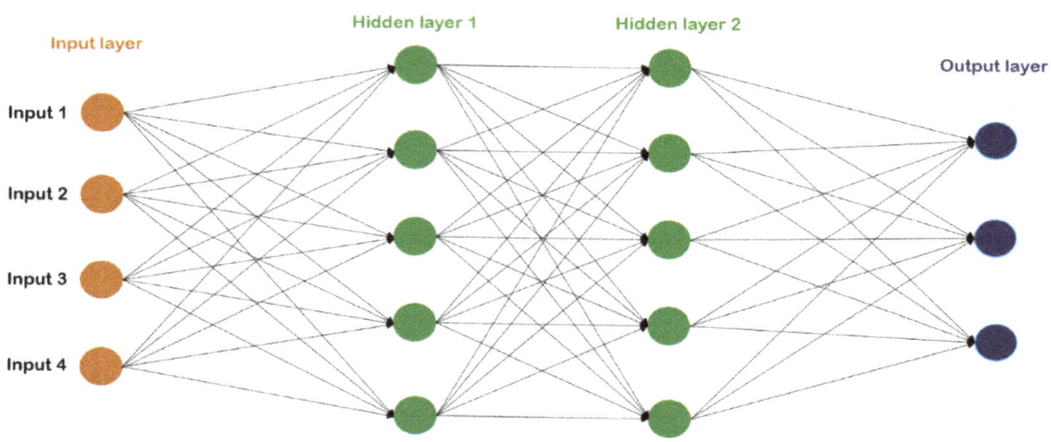

Figure 3. A schematic of neural network architecture, including input, hidden, and output layers.

The basic ANN mathematical formula can be expressed as

$$A_n = \sum_{j=1}^{n} (w_j \cdot I_j) + b \tag{1}$$

where A_n is the output, n is the number of input, w_j is the weight, I_j is the input, and b is the bias.

The output value varies with the activation function. The activation function, also known as the transfer function, is a mathematical equation determining neurons' output and can be divided into two types, namely, linear and nonlinear functions. A linear activation function generates the same linear result between the input and output layers. However, such a linear relationship is not enough for practical applications because the problems involve complex information and various parameters, such as image, video text, and sound. A neural network with a nonlinear activation function can tackle the limitations of the linear activation function. The commonly used activation functions are shown in Table 1. Note that the rectified linear unit (ReLU) and leaky ReLU are examples of nonlinear activation function because the slope is not constant for all values. Especially for ReLU, the slope is always either 0 for negative values or 1 for positive values.

Table 1. Activation functions of the neural network.

Activation Function	Equation	Plot
Linear	$f(x) = x$	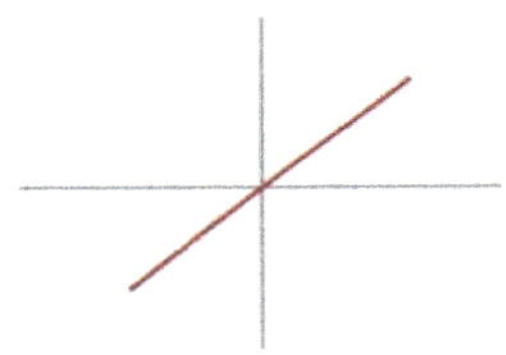
ReLU	$f(x) = \max(0, x)$	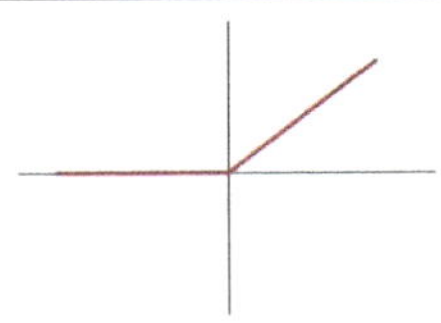
Leaky ReLU	$f(x) = \max(0.1 \cdot x, x)$	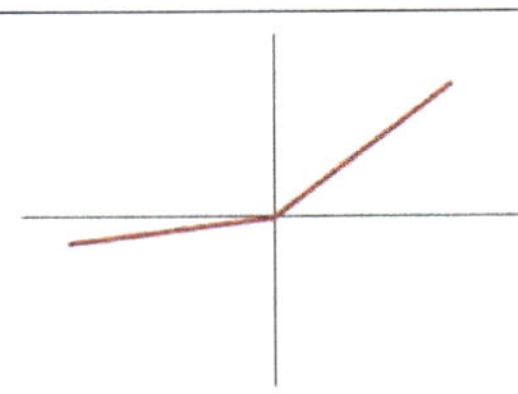
Tanh	$f(x) = \tanh(x)$	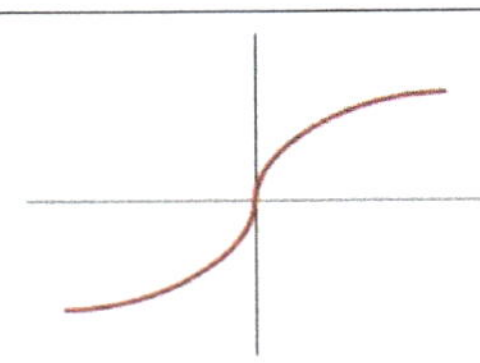
Sigmoid	$f(x) = \frac{1}{1+e^{-x}}$	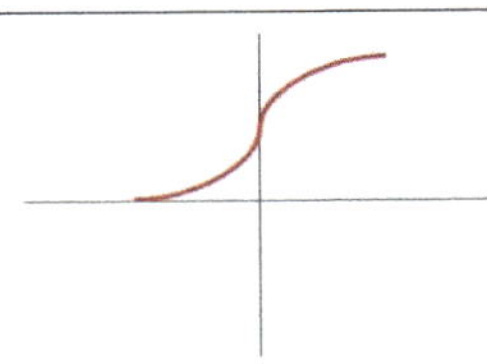

3.2. RNN

RNNs are specially designed for analyzing sequential data and have been successfully used in fields such as speech recognition, machine translation, and image captioning [19]. RNN processes sequence data by elements and preserves a state to represent the information at time steps [20]. A traditional neural network assumes that all units of the input vectors are independent. Consequently, the traditional neural network is ineffective for predicting using sequential data.

The architecture of RNN with three main components (input, hidden neuron, and activation function) is shown in Figure 4.

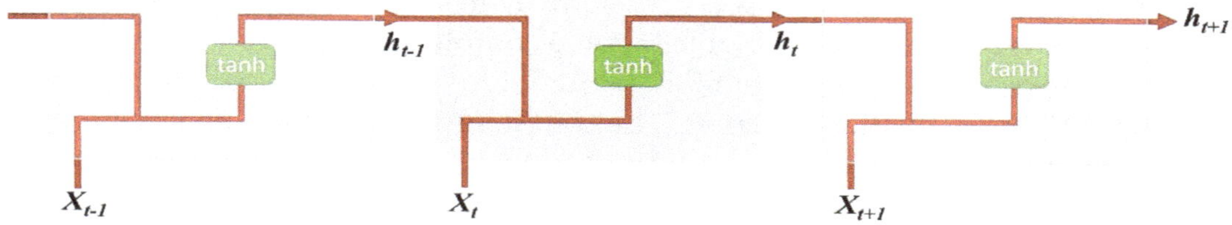

Figure 4. The structure of the recurrent neural network.

Previous hidden state (h_t) can be formulated as

$$h_t = \tanh(U \cdot x_t + W \cdot h_{t-1}) \tag{2}$$

where x_t is the input at time t, h_t is the hidden neuron at time t, U is the weight of the hidden layer, and W is the transition weights of the hidden layer. The input and previous hidden states are combined to produce information as the current and previous input go through the tanh function. Then, the output is the new hidden state, performing as the neural network memory because it holds information from the previous network.

Training regular RNNs can be challenging because of vanishing and exploding gradient problems. In the case of the exploding gradient, the problem can be solved after the backpropagation is closed at a certain point. However, the result is not optimal because all the weights are not updated. In the case of the vanishing gradient, it can be fixed by initializing the weights to reduce the possibility of vanishing gradient. However, an alternative treatment to solve the problem is to use LSTM, which we discuss later.

3.3. LSTM

LSTM is a time RNN proposed by Hochreiter et al. [21] to learn the long-term dependence of information. An LSTM has a similar flow as an RNN. The difference is the operation inside the cells. An LSTM unit comprises forget, input, and output gates (Figure 5). The forget gate categorizes the information that should be thrown away or kept. The input gate updated the cells, and the output gate decides the next hidden state. Furthermore, LSTM has an internal memory unit and gate mechanism to overcome both the vanishing gradient and explosion gradient problems in the training process of RNN [22].

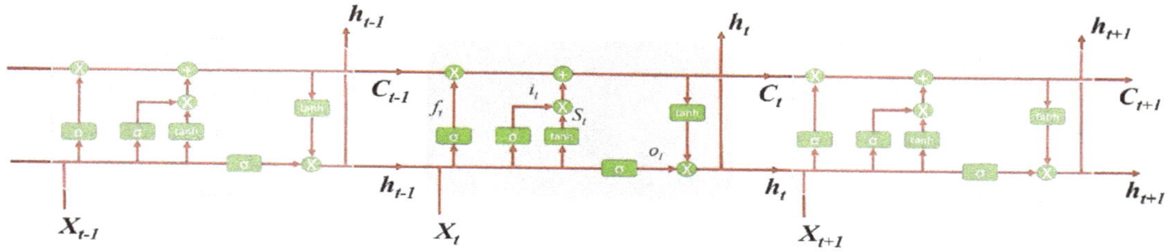

Figure 5. The structure of a long short-term memory network.

The calculation formulas related to the LSTM structure in Figure 4 are as follows:

$$f_t = \sigma\left(W_f \cdot X_t + U_f \cdot h_{t-1} + b_f\right) \tag{3}$$

$$i_t = \sigma(W_i \cdot X_t + U_i \cdot h_{t-1} + b_i) \tag{4}$$

$$S_t = \tanh(W_c \cdot X_t + U_c \cdot h_{t-1} + b_c) \tag{5}$$

$$C_t = i_t * S_t + f_t * S_{t-1} \tag{6}$$

$$o_t = \sigma(W_o \cdot X_t + U_o \cdot h_{t-1} + V_o \cdot C_t + b_o) \tag{7}$$

$$h_t = o_t * \tanh(C_t) \tag{8}$$

The mathematical symbols in the above equations are as follows:

1. X_t is the input vector to the memory cell at time t.
2. W_i, W_f, W_c, W_o, U_i, U_f, U_c, U_o, and V_o are weight matrices.
3. b_i, b_f, b_c, and b_o are bias vectors.
4. h_t is the value of the memory cell at time t.
5. S_t and C_t are the values of the candidate state of the memory cell and the state of the memory cell at time t, respectively.
6. σ and tanh are the activation functions.
7. i_t, f_t, and o_t are values of the input gate, the forget gate, and the output gate at time t.

The forget gate (f_t), input gate (i_t), and output gate (o_t) in Equations (3), (4) and (7) have values from 0 to 1 through the sigmoid function (σ). A value of one means that all input information passes through the gate, but a value of 0 shows that no input information passes [23]. The values of the candidate state of the memory cells in Equation (6) calculate the new information at time t, and its output through the tanh function has a value between −1 and 1. The state of the memory at the cell, controlled by the forget and input gates, is calculated as variable C_t of time t (Equation (7)). The selected values are converted into output by multiplying them by o_t and output becomes h_t (Equation (8)).

3.4. GRU

Cho et al. [24] first proposed GRU as a simpler RNN architecture than LSTM, resulting in easier computation and implementation. GRU is similar to LSTM in terms of remembering valuable information and capturing long-term dependencies. The strength of GRU is that the computational time is more efficient with less complexity because of fewer parameters than LSTM [25]. GRU also only has two gates, namely, a reset and an update gate. The update gate is the same as the forget and input gate in LSTM because it selects what information should be stored or erased. Meanwhile, the reset gate decides the amount of information that must be forgotten. Therefore, the training time of GRU is faster than LSTM.

The structure of GRU is shown in Figure 6, and the relationship between the input and output for GRU can be written as:

$$r_t = \sigma(W_r \cdot X_t + U_r \cdot h_{t-1} + b_r) \tag{9}$$

$$Z_t = \sigma(W_z \cdot X_t + U_z \cdot h_{t-1} + b_z) \tag{10}$$

$$A_t = \tanh(W_h \cdot X_t + U_h \cdot (r_t * h_{t-1}) + b_h) \tag{11}$$

$$h_t = (1 - Z_t) * h_{t-1} + Z_t * A_t \tag{12}$$

where r_t is the reset gate, Z_t is the update gate, A_t is the memory content, σ and tanh are the activation functions, and h_t is the final memory at the current time step. The reset (r_t) and update gate (Z_t) have values from 0 to 1 through the sigmoid function (σ) in Equations (9) and (10). Meanwhile, the memory content (A_t), using the rest gate to store the relevant information from the past, has a value between −1 and 1 through tanh.

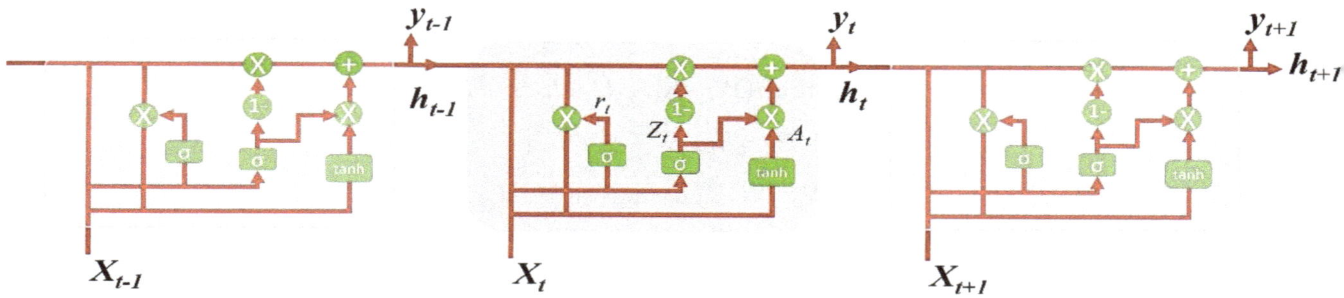

Figure 6. The structure of the gated recurrent unit network.

3.5. The Hybrid Model (CNN–LSTM)

CNN is a deep learning algorithm by considering spatial inputs. Identical to other neural networks, CNN neurons have learnable weights and biases. However, CNN is mainly used for processing data with a grid topology, giving it a specific characteristic of its architecture [26].

CNN is a feedforward network because information flow occurs in one direction only, that is, from their inputs to their outputs [27]. The CNN model uses three main layers, namely, the convolutional, pooling, and fully connected layers (Figure 7). The convolutional and pooling layers are used to reduce the computational complexity. Meanwhile, the fully connected layer is the flattened layer connected to the output. Various pooling techniques are available in the architecture of CNN. However, max pooling is mostly used in CNN layers, where the pooling window contains the maximum value from each element [28].

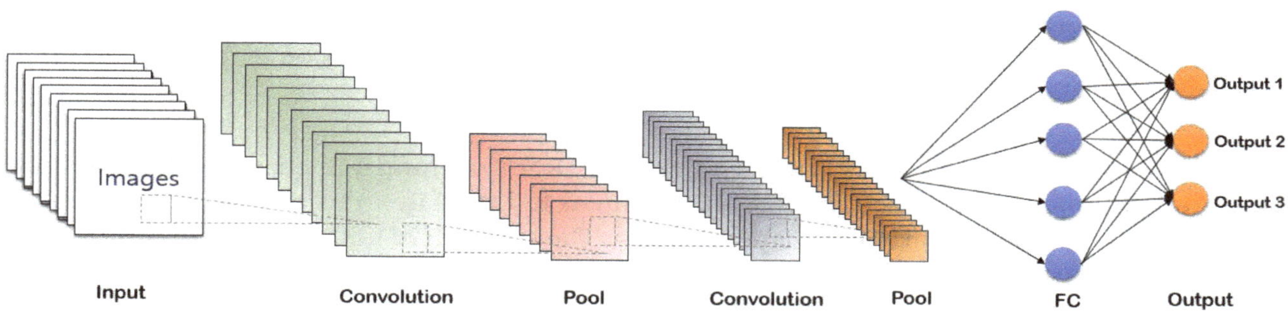

Figure 7. The structure of the convolutional neural network.

CNN–LSTM was developed for visual time series prediction problems and generating textual descriptions from the sequences of images. The CNN–LSTM architecture uses CNN layers for feature extraction on input data and combines with LSTM to support sequence prediction. Specifically, CNN extracts the features from spatial inputs and uses them in the LSTM architecture to output the caption. The architecture of the CNN–LSTM model is illustrated in Figure 8.

The applications of this hybrid model have been used to solve many problems, such as rod pumping [29], particulate matter [30], waterworks [31], and heart rate signals [32]. Studies have demonstrated promising results; for example, Xingjian et al. [33] predicted the future rainfall intensity in a local region over a relatively short period. The experiments show that the CNN–LSTM network captures spatiotemporal correlations better and consistently outperforms the fully connected LSTM (FC-LSTM) model for precipitation forecasting.

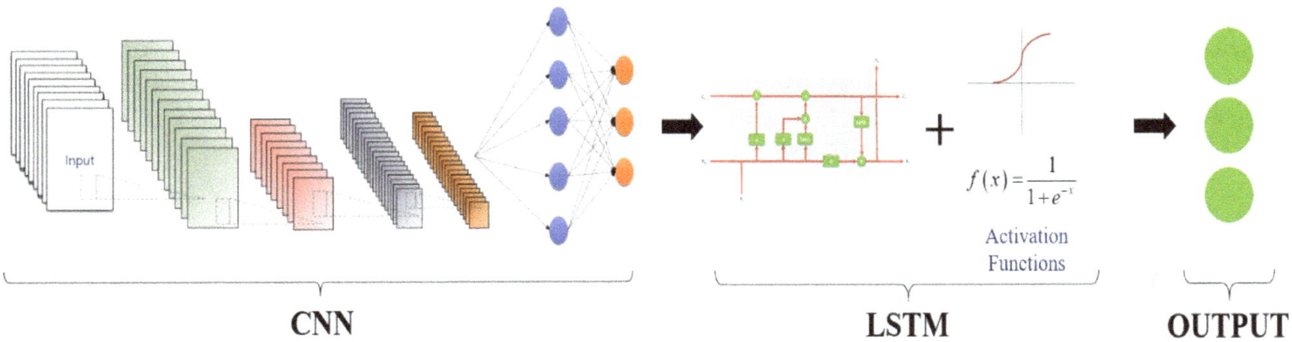

Figure 8. Illustration of convolutional neural network (CNN)–long short-term memory (LSTM) architecture.

4. Evaluation Metrics

Evaluation metrics are critical for explaining the forecast performance of deep learning models [34]. The metrics provide feedback regarding the accuracy of the forecasting to improve the models until a desirable accuracy is achieved. Various evaluation metrics are available for calculating the accuracy of prediction. The typical evaluation metrics for solar irradiance and PV power forecasting are summarized in Table 2. Here, P_{pred}, P_{meas}, and n represent the forecasted values at each time, the measured values at each time, and the number of sample data for the period, respectively.

Table 2. Evaluation metrics.

Evaluation Metric	Equation
Error	$P_{pred} - P_{meas}$
Mean absolute error (MAE)	$\frac{1}{n}\sum_{i=1}^{n}\lvert P_{pred} - P_{meas}\rvert$
Mean absolute percentage error (MAPE)	$\frac{1}{n}\sum_{i=1}^{n}\left\lvert\frac{P_{pred}-P_{meas}}{P_{meas}}\right\rvert \cdot 100$
Mean bias error (MBE)	$\frac{1}{n}\sum_{i=1}^{n}\left(P_{pred} - P_{meas}\right)$
Relative Mean bias error (rMBE)	$\frac{\sum_{i=1}^{n}\left(P_{pred}-P_{meas}\right)}{\sum_{i=1}^{n}P_{meas}} \cdot 100$
rRMSE	$\frac{\sqrt{\frac{1}{n}\sum_{i=1}^{n}\left(P_{pred}-P_{meas}\right)^2}}{\frac{1}{n}\sum_{i=1}^{n}P_{meas}} \cdot 100$
RMSE	$\sqrt{\frac{1}{n}\sum_{i=1}^{n}\left(P_{pred} - P_{meas}\right)^2}$
Forecasting skill	$1 - \frac{RMSE_{model}}{RMSE_{persistence}}$

Mean absolute error (MAE) measures the average magnitude of error in a set of predictions using the absolute value. If the absolute sign is removed, the evaluation metric becomes MBE, capturing the average bias in the prediction, such that positive and negative values represent overprediction and underprediction, respectively. However, root-mean-square error evaluation metric (RMSE) measures the deviation from the measurement, and thus, the smaller, the better. When mean values vary with location or system, a direct comparison of evaluation metrics could lead to a misunderstanding. In such cases, the percentage or relative metrics such as Mean absolute percentage error (MAPE) and relative root-mean-square error (rRMSE) are more useful. Forecasting skill measures the forecasting model over the persistence model regarding RMSE, where the persistence model assumes that the atmospheric conditions are stationary. A positive forecasting skill value means that the model is outperforming the persistence model. Note that measurement data include uncertainty by nature, and thus, the true values are unknown. Hence, researchers prefer to use differences, such as root-mean-square difference (rRMSD) and relative mean bias difference (rMBD) [35].

5. Analysis of Past Studies

In this section, we present notable findings in forecasting solar irradiance and PV power after analyzing the published studies based on RNN, LSTM, GRU, and CNN–LSTM hybrid models. In total, 35 papers from 2005 to 2020 were collected and plotted by the publication year (Figure 9). From 2005 to 2017, only a few papers were published because these deep learning models were unpopular then. However, the number of publications increased dramatically since 2018. The number of publications grew by 250%, from 4 in 2018 to 14 in 2019. Meanwhile, only the publications until early 2020 were collected, and more publications are expected by the end of the year.

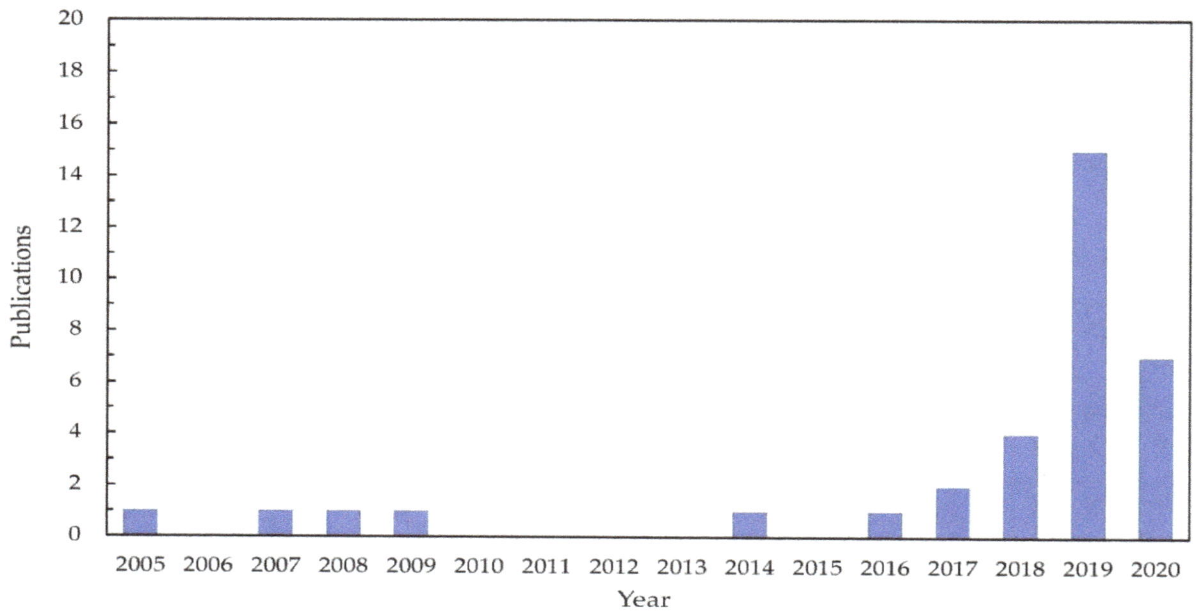

Figure 9. The number of publications that apply the deep learning models to predict solar irradiance and photovoltaic (PV) power from 2005 to 2020.

The proportion of publications by deep learning models for predicting solar irradiance and PV power individually is shown in Figure 10. In both cases, the LSTM accounts for most publications, with RNN and GRU following. Meanwhile, the CNN–LSTM hybrid model shows the lowest contribution. The popularity of the LSTM model is higher than that of the other standalone models because it provides promising accuracy in the case of solar energy forecasting. The CNN–LSTM also has a better performance than other models. However, the percentage of publications using this model is smaller than using other models because the CNN–LSTM is a new solar energy forecasting model.

It should be noted that we did an independent review for solar irradiance and PV power because the units and range values are different. Solar irradiance refers to the amount of solar radiation per unit area; meanwhile, PV power refers to the use of solar radiation as thermal energy through PV cells in the solar panel. The total solar radiation on the earth's atmosphere is approximately 1360 W/m^2, called the solar constant. This value is attenuated to the earth's surface through a complex series of reflections, absorptions, and remissions. Solar irradiance fluctuates because it is affected by several factors such as atmosphere condition, geographic location, season, and time of day. Although the amount of power generated by PV at a particular location depends on how much of the solar irradiance reaches it, the PV power output also relies on the solar panel's size and efficiency. Therefore, it is essential to describe the specification of the solar panel for getting accurate PV output information.

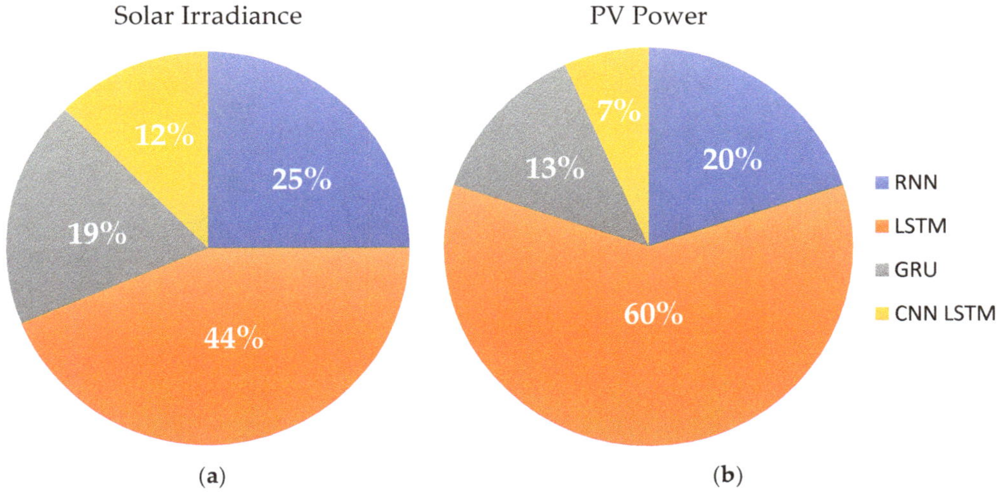

Figure 10. Distribution of studies using deep learning models to predict (**a**) solar irradiance and (**b**) PV power.

5.1. Accuracy

The prediction accuracy is the most critical factor in selecting a forecast model. We chose RMSE as the basic evaluation metric because it is most popular in solar energy forecasting. Because the mean value of solar irradiance or PV power differs by factors such as location and system size, rRMSE is a better measure to compare the accuracy between studies. Unfortunately, only a few studies presented rRMSE. We selected the best accuracy out of all the studied cases. Tables 3 and 4 list the accuracy of the papers, including forecast horizon, time interval, input parameters, and the size of PV systems in the case of PV forecasting.

Table 3. Solar irradiance forecasting.

Authors and Ref.	Forecast Horizon	Time Interval	Model	Input Parameter	Historical Data Description	RMSE (W/m^2)
Cao et al. [38]	1 day	hourly	RNN	-Solar irradiance	1995–2000 (2192 days)	44.326
Niu et al. [37]	10 min ahead 30 min ahead 1 h ahead	every 10 min	RNN	-Global solar radiation -Dry bulb temperature -Relative humidity -Dew point -Wind speed	22–29 May 2016 (7 days)	118 121 195
Qing et al. [39]	1 day ahead	hourly	LSTM	-Temperature -Dew Point -Humidity -Visibility -Wind Speed	March 2011–August 2012 January 2013–December 2013 (30 months)	76.245
Wang et al. [25]	1 day ahead	every 15 min	CNN–LSTM LSTM	-Solar irradiance	2008–2012 2014–2017 (3013 days)	32.411 33.294

Table 3. *Cont.*

Authors and Ref.	Forecast Horizon	Time Interval	Model	Input Parameter	Historical Data Description	RMSE (W/m^2)
Aslam et al. [40]	1 h ahead 1 day ahead	hourly	LSTM GRU RNN LSTM GRU RNN	-Solar irradiance	2007–2017 (10 years)	108.888 99.722 105.277 55.277 55.821 63.125
Ghimire et al. [36]	1 day ahead	every 30 min	CNN–LSTM	-Solar irradiance	January 2006–August 2018	8.189
Husein et al. [3]	1 day ahead	hourly	LSTM	-Temperature -Humidity -Wind speed -Wind direction -Precipitation -Cloud cover	January 2003–December 2017	60.310
Hui et al. [41]	1 day ahead	hourly	LSTM	-Temperature -Relative humidity -Cloud cover -Wind speed -Pressure	2006–2015 (10 years)	62.540
Byung-ki et al. [42]	1 day head	hourly	LSTM	-Temperature -Humidity -Wind speed -Sky cover -Precipitation -Irradiance	(1825 days)	30.210
Wojtkiewicz et al. [43]	1 h ahead	hourly	GRU LSTM	-GHI -Solar zenith angle -Relative humidity -Air Temperature	January 2004–December 2014	67.290 66.570
Yu et al. [44]	1 h ahead	hourly	LSTM	-GHI -Cloud type -Dew point -Temperature -Precipitation -Relative humidity -Solar Zenith Angle -Wind speed -Wind direction	2013–2017	41.370
Yan et al. [45]	5 min ahead 10 min ahead 20 min ahead 30 min ahead	every 1 min	LSTM GRU LSTM GRU LSTM GRU LSTM GRU	-Solar irradiance	2014	18.850 20.750 14.200 15.200 33.860 29.580 58.000 55.290

Regarding solar irradiance forecasting in Table 3, the forecast horizon ranges from 5 min to 1 day ahead. Meanwhile, the time interval data vary from every 5 min to 1 h. Generally, the CNN–LSTM as the hybrid model performs better to predict solar irradiance for one day-ahead forecasting. Ghimire et al. [36] have the best performances of all the studies with RMSE values of 8.189 W/m^2. For LSTM and GRU, the best performance was the prediction of solar irradiance for 10 min ahead at 5 min intervals. However, RNN tends to result in lower accuracy than other models for day-ahead hourly forecasting.

In this case, the main factor causing changes in solar irradiation is the presence of clouds. Unfortunately, only some studies explained the condition of the sky. For example, Wang et al. [25]

demonstrated that the CNN–LSTM hybrid outperformed the single LSTM to predict one day-ahead solar irradiation at 15 min intervals on sunny days, showing the small error where the RMSE value is less than 34 W/m^2. Meanwhile, Niu et al. [37] have the largest error in forecasting solar irradiance using the RNN model where the RMSE value is 195 W/m^2; however, the sky conditions are not mentioned in the present study. Hence, further study regarding deep learning models to predict solar irradiance is required to create a concrete solution.

The performances for PV power forecasting are listed in Table 4. Most publications focused on predicting PV power in intra-hour forecast horizons. The RMSE reveals a wide variation from 0.044 to 15,290 kW because PV power generation is proportional to system size. Accuracy comparison is more difficult in PV power forecasting than in solar irradiance forecasting, suggesting an increasing need for evaluation based on rRMSE. For the LSTM models in references (Zhang et al. [46]; Wang et al. [47]; Li et al. [48]), the RMSE value increases as the PV size increases. The values are 0.139, 0.398, and 0.885 kW for the PV size of 60.00, 153.48, and 199.16 m^2, respectively.

Table 4. PV power forecasting.

Authors and Ref.	**Forecast Horizon**	**Interval Data**	**Model**	**Input Variable**	**Historical Data Description**	**RMSE (kW)**	**PV Size**
Vishnu et al. [49]	1 h ahead 1 day ahead 1 w ahead	hourly	CNN–LSTM	-Irradiation -Wind speed -Temperature	March 2012–December 2018	0.053 0.051 0.045	N/A
Gensler et al. [50]	1 day ahead	hourly	LSTM	-PV power	(990 days)	0.044	N/A
Wang et al. [51]	1 h ahead	hourly	GRU	-Total column liquid water -Total column ice water -Surface pressure -Relative humidity -Total cloud cover -Wind speed -Temperature -Total precipitation -Total net solar radiation -Surface solar radiation -Surface thermal radiation	April 2012–May 2014	68.300	N/A
Zhang et al. [34]	1 min ahead	every 1 min	LSTM	-Sky images -PV Power	2006	0.139	10×6 m^2
Abdel-Nassar et al. [52]	1 h ahead	hourly	LSTM	-PV power	(12 months)	82.150	N/A
Lee et al. [53]	1 h ahead	hourly	LSTM	-PV power	June 2013–August 2016 (39 months)	0.563	N/A
Lee et al. [54]	1 h ahead	hourly	RNN	-Temperature -Relative humidity -Wind speed -Wind direction -Sky index -Precipitation -Solar altitude	June 2017–August 2018	0.160	N/A

Table 4. *Cont.*

Authors and Ref.	Forecast Horizon	Interval Data	Model	Input Variable	Historical Data Description	RMSE (kW)	PV Size
Li et al. [55]	15 min ahead 30 min ahead	N/A	RNN LSTM RNN LSTM	-PV power	January 2015–January 2016	6970 8700 15,290 15,570	N/A
Li et al. [48]	1 h ahead	every 5 min	LSTM GRU RNN	-PV power	June 2014–June 2016 (743 days)	0.885 0.847 0.888	199.16 m^2
Wang et al. [56]	5 min ahead	every 5 min	LSTM CNN–LSTM	-Current phase average -Wind speed -Temperature -Relative humidity -GHI -DHI	2014–2017 (4 years)	0.398 0.343	4×38.37 m^2
Wen et al. [57]	1 h ahead	hourly	LSTM	-Wind direction -Temperature -Humidity -Wind speed -GHI -DHI	1 January–1 February 2018	7.536	N/A
Sharadga et al. [58]	1 h ahead 2 h ahead 3 h ahead	hourly	LSTM	-PV power	January–October 2010	841 1102 1824	N/A

5.2. Types of Input Data

The type of input data, endogenous and exogenous, can classify forecasting models. In the endogenous model, the type of input and output data is identical. That is, historical PV power data are used for forecasting future PV power. However, the exogenous model uses other data types, such as ambient temperature, humidity, wind speed, wind direction, and sun position, besides the type of output data. It should be noted that the cited references used historical data prediction as input rather than numerical weather prediction data. The time period of input data is also summarized in Tables 3 and 4. The number of publications by exogenous and endogenous inputs, and the forecast horizon is illustrated in Figure 11. For the intra-hour and day-ahead forecast horizons, the endogenous model outnumbers the exogenous model.

Li et al. [55] presented RNN and LSTM models using endogenous inputs for predicting PV power output in the very short term. They used both PV power data from the previous day and previous forecasting data as the input. The results demonstrated that the average RMSE, MAPE, and MAE outperform the other models such as SVM, radial basis function (RBF), back propagation neural network (BPNN), and persistence in the 15 and 30 min forecasting horizons.

Husein et al. [3] proposed an LSTM model using exogenous inputs to forecast day-ahead solar irradiance. The models used dry bulb temperature, dewpoint temperature, and relative humidity as input from six locations. The results showed that LSTM outperforms the feedforward neural network (FFNN) for data from all locations. They also simulated a one-year operation of a commercial building microgrid using the actual and forecasted solar irradiance, and the results showed that using the forecasting approach increases the annual energy savings by 2% compared with FFNN.

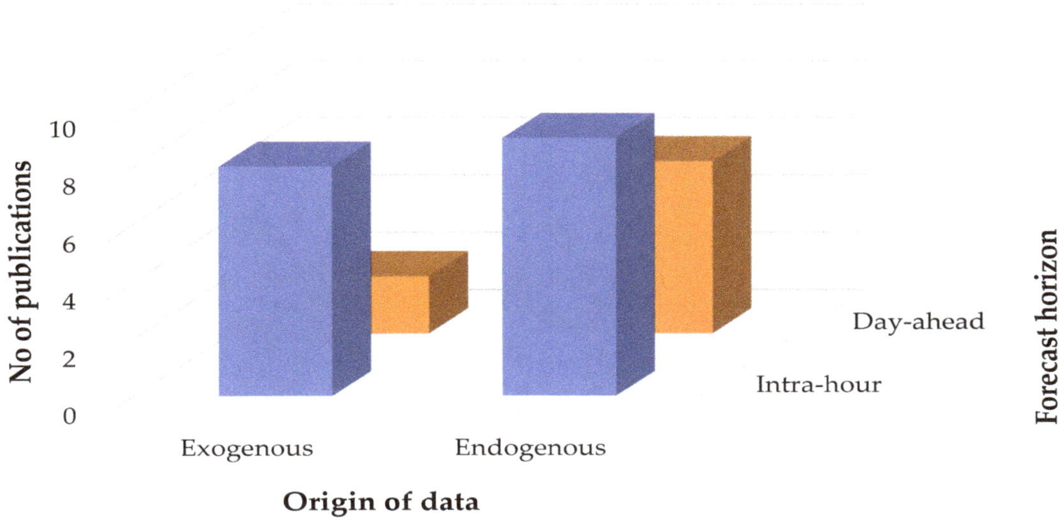

Figure 11. The number of publications based on the origin of data and forecast horizon.

We investigated the exogenous input parameters from 25 publications (Figure 12). The less frequently used parameters include visibility, sky images, column ice water, column liquid water, and pressure. However, it shows that temperature, humidity, and wind speed are used more frequently because they are easier to collect than other parameters.

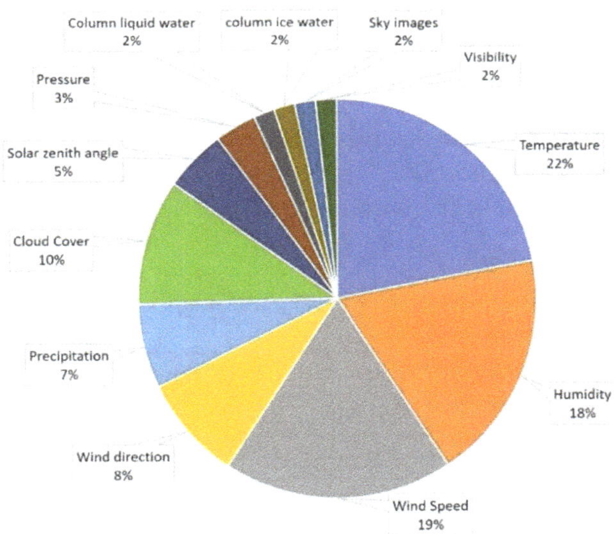

Input Parameter	Number of parameter
Temperature	13
Humidity	11
Wind Speed	11
Wind direction	5
Precipitation	4
Cloud Cover	6
Solar zenith angle	3
Pressure	2
Column liquid water	1
column ice water	1
Sky images	1
Visibility	1

Figure 12. Percentage of the input parameters used to predict solar irradiance and PV power.

The period of historical data also affects the performance of the prediction. If the time series is too short, it will lead to a lack of information for learning, and if the time series is too long, it will increase the complexity of the algorithm. However, the increase in time series data does not guarantee better performance. Wang et al. [47] compared the errors of the LSTM and CNN–LSTM according to input sequences (Table 5). The results show that errors increase from 0.5 to 2 years of input time series data. The best accuracy was observed when three years of input sequences were used. Historical data of more than three years degraded performances, implying that the period of input data must be optimized.

Table 5. Root-mean-square error evaluation metric (RMSE) of LSTM vs. CNN–LSTM based on the input sequence.

Input Sequence (Years)	LSTM		CNN–LSTM (kW)	
	RMSE (kW)	MAE (kW)	RMSE (kW)	MAE (kW)
0.5	1.244	0.654	1.161	0.559
1	1.393	0.616	1.434	0.628
1.5	1.533	0.599	1.248	0.529
2	1.320	0.457	0.941	0.397
2.5	0.945	0.389	0.426	0.198
3	0.398	0.181	0.343	0.126
3.5	1.150	0.455	0.991	0.384
4	1.465	0.565	0.886	0.405

5.3. *Forecast Horizon*

The forecast horizon is the length of time into the future for which a model can predict, and it strongly influences the performances and characteristics of the forecast. Forecasting horizons can be divided into four types [59]:

1. Very short-term forecasting (1 min to several minutes ahead).
2. Short-term forecasting (1 h or several hours ahead to 1 day or 1 week ahead).
3. Medium-term forecasting (1 month to 1 year ahead).
4. Long-term forecasting (1 to 10 years ahead).

Yan et al. [45] studied LSTM and GRU's performance to predict solar irradiance for very short-term forecasting. The experiments were conducted in time ranges of 5, 10, 20, and 30 min and the four seasons. The RMSE in Table 6 shows that LSTM and GRU's smallest error occurred in winter for the 10 min forecast horizon. Generally, the two models showed a gradual increase of errors for the 10 to 30 min solar irradiance forecasting.

Table 6. RMSE of LSTM and gated recurrent unit (GRU) for very short-term forecasting.

Forecast Horizon (min)	Model	Spring		Summer		Autumn		Winter	
		RMSE	MAE	RMSE	MAE	RMSE	MAE	RMSE	MAE
		(W/m^2)		(W/m^2)		(W/m^2)		(W/m^2)	
5	LSTM	36.67	26.95	89.91	59.20	18.85	13.13	44.24	21.58
	GRU	36.82	27.18	89.77	59.70	20.75	16.03	43.66	23.60
10	LSTM	41.02	29.65	42.23	32.96	53.01	33.62	14.20	11.09
	GRU	44.71	34.42	41.08	30.92	55.00	38.35	15.20	12.83
20	LSTM	56.22	49.09	46.31	40.58	33.86	28.11	43.54	39.10
	GRU	45.23	36.78	53.97	47.44	29.58	24.55	41.03	37.09
30	LSTM	58.77	47.54	58.00	47.82	81.75	59.08	61.68	52.29
	GRU	60.42	49.65	55.29	50.52	82.12	60.71	62.33	54.13

Ghimire et al. [36] used the CNN–LSTM hybrid model and three standalone models for solar irradiance forecasting. The forecasting errors for those models are listed in Table 7. CNN–LSTM was used using 30 min interval data to predict solar irradiance for a 1 day up to 1-month forecast horizon, as measured by RMSE and MAE. Compared with RNN, LSTM, and GRU, the hybrid model outperforms other models to predict solar irradiance in all forecasting horizons.

Table 7. Evaluation of predictive model (RMSE and MAE) over multiple forecast horizons.

Model	RMSE (W/m^2)				MAE (W/m^2)			
	1 Day	1 Week	2 Weeks	1 Month	1 Day	1 Week	2 Weeks	1 Month
CNN–LSTM	8.189	16.011	14.295	32.872	6.666	9.804	8.238	13.131
LSTM	21.055	18.879	16.327	33.387	18.339	11.275	10.750	14.307
RNN	20.177	18.113	15.494	41.511	18.206	11.387	10.492	26.858
GRU	14.289	21.464	19.207	57.589	11.320	15.658	14.005	39.716

5.4. Type of Season and Weather

The accuracy of deep learning models has been proven for the different seasons and types of weather. For example, Li et al. [48] used RNN, LSTM, and GRU models for short-term PV power forecasting. The performance evaluations of the different models for each season and types of weather are presented in Table 8. Except for winter, generally, the LSTM model outperforms the two other models. In winter, the GRU model is better than LSTM and RNN for all types of weather.

Table 8. RMSE of PV power prediction based on the type of weather (sunny, cloudy, and rainy).

Season	Type of Weather	LSTM (kW)	GRU (kW)	RNN (kW)
Winter	Sunny	1.2541	1.2399	1.2468
	Cloudy	1.1279	0.2206	0.2867
	Rainy	2.2336	2.0876	2.1223
Spring	Sunny	0.1643	0.2456	0.3431
	Cloudy	0.2759	0.6452	0.4222
	Rainy	0.8107	1.0036	0.8604
Summer	Sunny	0.9701	1.0748	0.8514
	Cloudy	0.8398	0.9323	0.8812
	Rainy	0.3009	0.5805	0.4993
Autumn	Sunny	0.7395	0.8029	0.7778
	Cloudy	1.0540	1.2110	1.1365
	Rainy	2.4216	2.3687	2.4275

5.5. Training Time

Deep learning with many parameters requires distributed training, where training time is critical [60]. Training time is the product of the deep learning models that must be performed to reach the desired level of accuracy. Each deep learning model has a different training time to reach the best performance. Hence, in this section, we compared the training time for each model to know which model is more efficient in forecasting solar irradiation and PV power.

Wang et al. [25] conducted forecasting experiments for LSTM and GRU to experience information training time in the best, worst, and average cases. LSTM and GRU's training times in the three cases are presented in Table 9, proving that GRU is better than LSTM because the longest training time of GRU is shorter than the best training time of the LSTM.

Table 9. The training time of LSTM vs. GRU.

Model	The Best Case/s	The Worst Case/s	The Average Case/s
LSTM	393.01	400.57	396.27
GRU	354.92	379.57	365.40

The training times of LSTM and GRU in Seoul and Busan in 2016 and 2017 are shown in Table 10. Aslam et al. [40] measured LSTM and GRU's performance as training time in a system with an AMD Ryzen Thread ripper 2950X and 64 GB RAM. The mean values taken from 10 runs for accurate results were taken. GRU performed faster than LSTM in hourly and daily forecasting. For hourly prediction, the difference in training time is more than 200 s, and for daily forecasting, it is less than 20 s in each year and region.

Table 10. The training time of LSTM vs. GRU.

Region	Year	Hourly		Daily	
		LSTM (s)	GRU (s)	LSTM (s)	GRU (s)
Seoul	2017	1251.23	1004.15	88.35	72.56
	2016	1060.82	832.63	77.98	64.12
Busan	2017	1269.21	1028.43	90.42	75.44
	2016	1023.27	830.54	75.99	64.29

The training times of LSTM and CNN–LSTM are compared in Table 11. The hybrid model has a longer training time than LSTM, which is 983.71 s because the hybrid model must extract both the temporal and spatial features of the data [47].

Table 11. Performance of LSTM vs. CNN–LSTM.

Model	LSTM (s)	CNN–LSTM (s)
Training time	70.490	983.701

5.6. Comparison with Other Models

In this section, RNN, LSTM, GRU, and CNN–LSTM are compared in Tables 11–15 with other machine learning models and deep learning models. Pang et al. [61] proposed a hierarchical approach to predict solar irradiance using ANN and RNN in Tuscaloosa, Alabama, USA. The data from 22 to 28 May 2016 were used to predict solar irradiance into three interval times (10, 30, and 60 min) (Table 12). Consequently, RNN outperformed ANN in forecasting horizon data sampling, and the best result for the forecast horizon is when the shortest horizon was considered.

Table 12. Comparison between Recurrent neural network (RNN) and artificial neural network (ANN).

Forecast Horizon (min)	RMSE (W/m^2)	
	ANN	RNN
10	55.7	41.2
30	63.3	53.3
60	170.9	58.1

Table 13. Comparison between LSTM and other models.

Model	FFNN (W/m^2)	SVR (W/m^2)	LSTM (W/m^2)
RMSE	0.160	0.110	0.086

Table 14. Comparison between CNN–LSTM and other models.

Model	RMSE (kW)	
	With Weather Data	Without Weather Data
RFR	0.178	0.191
SVR	0.122	0.126
CNN–LSTM	0.098	0.140

Table 15. Comparison between GRU and multilayer perceptron (MLP).

Season	RMSE (kW)				
	Winter	Spring	Summer	Autumn	Average
GRU	847	917	1238	1074	1035
MLP	916	1069	1263	1061	1086

In the case of LSTM, Alzahrani et al. [62] studied a machine learning approach in Chicago, USA, where the study compared various prediction models on a test dataset using FFNN and support vector regression (SVR). The data were collected over four days (24 March, 8 February, 8 October, and 12 August) and split into three parts, namely, data training, data testing, and data validation in portions of 70%, 15%, and 15%, respectively. The results in Table 13 show that LSTM has a better performance in achieving an RMSE of 0.086 W/m^2, whereas FFNN and SVR obtained 0.160 and 0.110 W/m^2, respectively.

Lee et al. [63] compared CNN–LSTM, random forest regression (RFR), and SVR. They considered two sets of inputs: a PV measurement incorporated with weather values from a nearby meteorological station and only past PV measurements. They collected 18,620 hourly data time series and divided the data training and data testing as 75% and 25%, respectively. The results in the first case show that CNN–LSTM has the lowest error metrics with an RMSE of 0.098 kW. However, the excluded weather data case shows that SVR is better than RFR and CNN–LSTM with 0.126 kW (Table 14).

Li et al. [48] presented their results for the comparison between GRU and multilayer perceptron (MLP) in a different season (Table 15). They collected the data in DKASC, Alice Springs, Australia. The generation of the PV system is 26.500 kW. In this case, they used the data from 1 June 2014 to 31 May 2015, as the training dataset, and the data for 12 June 2016, were used as the testing dataset. The results show that MLP has a better performance than GRU only in the autumn, with a value of 1061 kW. However, the average value of GRU has the most representative attributes of all the seasons compared with MLP.

6. Conclusions

This paper introduced deep learning models as techniques to predict solar irradiance and PV power generation. To represent a complete review of the models, the evaluation for the PV power and solar irradiance forecasting is made apart. The main reason because the output is different in terms of function, unit, and range value. Although the value fluctuates because it is affected by several factors such as atmosphere condition, geographic location, season, and time of day, the evaluation of solar irradiance forecasting is uncomplicated to understand. Radiation from the sun that reaches the earth's surface is measured in power per unit area; hence it is possible to compare the solar irradiance at different locations. However, apart from the main factors, the PV power output relies on the solar panel's size and efficiency. Therefore, in part of the review of PV power forecasting, the solar panel size is described to obtain accurate PV output information.

RNN, LSTM, GRU, and CNN–LSTM have become the topics of research interest because of their popularity in predicting solar energy. They also offer many advantages over other machine learning models, especially regarding time series data forecasting. Each model has its strengths and limitations to predict solar irradiance and PV power; therefore, it is challenging to decide which is the best among all the models. However, from the studies reviewed in this paper, we propose the following conclusions:

- In the case of the single model, most studies explain that LSTM and GRU show better performance than RNN in all conditions because LSTM and GRU have internal memory to overcome the vanishing gradient problems occurring in the RNN.

- The hybrid model (CNN–LSTM) outperforms the three standalone models in predicting solar irradiance. More specifically, the evaluation metrics for this hybrid model are substantially smaller than those of the standalone models. However, the CNN–LSTM model requires complex input data, such as images, because it has a CNN layer inside.
- The training time should be considered to recognize the performance of the models. This work reveals that the statistics of GRU are more efficient than that of LSTM in the case of computational time because the average time for LSTM to train the data is relatively longer than that for GRU. Therefore, considering training time and forecasting accuracy, the GRU model can generate a satisfactory result for forecasting PV power and solar irradiance.
- Comparisons between the deep learning models and other machine learning models conclude that these models were better used in predicting solar irradiance and PV power (Section 5.6). Most studies show that the accuracy of the proposed models is better than other models, such as ANN, FFNN, SVR, RFR, and MLP.

Author Contributions: R.A.R.: conceptualization, methodology, formal analysis, writing—original draft; R.A.A.R.: conceptualization, investigation; H.-J.L.: writing—review and editing, supervision, funding acquisition. All authors have read and agreed to the published version of the manuscript.

Abbreviation

ANN	Artificial neural network
BPNN	Back propagation neural network
CNN	Convolutional neural network
DHI	Diffuse horizontal irradiance
FFNN	Feedforward neural network
GHI	Global horizontal irradiance
GRU	Gated recurrent unit
LSTM	Long short-term memory
MAE	Mean absolute error
MAPE	Mean absolute percentage error
MLP	Multilayer perceptron
PV	Photovoltaic
RBF	Radial basis function
ReLU	Rectified linear unit
RFR	Random forest regression
RMSE	Root-mean-square error
RNN	Recurrent neural network
rRMSE	Relative root-mean-square error
SVR	Support vector regression

References

1. Mohanty, S.; Patra, P.K.; Sahoo, S.S.; Mohanty, A. Forecasting of solar energy with application for a growing economy like India: Survey and implication. *Renew. Sustain. Energy Rev.* **2017**, *78*, 539–553. [CrossRef]
2. Das, U.K.; Tey, K.S.; Seyedmahmoudian, M.; Mekhilef, S.; Idris, M.Y.I.; Van Deventer, W.; Horan, B.; Stojcevski, A. Forecasting of photovoltaic power generation and model optimization: A review. *Renew. Sustain. Energy Rev.* **2018**, *81*, 912–928. [CrossRef]
3. Husein, M.; Chung, I.-Y. Day-Ahead Solar Irradiance Forecasting for Microgrids Using a Long Short-Term Memory Recurrent Neural Network: A Deep Learning Approach. *Energies* **2019**, *12*, 1856. [CrossRef]
4. Lappalainen, K.; Wang, G.C.; Kleisslid, J. Estimation of the largest expected photovoltaic power ramp rates. *Appl. Energy* **2020**, *278*, 115636. [CrossRef]
5. Sobri, S.; Koohi-Kamali, S.; Rahim, N.A. Solar photovoltaic generation forecasting methods: A review. *Energy Convers. Manag.* **2018**, *156*, 459–497. [CrossRef]

6. Ng, A. Machine Learning Yearning: Techincal Strategy for AI Engineers, in the Era of Deep Learning. 2018. Available online: https://www.deeplearning.ai/machine-learning-yearning (accessed on 30 November 2020).
7. Zang, H.; Liu, L.; Sun, L.; Cheng, L.; Wei, Z.; Sun, G. Short-term global horizontal irradiance forecasting based on a hybrid CNN-LSTM model with spatiotemporal correlations. *Renew. Energy* **2020**, *160*, 26–41. [CrossRef]
8. Shuai, Y.; Notton, G.; Kalogirou, S.; Nivet, M.-L.; Paoli, C.; Motte, F.; Fouilloy, A. Machine learning methods for solar radiation forecasting: A review. *Renew. Energy* **2017**, *105*, 569–582. [CrossRef]
9. Carrera, B.; Kim, K. Comparison Analysis of Machine Learning Techniques for Photovoltaic Prediction Using Weather Sensor Data. *Sensors* **2020**, *20*, 3129. [CrossRef]
10. Caldas, M.; Alonso-Suárez, R. Very short-term solar irradiance forecast using all-sky imaging and real-time irradiance measurements. *Renew. Energy* **2019**, *143*, 1643–1658. [CrossRef]
11. Miller, S.D.; Rogers, M.A.; Haynes, J.M.; Sengupta, M.; Heidinger, A.K. Short-term solar irradiance forecasting via satellite/model coupling. *Sol. Energy* **2018**, *168*, 102–117. [CrossRef]
12. Xie, T.; Zhang, G.; Liu, H.; Liu, F.; Du, P. A Hybrid Forecasting Method for Solar Output Power Based on Variational Mode Decomposition, Deep Belief Networks and Auto-Regressive Moving Average. *Appl. Sci.* **2018**, *8*, 1901. [CrossRef]
13. Wang, K.; Li, K.; Zhou, L.; Hu, Y.; Cheng, Z.; Liu, J.; Chen, C. Multiple convolutional neural networks for multivariate time series prediction. *Neurocomputing* **2019**, *360*, 107–119. [CrossRef]
14. Chaouachi, A.; Kamel, R.M.; Nagasaka, K. Neural Network Ensemble-Based Solar Power Generation Short-Term Forecasting. *J. Adv. Comput. Intell. Intell. Inform.* **2010**, *14*, 69–75. [CrossRef]
15. Yadav, A.K.; Chandel, S. Solar radiation prediction using Artificial Neural Network techniques: A review. *Renew. Sustain. Energy Rev.* **2014**, *33*, 772–781. [CrossRef]
16. Mellit, A.; Kalogirou, S.A. Artificial intelligence techniques for photovoltaic applications: A review. *Prog. Energy Combust. Sci.* **2008**, *34*, 574–632. [CrossRef]
17. Cheon, J.; Lee, J.T.; Kim, H.G.; Kang, Y.H.; Yun, C.Y.; Kim, C.K.; Kim, B.Y.; Kim, J.Y.; Park, Y.Y.; Jo, H.N.; et al. Trend Review of Solar Energy Forecasting Technique. *J. Korean Sol. Energy Soc.* **2019**, *39*, 41–54. [CrossRef]
18. Hameed, W.I.; Sawadi, B.A.; Al-Kamil, S.J.; Al-Radhi, M.S.; Al-Yasir, Y.I.; Saleh, A.L.; Abd-Alhameed, R.A. Prediction of Solar Irradiance Based on Artificial Neural Networks. *Inventions* **2019**, *4*, 45. [CrossRef]
19. Fan, C.; Wang, J.; Gang, W.; Li, S. Assessment of deep recurrent neural network-based strategies for short-term building energy predictions. *Appl. Energy* **2019**, *236*, 700–710. [CrossRef]
20. Chollet, F.; Allaire, J. *Deep Learning with R*; Manning Publications: Shelter Island, NY, USA, 2018.
21. Hochreiter, S.; Schmidhuber, J. Long Short-Term Memory. *Neural Comput.* **1997**, *9*, 1735–1780. [CrossRef]
22. Liu, Y.; Guan, L.; Hou, C.; Han, H.; Liu, Z.; Sun, Y.; Zheng, M. Wind Power Short-Term Prediction Based on LSTM and Discrete Wavelet Transform. *Appl. Sci.* **2019**, *9*, 1108. [CrossRef]
23. Kim, H.Y.; Won, C.H. Forecasting the volatility of stock price index: A hybrid model integrating LSTM with multiple GARCH-type models. *Expert Syst. Appl.* **2018**, *103*, 25–37. [CrossRef]
24. Cho, K.; Van Merrienboer, B.; Gulcehre, C.; Bahdanau, D.; Bougares, F.; Schwenk, H.; Bengio, Y. Learning phrase representations using RNN encoder-decoder for statistical machine translation. In Proceedings of the 2014 Conference on Empirical Methods in Natural Language Processing (EMNLP), Doha, Qatar, 14–21 October 2014; pp. 1724–1734. [CrossRef]
25. Wang, F.; Yu, Y.; Zhang, Z.; Li, J.; Zhen, Z.; Li, K. Wavelet Decomposition and Convolutional LSTM Networks Based Improved Deep Learning Model for Solar Irradiance Forecasting. *Appl. Sci.* **2018**, *8*, 1286. [CrossRef]
26. Yann, L.; Yoshua, B. Convolutional Networks for Images, Speech, and Time-Series. *Handb. Brain Theory Neural Netw.* **1995**, *10*, 2571–2575.
27. Rawat, W.; Wang, Z. Deep Convolutional Neural Networks for Image Classification: A Comprehensive Review. *Neural Comput.* **2017**, *29*, 2352–2449. [CrossRef]
28. Rehman, A.U.; Malik, A.K.; Raza, B.; Ali, W. A Hybrid CNN-LSTM Model for Improving Accuracy of Movie Reviews Sentiment Analysis. *Multimed. Tools Appl.* **2019**, *78*, 26597–26613. [CrossRef]
29. He, Y.; Liu, Y.; Shao, S.; Zhao, X.; Liu, G.; Kong, X.; Liu, L. Application of CNN-LSTM in Gradual Changing Fault Diagnosis of Rod Pumping System. *Math. Probl. Eng.* **2019**, *2019*, 4203821. [CrossRef]
30. Huang, C.-J.; Kuo, P.-H. A Deep CNN-LSTM Model for Particulate Matter (PM2.5) Forecasting in Smart Cities. *Sensors* **2018**, *18*, 2220. [CrossRef]
31. Cao, K.; Kim, H.; Hwang, C.; Jung, H. CNN-LSTM coupled model for prediction of waterworks operation data. *J. Inf. Process. Syst.* **2018**, *14*, 1508–1520. [CrossRef]

32. Swapna, G.; Soman, K.P.; Vinayakumar, R. Automated detection of diabetes using CNN and CNN-LSTM network and heart rate signals. *Procedia Comput. Sci.* **2018**, *132*, 1253–1262. [CrossRef]
33. Shi, X.; Chen, Z.; Wang, H.; Yeung, D.Y.; Wong, W.K.; Woo, W.C. Convolutional LSTM network: A machine learning approach for precipitation nowcasting. *Adv. Neural Inf. Process. Syst.* **2015**, *2015*, 802–810.
34. Zhang, J.; Florita, A.; Hodge, B.-M.; Lu, S.; Hamann, H.F.; Banunarayanan, V.; Brockway, A.M. A suite of metrics for assessing the performance of solar power forecasting. *Sol. Energy* **2015**, *111*, 157–175. [CrossRef]
35. Lave, M.; Hayes, W.; Pohl, A.; Hansen, C.W. Evaluation of Global Horizontal Irradiance to Plane-of-Array Irradiance Models at Locations across the United States. *IEEE J. Photovolt.* **2015**, *5*, 597–606. [CrossRef]
36. Ghimire, S.; Deo, R.C.; Raj, N.; Mi, J. Deep solar radiation forecasting with convolutional neural network and long short-term memory network algorithms. *Appl. Energy* **2019**, *253*, 113541. [CrossRef]
37. Niu, F.; O'Neill, Z. Recurrent Neural Network based Deep Learning for Solar Radiation Prediction. In Proceedings of the 15th IBPSA Conference, San Francisco, CA, USA, 7–9 August 2017; pp. 1890–1897.
38. Cao, S.; Cao, J. Forecast of solar irradiance using recurrent neural networks combined with wavelet analysis. *Appl. Therm. Eng.* **2005**, *25*, 161–172. [CrossRef]
39. Qing, X.; Niu, Y. Hourly day-ahead solar irradiance prediction using weather forecasts by LSTM. *Energy* **2018**, *148*, 461–468. [CrossRef]
40. Aslam, M.; Lee, J.-M.; Kim, H.-S.; Lee, S.-J.; Hong, S. Deep Learning Models for Long-Term Solar Radiation Forecasting Considering Microgrid Installation: A Comparative Study. *Energies* **2019**, *13*, 147. [CrossRef]
41. He, H.; Lu, N.; Jie, Y.; Chen, B.; Jiao, R. Probabilistic solar irradiance forecasting via a deep learning-based hybrid approach. *IEEJ Trans. Electr. Electron. Eng.* **2020**, *15*, 1604–1612. [CrossRef]
42. Jeon, B.-K.; Kim, E.-J. Next-Day Prediction of Hourly Solar Irradiance Using Local Weather Forecasts and LSTM Trained with Non-Local Data. *Energies* **2020**, *13*, 5258. [CrossRef]
43. Wojtkiewicz, J.; Hosseini, M.; Gottumukkala, R.; Chambers, T. Hour-Ahead Solar Irradiance Forecasting Using Multivariate Gated Recurrent Units. *Energies* **2019**, *12*, 4055. [CrossRef]
44. Yu, Y.; Cao, J.; Zhu, J. An LSTM Short-Term Solar Irradiance Forecasting Under Complicated Weather Conditions. *IEEE Access* **2019**, *7*, 145651–145666. [CrossRef]
45. Yan, K.; Shen, H.; Wang, L.; Zhou, H.; Xu, M.; Mo, Y. Short-Term Solar Irradiance Forecasting Based on a Hybrid Deep Learning Methodology. *Information* **2020**, *11*, 32. [CrossRef]
46. Zhang, J.; Verschae, R.; Nobuhara, S.; LaLonde, J.-F. Deep photovoltaic nowcasting. *Sol. Energy* **2018**, *176*, 267–276. [CrossRef]
47. Wang, K.; Qi, X.; Liu, H. Photovoltaic power forecasting based LSTM-Convolutional Network. *Energy* **2019**, *189*, 116225. [CrossRef]
48. Li, P.; Zhou, K.; Lu, X.; Yang, S. A hybrid deep learning model for short-term PV power forecasting. *Appl. Energy* **2020**, *259*, 114216. [CrossRef]
49. Suresh, V.; Janik, P.; Rezmer, J.; Leonowicz, Z. Forecasting Solar PV Output Using Convolutional Neural Networks with a Sliding Window Algorithm. *Energies* **2020**, *13*, 723. [CrossRef]
50. Gensler, A.; Henze, J.; Sick, B.; Raabe, N. Deep Learning for solar power forecasting—An approach using AutoEncoder and LSTM Neural Networks. In Proceedings of the 2016 IEEE International Conference on Systems, Man, and Cybernetics (SMC), Budapest, Hungary, 9–12 October 2016; pp. 002858–002865.
51. Wang, Y.; Liao, W.; Chang, Y. Gated Recurrent Unit Network-Based Short-Term Photovoltaic Forecasting. *Energies* **2018**, *11*, 2163. [CrossRef]
52. Abdel-Nasser, M.; Mahmoud, K. Accurate photovoltaic power forecasting models using deep LSTM-RNN. *Neural Comput. Appl.* **2019**, *31*, 2727–2740. [CrossRef]
53. Lee, D.; Kim, K. Recurrent Neural Network-Based Hourly Prediction of Photovoltaic Power Output Using Meteorological Information. *Energies* **2019**, *12*, 215. [CrossRef]
54. Lee, D.; Jeong, J.; Yoon, S.H.; Chae, Y.T. Improvement of Short-Term BIPV Power Predictions Using Feature Engineering and a Recurrent Neural Network. *Energies* **2019**, *12*, 3247. [CrossRef]
55. Li, G.; Wang, H.; Zhang, S.; Xin, J.; Liu, H. Recurrent Neural Networks Based Photovoltaic Power Forecasting Approach. *Energies* **2019**, *12*, 2538. [CrossRef]
56. Wang, K.; Qi, X.; Liu, H. A comparison of day-ahead photovoltaic power forecasting models based on deep learning neural network. *Appl. Energy* **2019**, *251*, 113315. [CrossRef]
57. Wen, L.; Zhou, K.; Yang, S.; Lu, X. Optimal load dispatch of community microgrid with deep learning based solar power and load forecasting. *Energy* **2019**, *171*, 1053–1065. [CrossRef]

58. Sharadga, H.; Hajimirza, S.; Balog, R.S. Time series forecasting of solar power generation for large-scale photovoltaic plants. *Renew. Energy* **2020**, *150*, 797–807. [CrossRef]
59. Raza, M.Q.; Nadarajah, M.; Ekanayake, C. On recent advances in PV output power forecast. *Sol. Energy* **2016**, *136*, 125–144. [CrossRef]
60. Gupta, S.; Zhang, W.; Wang, F. Model accuracy and runtime tradeoff in distributed deep learning: A systematic study. In Proceedings of the 2016 IEEE 16th International Conference on Data Mining (ICDM), Barcelona, Spain, 12–15 December 2016. [CrossRef]
61. Pang, Z.; Niu, F.; O'Neill, Z. Solar radiation prediction using recurrent neural network and artificial neural network: A case study with comparisons. *Renew. Energy* **2020**, *156*, 279–289. [CrossRef]
62. Alzahrani, A.; Shamsi, P.; Dagli, C.; Ferdowsi, M. Solar Irradiance Forecasting Using Deep Neural Networks. *Procedia Comput. Sci.* **2017**, *114*, 304–313. [CrossRef]
63. Lee, W.; Kim, K.; Park, J.; Kim, J.; Kim, Y. Forecasting Solar Power Using Long-Short Term Memory and Convolutional Neural Networks. *IEEE Access* **2018**, *6*, 73068–73080. [CrossRef]

2

Renewable Energy Sources in Transport on the Example of Methyl Esters and Bioethanol

Adam Kupczyk, Joanna Mączyńska-Sęczek, Ewa Golisz and Piotr F. Borowski *

Institute of Mechanical Engineering, Warsaw University of Life Sciences, 166 Nowoursynowska Str., 02-787 Warsaw, Poland; adam_kupczyk@sggw.edu.pl (A.K.); joanna_maczynska@sggw.edu.pl (J.M.-S.); ewa_golisz@sggw.edu.pl (E.G.)

* Correspondence: piotr_borowski@sggw.edu.pl or pborowski@autograf.pl

Abstract: The aim of this article is to show how the situation of domestic biofuels in Poland and the alternative fuels sector is changing. On the basis of our own research and the available literature, changes in the production potential of Poland in the field of alternative fuels/biofuels are examined. Moreover, the reasons for changes in the production structure are analyzed. The flexibility of production potential to changes in legal conditions is assessed, and the value of sectors and their future are determined. This paper discusses legal and market aspects related to selected renewable energy sources used in Polish transport. Based on a review of literature and statistical data, the production and use of methyl esters and bioethanol are characterized in detail. Analysis of legal regulations enables the presentation of targets regarding renewable energy source (RES) use in transport by 2030. The results of studies conducted in 2020 are discussed with regard to the value (attractiveness) of methyl esters, bioethanol, electromobility and liquid bio-hydrocarbons, and these are compared and contrasted with the result from earlier years since 2010.

Keywords: renewable energy sources; esters; processes; eco-mobility; CO_2 emission reduction, mechanical engineering

---

1. Introduction

The transport sector plays a vital role in the economy. At the same time, it has a negative effect on the natural environment [1]. The energy consumed for transport purposes comes in a majority of cases from petroleum, which leads to a decline in natural fuel resources and to the emission of greenhouse gases in significant amounts [2–4]. Transport-related emissions come first and foremost from the direct combustion of fuels used in road transport, non-electrified railway transport and inland sailing [5]. Reducing emissions in road transport can also be implemented in parallel by using appropriate road surfaces [6,7]. Limiting the negative effects of transport is a significant component of the policy of the European Union (UE) and one of the chief actions taken in this area is the imposition among member states to use renewable energy sources (RES) for transport purposes [8]. The research undertaken in this study aimed to identify areas of development and dysfunction in the field of domestic alternative fuels. The research was carried out as a continuous monitoring of the sector, mainly on the basis of the point value (attractiveness) method of sectors, according to Porter's five forces [9]. After joining the EU in 2004, Poland planned a strong expansion of its own transport biofuels into the various EU markets. Conventional transport biofuels were expected to have many advantages, including reducing the emission of greenhouse gases. However, aspects that undermine their ecological character, such as the real reduction of greenhouse gas emissions and problems with achieving the assumed target (the national index target (NIT)), caused a significant increase in the importance of other alternative fuels as well. In Poland, the obligation has been fulfilled mostly by using methyl esters and bioethanol [10]. Before World War II, Poland used industrial ethanol as an additive to gasoline (approx. 10 million liters). In the 1970s, the technology for the production of lignocellulosic ethanol from wood waste was developed, whereas in the 1990s it was planned to implement bioethanol and gasoline blending technologies [11]. After Poland joined the EU, the industry was built from scratch, using the existing capacity for bioethanol and biodiesel production. Despite the expansion of plants in this sector, Poland's position was not favorable because of problems with the achievement of index targets and use of the existing production capacity began to occur. Production was based on imported technologies, but unfortunately these technologies were already in the mature phase, which resulted in a quick transition to a declining phase. As a result of investing in declining technologies, the benefits of their exploitation systematically decreased, and it provided less opportunities for development or meeting increasingly restrictive legal conditions in these sectors [12].

2. Literature Review

2.1. Legal Grounds at the European Union Level

The EU legislature, in the area of RES, has evolved substantially over recent years. Currently, the most important legal regulations stimulating the increase in use of RES includes Directives 2009/28/EC (referred to as RED) [13], 2015/1513 [14] and 2018/2001 (referred to as RED II) [15]. These directives bind the member states in terms of targets to be met, leaving them with freedom of choice as regards to the form and means of their accomplishment. The specification of the most important obligations resulting from these directives is presented in Table 1.

Table 1. Specification of the most important obligations resulting from the directives in the scope of use of renewable energy sources (RES) in transport for member states of the European Union.

Directive Name	Target for RES Use in Transport
Directive 2009/28/EC (RED)	Min. 10% share of renewable energy in end use of energy in transport in 2020
Directive 2015/1513	Max. 7% share of conventional biofuels in the target set for 2020
Directive 2018/2001 (RED II)	Min. 14% share of renewable energy in end use of energy in transport in 2030 (including but not limited to 3.5% advanced biofuels, max. 7% conventional biofuels)

Source: own study based on [13–15].

RED set a target ensuring a 10% share of energy from renewable sources in transport by 2020. Initially, biofuels (RED defines biofuels as liquid or gaseous fuels for transport produced from biomass) produced from food and feed plans—i.e., the so-called conventional biofuels—were mostly used for this purpose. For a long time, fuels of this sort were regarded as an effective means of emission reduction [16]. However, their perception changed over time and analyses showed that the production of such biofuels may entail indirect negative consequences as well [4]. The current results of studies on the value of biofuel sectors in Poland do not encourage investment in biofuels of this generation, i.e., conventional biofuels [17]. It has been observed that the growing demand for raw materials for the production of methyl esters and bioethanol (the most widely used biofuels), such as rapeseed or corn, is a form of competition for foodstuff and feedstuff markets [18]. If grasslands or farmlands previously used for the needs of those markets are adapted for the production of biofuels, it may be necessary to move the first type of production (for food purposes) to other lands, which are not farmlands. This phenomenon is known as indirect land use change (ILUC) which, when related to the transformation of lands rich in coal, can lead to considerable greenhouse gas emissions [14]. In order to minimize the effects of indirect land use change, in the target set for 2020 it was proposed to reduce the share of the biofuels produced from oleaginous, sugar and cereal crops, as well as other high-starch crops to max. 7%. This modification was implemented with Directive 2015/1513. Concurrently, the directive commenced a period of promotion of the so-called advanced biofuels (advanced biofuels are defined as biofuels produced from the raw materials listed in Part A of Annex IX to Directive 2015/1513/EC or the raw materials listed in the national legislation), i.e., biofuels not competing with food producing crops (made of waste or algae, for instance) [19]. Such biofuels allow for the reduction of greenhouse gas emissions and, at the same time, are characterized by a low risk of indirect land use change [20].

The last of the mentioned directives, RED II, obliged the member states to ensure at least a 14% share of renewable energy in the end energy consumption in transport by 2030, provided that the share of advanced biofuels does not fall below 3.5%. Maintaining the limit imposed by Directive 2015/1513, RED II stipulates that the share of conventional biofuels in the target set for 2030 in the given member states cannot exceed their share in 2020 by more than 1%, and that the maximum share can be 7%.

2.2. National Legislature in the Scope of RES in Transport

The member states are obliged to transpose (implement) the said directives into their internal legal systems. In Poland, the provisions of RED and Directive 2015/1513 are transposed into the national legislation mostly under the relevant act relating to biocomponents and liquid biofuels, which is amended on a regular basis [21]. RED II is to be transposed in 30 June 2021.

For the purposes of further discussion, it is necessary to quote the statutory definitions of "liquid biofuels" and "biocomponents". Liquid biofuels are defined as "engine gasoline containing over 10% of biocomponents in volume; diesel oil containing over 7% of biocomponents in volume; bioethanol, ester, liquid bio-hydrocarbons—being independent fuels." The notion of biocomponents encompasses such substances as bioethanol, ester and liquid bio-hydrocarbons, which are "produced from biomass for the purpose of producing fuels, excluding other renewable fuels" [22].

To enable the accomplishment of the target set for 2020 in the scope of the share of renewable energy in transport, the so-called National Index Target (NIT) was established. It aimed to increase RES use in transport on a gradual, systematic basis. Under the act, NIT is defined as "the minimum share of other renewable fuels and biocomponents found in fuels used in all types of transport in the overall amount of liquid fuels and liquid biofuels consumed during a calendar year in road and railway transport, calculated according to the calorific value" [21]. For 2020, the NIT was set at 8.5% [22]. The NIT obligation is one of the mains instruments of realization of the target set by RED. In addition to this obligation, the target is being accomplished mostly by use of renewable electric energy in transport [23].

Implementing the provisions of Directive 2015/1513, the discussed act stipulates that the share of conventional biocomponents, i.e., those made of high-starch, sugar and oleaginous crops and crops cultivated for energy purposes on farmlands (as main crops), in the fuels used in transport cannot exceed 7.0% of the overall quantity of liquid fuels and biofuels used in road and railway transport in 2020. As mentioned in the previous subsection, the said directive led to a period of advanced biofuel promotion. In some countries, including Poland, such biofuels—as well as biofuels produced from used cooking oil or animal fats—are double-counted for realization of the targets set with regard to the use of renewable energy in transport [24]. Under the national legislation, the share of advanced biocomponents in fuels in 2020 cannot be less than 0.1% of the quantity of liquid fuels and biofuels used in road and railway transport [21].

3. Materials and Methods

3.1. Production and Use of Biocomponents in Poland

As mentioned above, the EU-imposed obligation regarding the minimum share of renewable energy in Polish transport in 2020 is satisfied mainly by way of the use of methyl esters and bioethanol [10]. Liquid bio-hydrocarbons are also used as biocomponents, but in considerably lower quantities.

Fatty acid methyl esters (FAMEs) are produced based on vegetable oils, e.g., rapeseed oil or used cooking oil [25]. Bioethanol is a dehydrated ethyl alcohol produced from biomass or waste, containing monosaccharides or polysaccharides [20]. Liquid bio-hydrocarbons are liquid hydrocarbons produced from biomass, including hydro-refined oils, e.g., fuels of the hydrogenated vegetable oil (HVO) type, [21], which are raw materials of great potential [26].

In Poland, the basic raw material used to produce methyl esters is rapeseed oil. In 2019, it was used in a quantity of 869 kt (1 t = 1000 kg) (90% of all raw materials). The production of bioethanol is dominated by corn, which was used in a quantity of 512 kt (48%) [27]. Both rapeseed oil and corn are foodstuff raw materials that are in competition for foodstuff and feedstuff markets; thus, the biocomponents produced from them are not eligible for double-counting in terms of NIT realization. However, Polish producers have been using more waste raw materials (eligible raw materials) since 2018 [28]. In 2019, the share of that sort of raw materials in the overall quantity of raw materials used to produce methyl esters was 11%, and it was 28% in the overall quantity of raw materials used to produce bioethanol [27]. In the production of liquid bio-hydrocarbons in Poland, the prevalent raw material used is ethyl alcohol of agricultural origin (58% of all raw materials), which makes the biocomponents produced from it eligible for double-counting for the purposes of NIT realization.

The overall annual performance of the installations used to produce methyl esters in Poland as of 25 September 2020 was 1.52 k(m^3), to produce bioethanol—0.9 k(m^3), and bio-hydrocarbons—0.25 k(m^3) [29]. Having recalculated the declared production capacities expressed in volumetric units per mass units (using the biocomponent density, calculated on the basis of data regarding calorific values specified in the Regulation of the Minister of Economy of 21 October 2014 on calorific value of specific biocomponents and liquid fuels (Journal of Laws of 2014, item 1517)) and having analyzed the data on the quantities of produced biocomponents (Table 2), it turns out that the production capacities of Polish producers are not utilized at 100%, which a phenomenon that has been observed in Poland for many years [30]. In the case of bioethanol producers in the period from 2010 to 2018, their production capacities were utilized at 28% on average; in the case of methyl ester producers it was 62% [31].

Table 2. Quantity of raw materials used and biocomponents produced using them in Poland in 2019 (kt).

Type	Raw Material	Quantity of Raw Material Used (kt)		Quantity of Biocomponent Produced (kt)	
		(1)	(2)	(1)	(2)
Methyl ester	Rapeseed oil	869.36		877.14	83.00
	other	10.49	85.02		
Bioethanol	corn	511.50		213.04	12.41
	Soybean waste suspension (3)		276.59		
	other	228.59	16.83		
Liquid bio-hydrocarbons	Ethyl alcohol of agricultural origin (4)		0.07	0.01	0.05
	other	0.01	0.04		

(1) Raw material/biocomponent not eligible for double-counting for national index target (NIT) realization; (2) raw material/biocomponent eligible for double-counting for NIT realization; (3) waste code 02 03 80; (4) waste code 02 06 99. Source: own study based on [27].

Table 3 illustrates the balance of methyl esters and bioethanol in Poland from 2010 to 2018. According to the data published by Statistics Poland, at the end of the analyzed period the production of methyl esters (888.9 kt) was more than twice as high compared to the beginning of the period (379.8 kt). The national consumption grew slightly by 33.5 kt. The volume of import of that biocomponent declined by 80.1 kt. The comparison of import and domestic production is notable. Initially, in 2010, more methyl esters were imported than were produced domestically. At the end of the analyzed period, imports accounted for 36% of their production. Within the analyzed period, exports of methyl esters increased significantly year-to-year and reached 797.3 kt in 2016, which is over sixty times higher compared to 2010. Production of bioethanol in 2010–2018 increased as well, but the increase was not as significant as in the case of methyl esters (i.e., by 33.5 kt). Imports of this biocomponent decreased almost twice (from 118.0 thousand t in 2010 down to 59.96 kt in 2018) and its consumption decreased as well, by 23.3 kt.

Table 3. Balance of methyl esters and bioethanol in Poland in 2010–2018 (kt).

	2010	2011	2012	2013	2014	2015	2016	2017	2018
				Methyl Esters					
Domestic production	379.80	363.89	605.39	642.37	739.01	788.24	900.51	899.31	888.90
Import	397.69	495.63	181.39	156.22	129.88	189.30	216.65	310.15	317.50
Export	13.06	37.44	48.51	157.98	235.04	420.17	797.27	753.32	393.19
Inventories change	−3.47	1.37	−8.37	19.92	−2.90	−11.93	−7.98	−0.62	−18.81
Domestic use	760.96	823.45	729.90	660.52	630.95	545.44	311.90	455.52	794.41
				Bioethanol					
Domestic production	152.80	136.58	172.53	190.92	142.79	173.93	195.37	190.63	186.32
Import	118.03	120.06	43.83	41.16	64.20	65.68	69.02	70.10	59.96
Export	1.45		1.9	2.73	2.19	2.33	2.95	10.15	1.74
Inventories change	−3.09	−4.83	2.48	−3.04	1.21	0.72	−1.44	−3.24	−1.28
Domestic use	266.29	251.81	216.94	226.31	206.01	238.00	260.00	247.33	243.26

Source: [32–34].

3.2. *Assessment of the Level of Fulfilment of EU Obligations and Challenges of the Near Future*

Under RED, the member states, including Poland, were obliged to achieve at least a 10% share of renewable energy in end energy consumption in transport in 2020. In 2018, Poland demonstrated a share of 5.6%, which was lower by 1 percentage point than in 2010 (Table 4).

Table 4. Share of renewable energy in Polish transport in 2010–2018 (%).

Year	**2010**	**2011**	**2012**	**2013**	**2014**	**2015**	**2016**	**2017**	**2018**
Share (%)	6.6	6.8	6.5	6.6	6.2	5.6	3.9	4.2	5.6

Source: own study based on [35].

To a great extent, the reduction in the share of renewable energy in Polish transport—apparent in 2015 and considerably deepening in 2016—resulted from a significant increase in fuel consumption, which was a consequence of activities aiming at limitation of the functioning of the so-called gray market. There resulted, inter alia, an increase in the official use of diesel oil, from 11.5 million t in 2015 to 17.2 million t in 2018 and an increase in its imports from 1 million t to 4.7 million t [36]. Along with the limitation of the gray market and the increase in the use of conventional fuels, the issue of ensuring proper quantities of biocomponents intensified. Their share in 2020, according to the set level of the national index target, was to be 8.5%.

According to the list kept by the President of the Energy Regulatory Office (status as of 4 October 2020), 16 entities in Poland are obliged to meet NIT [37]. Due to the fact that the sector of liquid fuel production and liquid fuel imports or intra-community purchase is characterized by high concentrations, an analogous situation occurs in the entities pursuing the national index target. As a result, PKN Orlen S.A. (Płock/Warsaw, Poland) and Grupa Lotos S.A. (Gdańsk, Poland) are the entities responsible for its realization, with a contribution of ca. 85% [36].

Although using fatty acid methyl esters and bioethanol are the most common method to meet the commitments resulting from RED in Poland, due to numerous limitations (including technical barriers, the so-called blending wall, which limit their use in liquid fuels down to standard E5/E10 and B7 [23]) and the current market situation, the NIT level that can be achieved with them does not exceed 6% [23]. Therefore, it is estimated that in order to reach the level of 8.5%, it is necessary to use ca. 700 kt of liquid bio-hydrocarbons in transport (ca. 500 kt produced from foodstuff raw materials and 170–190 kt produced from raw materials eligible for double-counting). In the case of bio-hydrocarbons, there is no blending wall issue, and they can be added to liquid fuels in excess of the standards effective for esters and bioethanol [23].

In addition to the most commonly used esters and bioethanol, the NIT can also include bio-hydrogen produced using biomethane, as well as products of co-hydrogenation [38]. This process consists in co-hydrogenation of petroleum derivative fractions with vegetable oils or animal fat, allowing the output of a product containing a high-quality biocomponent.

At present, biomethane utilization for fulfilment of the obligations regarding RES in transport is low (below 1% of all biocomponents). However, it is seen as a promising direction [23], taking into account the potential of the raw materials from which it can be produced and which are eligible for assignment to the advanced biofuel category [15]. It is estimated that the demand for biomethane for biohydrogen production may reach the level of ca. 700 million m^3 [23]. Co-hydrogenation is a process which consists in simultaneous hydro-refining of biomass and the fractions from petroleum refining and which results in the formation of liquid bio-hydrocarbons and bio-propane in a mixture with hydrocarbons from petroleum refining [38].

3.3. *Sector Value Studies*

In this study, we used a method of number scoring to determine the value of a sector [39]. This method is classified as a qualitative method [40]. It was based on the assumption that properly selected criteria (factors) determine the attractiveness of the assessed economic field [41,42]. The scoring method was used because it enables a multi-criteria, weighted comparative assessment of several sectors and is a method verified by the authors [17]. It makes it possible to assess the development/decline of sectors over the years and allows for analysis in the future. By assessing the value of a sector, we can determine its structural features. Selected criteria, properly adapted to the analyzed sectors, are evaluated on a 0–3 scale (depending on the intensity of their occurrence). Each factor is given the appropriate weight. The use of weighted assessments occurs due to the different importance of individual factors, as some of them may have a higher or lower influence on the assessment of the sector. Using an identical set of criteria, one can compare and contrast any number of sectors. In the studies considered, each sector was evaluated by means of the so-called grade matrix, containing 15 identical factors each from the group of market, financial, technical and organizational criteria (such as sector size and its expected growth rate, raw material availability, price level, competition strength, risk of substitution or favoring of European Union policy). The result of assessing the sectors is a ranked list, presenting the sectors from the most valuable to the least valuable. The assessment was carried out by sector experts, i.e., persons having extensive knowledge in this area of study.

The value of the sector of biocomponents, such as methyl esters and conventional bioethanol (produced from raw foodstuff materials) has been studied since 2007. Moreover, in 2013–2018 the scope of studies covered advanced bioethanol, produced from lignocellulose. As of 2018, evaluation has also been carried out on the sector of electric energy used in transport (electromobility) and since 2019—the sector of liquid bio-hydrocarbons.

Our studies omitted the sector of bioethanol (a subsector) based on ethyl alcohol of agricultural origin (two-phase production) produced in Polish agricultural distilleries. Since joining the EU, the significance of those sectors has been marginalized due to low price competition, the lack of economies of scale and environmental aspects (no CO_2 emission reduction). A series of factors led to a decrease in the number of agricultural distilleries from ca. 1000 operating after World War II to several dozen operating nowadays, with organizations representing them currently winding up operations, e.g., the Association of Polish Distilleries in 2020.

The table below (Table 5) presents the 2020 results, compared with 2010–2019 results.

At the beginning of the studied period, the sectors of conventional methyl esters and bioethanol were characterized by relatively high attractiveness. Producers could use different forms of investment support or excise duty reliefs—the market was soaking products up and growing. At the time, the value of those sectors was under the influence of domestic and EU demand (opportunities for exporting Polish biocomponents), favorable legal conditions in the EU and beneficial forecasts, as well as legal conditions and political support at the domestic level. Over time, however, the attractiveness of those sectors declined, which was a result of different economic, political, legal, technological or environmental factors. 2011 witnessed the disappearance of beneficial excise duty reliefs and it turned out that, according to the adopted CO_2 emission reduction methodology, the achieved value was not sufficient with regard to these fuels. In consequence, the need arose (entailing costs) to properly adapt the operating installations and technologies for the production of these biocomponents, so that the CO_2 emission reduction could be boosted. The current capacities, according to the studies, in terms of esters and bioethanol will be tolerated by ca. 2030; after 2035 they will either be transformed (direction seeking) or enter a clear decline phase.

Table 5. Obtained study results regarding the values of selected Polish sectors of biocomponents and electric energy used in transport in 2010–2020 (%).

Assessed Sector	Study Year										
	2010	2011	2012	2013	2014	2015	2016	2017	2018	2019	2020
Convention methyl esters	58.3	53.3	43.2	41.7	34.5	33.5	33.4	32.1	31.6	31.5	32.0
Conventional bioethanol	44.2	43.3	33.3	37.5	27.3	26.7	26.6	22.8	22.3	24.0	25.0
Lignocellulosic bioethanol				56.9	66.2	66.2	69.5	70.6	71.1		
Electrical energy									70.6	72.5	65.0
Liquid bio-hydrocarbons										58.3	61.2

Source: Own study. Results from prior years published in [17,43–46].

Lignocellulosic bioethanol was intended to come to the rescue of the market needs and requirements regarding CO_2 emission reductions—for a short time, the sector was seen as very promising, thanks to which it achieved the peak value. However, the situation was reversed. Studies on the production of lignocellulosic bioethanol usually ended at the pilot stage and the costs of its production were not accepted by the market. The production of lignocellulosic bioethanol is quite complicated due to the complex structure of raw materials. The best effects are achieved when wooden raw material is used with specific repeatable features, i.e., raw material from crops (crops mean increased energy expenditures and limited CO_2 emission reduction, as well as the need for available farmland). With relatively high CO_2 emission reductions, production costs were the main barrier. Currently, interest in this biocomponent is marginal.

Electromobility is a new sector with has given rise to hopes and an unquenched demand in Europe, including Poland [47]. Electric vehicles are being refined so as to bring their features (e.g., reach or driving time as well as mechanical strength) closer to conventional vehicles [48]. The issue of charging battery-powered vehicles is significant as well [49]. Electric vehicles are often referred to as zero-emission vehicles, although this is sometimes challenged. Taking into account the source of electric energy, often with high emissions (a network of "carbon" energy), the actual CO_2 emission reduction for the electric car may not be so high. According to political announcements, the sector will be developed intensively and in 2030 electromobility is to account for a considerable portion of renewable energy in transport. Electromobility makes sense and has a positive effect on the environment when it is based on electric energy from RES [50].

Therefore, it is important that energy production does not contribute to the increase in pollutant emissions [51,52]. The problems of climate change and air pollution necessitate the search for new, ecological, widely available and efficient methods of energy production. Leading the way in this respect is renewable energy using wind, sun, water and biomass resources. Therefore, alternative raw materials and energy materials to coal should be sought. One of these solutions may be the use of renewable biomass from bamboo instead of the previously used wood biomass [53,54]. Another solution may be technologies for converting waste into energy, from the use of digestate from a biogas plant cooperating with a sewage treatment plant [55].

The sector of liquid hydrocarbons is a new sector, studied since 2019. Hydrocarbons, the substrates of which are organic waste products, demonstrate high CO_2 emission reductions and can significantly contribute to meeting the targets set by RED and RED II. The sector is characterized by value growth, and the biocomponent in question is gaining ground, but is nonetheless of little significance as yet. The sector of liquid bio-hydrocarbons enjoys legal support, a market pull-selling phenomenon and very good forecasts.

4. Results

Poland has a long tradition in the use of alternative fuels. After World War I it used industrial ethanol for mixing with gasoline, in the 1970s wood waste was utilized to form lignocellulosic ethanol, and in the 1990s, technologies for blending bioethanol with gasoline were developed.

As regards the target of achieving at least 10% renewable energy in end energy consumption in transport in 2020 (which is to grow to 14% in 2030), as well as to limit the share of conventional biofuels (max. 7%), Poland is facing quite a challenge.

There are high penalties for failing to achieve the required indicators. Moreover, the methods of calculating indicators have been modified to their detriment (the support for advanced technologies was reduced). The interest of the EU in promoting bioethanol and esters will actually end after 2030, when all support and promotion measures for this sector will be abolished.

According to the position of the Ministry of Energy (ME) from 2016, the achievement of 8.5% NIT was to allow realization of the EU commitment and the electric energy used in transport was to supplement it [56]. In reality, however, the entities pursuing NIT have managed to achieve only ca. 5.5–6% [57] and electromobility, due to insufficiently widespread infrastructure, is still a thing of the future [58].

Currently, Poland does not have any RES sector for transport with major market significance, high production capacities and high CO_2 emission reduction potential. Admittedly, sectors of conventional biocomponents offer high production capacities, but their potential is not utilized to the fullest due to technical limitations. The value and competitiveness of these sectors is steadily on the fall, considering the EU policy that is unfavorable to raw materials used in their production, as well as low CO_2 emission reductions. The sector of lignocellulosic bioethanol, which initially gave rise to high hopes, does not function in the market reality.

5. Conclusions

After Poland's accession to the EU, it turned out that the bioethanol industry did not play its intended role, and investments in bioethanol were initiated (single-phase plants). Despite playing a significant role in the initial phase (Poland initially met relatively low requirements, including Directive 2003/30/EC), the departure of the EU from the biofuel support policy in 2011 significantly disturbed the functioning of this sector. The EU partially resigned from the use of bioethanol. The promotion of biofuels, esters and bioethanol is to be completed by 2030.

Chances of meeting the targets set by the European Union are currently being sought mostly in products resulting from the process of co-hydrogenation and in biohydrogen produced using biomethane. Liquid bio-hydrocarbons formed as a result of co-hydrogenation have no limitations as to their mixing with conventional fuels (unlike methyl esters and bioethanol) and domestic fuel producers possess potential that could be employed in terms of launching HVO production using local installations intended to refine petroleum [36]. The liquid bio-hydrocarbon sector is one with great perspectives for the future, and is the only one with a rising value. However, its production capacity is currently at the initial stage of development. It is thus similar to biomethane, which is currently used in marginal quantities. However, taking into account the potential of the raw materials from which it can be produced and which are eligible for assignment to the advanced biofuel category, it gives rise to high hopes.

The results of this research may be used primarily by politicians and producers to ensure the achievement of the indicators set by the EU. Therefore, it is important to monitor the market on an ongoing basis and draw conclusions about the effect of legal and political conditions on the achieved indicators. This allows people to verify current legislation and modify or implement new solutions, which will allow them to achieve the set goals and avoid paying penalties. For scientists, this is an indication of what methods have been used to calculate the value of the sector and may be an impulse for possible modifications of the methods used.

6. Limitations

Limitations, in terms of development of the biocomponent sector, result mostly from legal and environmental (ecological) regulations. The emergence of new legal conditions 2009/28/EC and a maximum of 7% in the fuel structure for bioethanol and esters are the main causes of problems in the Polish fuel industry. The legal and political landscape decides whether or not to support individual biofuel sectors. CO_2 emission reduction on the basis of biocomponents is low in Poland; hence, there are problems with meeting the requirements stemming from NIT. However, the electromobility sector will continue to develop, using also the mechanical engineering knowledge to construct innovative engines with new mechanical and technical parameters, and all those activities will have an effect on CO_2 emission reductions if electric energy is generated with raw materials produced from renewable (zero-emission) sources, rather than with fossil fuels (e.g., coal).

Author Contributions: Conceptualization, A.K., J.M.-S., and P.F.B.; Methodology, A.K. and J.M.-S.; Software, E.G.; Validation, P.F.B., A.K. and E.G.; Formal Analysis, A.K. and P.F.B.; Investigation, A.K., P.F.B., E.G. and J.M.-S.; Resources, P.F.B., E.G. and J.M.-S.; Data Curation, A.K.; Writing–Original Draft Preparation, J.M.-S. and P.F.B.; Writing–Review & Editing, P.F.B.; E.G.; Visualization, P.F.B.; Supervision, A.K. and P.F.B.; Project Administration, A.K. and P.F.B. All authors have read and agreed to the published version of the manuscript.

References

1. Masjuki, H.M.; Kalam, M.A. An Overview of Biofuel as a Renewable Energy Source: Development and Challenges. *Procedia Eng.* **2013**, *56*, 53. [CrossRef]
2. Pietrzak, K.; Pietrzak, O. Environmental Effects of Electromobility in a Sustainable Urban Public Transport. *Sustainability* **2020**, *12*, 1052. [CrossRef]
3. Susmozas, A.; Martín-Sampedro, R.; Ibarra, D.; Eugenio, M.E.; Iglesias, R.; Manzanares, P.; Moreno, A.D. Process Strategies for the Transition of 1G to Advanced Bioethanol Production. *Processes* **2020**, *8*, 1310. [CrossRef]
4. European Environment Agency (EEA). Transport. Available online: www.eea.europa.eu/pl/themes/transport/intro (accessed on 3 October 2020).
5. Klimat dla Polski Polska dla Klimatu 1988–2018–2050, R. Jeszke, Ed. (Climate for Poland Poland for the Climate 1988–2018–2050). Available online: https://cop24.gov.pl/fileadmin/user_upload/files/1._Klimat-dla-Polski-Polska-dla-Klimatu_PL.pdf (accessed on 27 November 2020).
6. Giunta, M.; Lo Bosco, D.; Leonardi, G.; Scopelliti, F. Estimation of Gas and Dust Emissions in Construction Sites of a Motorway Project. *Sustainability* **2019**, *11*, 7218. [CrossRef]
7. Praticò, F.G.; Giunta, M.; Mistretta, M.; Gulotta, T.M. Energy and Environmental Life Cycle Assessment of Sustainable Pavement Materials and Technologies for Urban Roads. *Sustainability* **2020**, *12*, 704. [CrossRef]
8. Ciucci, M. Energia ze źródeł Odnawialnych. (Energy from renewable sources). Available online: www.europarl.europa.eu/factsheets/pl/sheet/70/energia-ze-zrodel-odnawialnych (accessed on 3 October 2020).
9. Porter, M.E. Competitive strategy. *Meas. Bus. Excell.* **1997**, *1*, 12–17. [CrossRef]
10. Gradziuk, P.; Jendrzejewski, B. Wyzwania dla sektora biopaliw w kontekście polityki klimatyczno-energetycznej Unii Europejskiej. (Challenges for the biofuel sector in the context of the European Union's climate and energy policy). *Rocz. Nauk. Stowarzyszenia Ekon. Rol. Agrobiz.* **2007**, *2*, 58–65.
11. Kupczyk, A.; Borowski, P.; Powałka, M.; Ruciński, D. *Biopaliwa Transportowe w Polsce. Stan Aktualny i Perspektywy*; Transport Biofuels in Poland. Current Status And Prospects; Oficyna Wydawniczo-Poligraficzna ADAM: Warsaw Poland, 2011.
12. Borowski, P.; Gawron, J.; Golisz, E.; Kupczyk, A.; Piechocki, J.; Powałka, M.; Tucki, K. *Wpływ Redukcji Emisji CO_2 na Funkcjonowanie Sektorów Biopaliw Transportowych w Polsce. Wyd*; Impact of CO_2 Emission Reduction on the Operation of the Transport Biofuels Sector in Poland; WEMA: Warsaw, Poland, 2014.
13. EC. *Directive 2009/28/EC of the European Parliament and of the Council of 23 April 2009 on the Promotion of the Use of Energy from Renewable Sources and Amending and Subsequently Repealing Directives 2001/77/EC and 2003/30/EC*; European Commission: Brussels, Belgium, 2009.

14. EC. *Directive (EU) 2015/1513 of the European Parliament and of the Council of 9 September 2015 Amending Directive 98/70/EC Relating to the Quality of Petrol and Diesel Fuels and Amending Directive 2009/28/EC on the Promotion of the Use of Energy from Renewable Sources*; European Commission: Brussels, Belgium, 2015.
15. EC. *Directive (EU) 2018/2001 of the European Parliament and of the Council of 11 December 2018 on the Promotion of the Use of Energy from Renewable Sources*; European Commission: Brussels, Belgium, 2018.
16. Subramaniam, Y.; Masron, T.A.; Azman, N.H.N. Biofuels, environmental sustainability, and food security: A review of 51 countries. *Energy Res. Soc. Sci.* **2020**, *68*, 101549. [CrossRef]
17. Mączyńska, J.; Krzywonos, M.; Kupczyk, A.; Tucki, K.; Sikora, M.; Pińkowska, H.; Bączyk, A.; Wielewska, I. Production and use of biofuels for transport in Poland and Brazil—The case of bioethanol. *Fuel* **2019**, *241*, 989–996. [CrossRef]
18. Koizumi, T. Biofuels and food security. *Renew. Sustain. Energy Rev.* **2015**, *52*, 829–841. [CrossRef]
19. Kim, E.J.; Kim, S.; Choi, H.-G.; Han, S.J. Co-production of biodiesel and bioethanol using psychrophilic microalga Chlamydomonas sp. KNM0029C isolated from Arctic sea ice. *Biotechnol. Biofuels* **2020**, *13*, 1–13. [CrossRef]
20. Zalega, K. Znaczenie biopaliw i systemu ich certyfikacji—Polityka energetyczna i klimatyczna Unii Europejskiej (Importance of biofuels and their certification system - European Union energy and climate policy). *Kontrola Państwowa* **2017**, *62*, 40–60.
21. Ustawa z Dnia 25 Sierpnia 2006 r. o Biokomponentach i Biopaliwach Ciekłych (Dz.U. 2006 nr 169 poz. 1199, z późn. zm). (The Act of August 25, 2006 on Biocomponents and Liquid Biofuels). Available online: http://isap.sejm.gov.pl/isap.nsf/DocDetails.xsp?id=wdu20061691199 (accessed on 3 October 2020).
22. Energy Regulatory Office. Wysokość Narodowych Celów Wskaźnikowych Ustalona na Lata 2017–2020 (The amount of the National Indicative Targets Established for the Years 2017–2020). Available online: https://www.ure.gov.pl/pl/urzad/informacje-ogolne/aktualnosci/7017,Wysokosc-Narodowych-Celow-Wskaznikowych-ustalona-na-lata-2017-2020.html (accessed on 3 October 2020).
23. Uzasadnienie do Projektu Ustawy o Zmianie Ustawy o Biokomponentach i Biopaliwach Ciekłych Oraz Uzasadnienie do Projektu Niektórych Innych Ustaw z 2019 r. (DKPL.WK.10.2.77.2019.AMU(8)) (Justification to the Draft Act Amending the Act on Biocomponents and Liquid Biofuels and Justification to the Draft Certain Other Acts of 2019). Available online: http://orka.sejm.gov.pl/Druki8ka.nsf/Projekty/8-020-1392-2019/$file/8-020-1392-2019.pdf (accessed on 3 October 2020).
24. Giuntoli, J. Advanced Biofuel Policies in Select UE Member States: 2018 Update. International council on Clean Transportation. Available online: https://theicct.org/sites/default/files/publications/Advanced_biofuel_policy_eu_update_20181130.pdf (accessed on 6 November 2019).
25. Energia z Zasobów Odnawialnych. Gdańska Infrastruktura Wodociągowo-Kanalizacyjna Sp. z o.o. (Energy from Renewable Sources). Available online: http://www.giwk.pl/files/21/219/245/podrecznik.pdf (accessed on 4 October 2020).
26. Rimkus, A.; Stravinskas, S.; Matijošius, J. Comparative Study on the Energetic and Ecologic Parameters of Dual Fuels (Diesel–NG and HVO–Biogas) and Conventional Diesel Fuel in a CI Engine. *Appl. Sci.* **2020**, *10*, 359. [CrossRef]
27. Krajowy Ośrodek Wsparcia Rolnictwa. Dane Dotyczące Ilości Biokomponentów Wytworzonych, a Następnie Rozporządzonych Przez Dokonanie Jakiejkolwiek Czynności Prawnej lub Faktycznej, Lub Przeznaczonych do Wytworzenia Paliw Ciekłych lub Biopaliw Ciekłych Przez Wytwórców, z Uwzględnieniem Podstawowych Surowców Użytych do ich Wytworzenia (I-IV Kwartał 2019). National Agricultural Support Center. Data on the amount of Biocomponents Produced and then Disposed of by any Legal or Actual Act, or Intended for the Production of Liquid Fuels or Liquid Biofuels by Producers, including the Basic Raw Materials Used for Their Production (Q1–Q4 2019). 2020. Available online: https://bip.kowr.gov.pl/informacje-publiczne/odnawialne-zrodla-energii/informacje-dotyczace-rynku-biokomponentow (accessed on 4 October 2020).
28. KOWR. Dane Dotyczące Wytwarzania Biokomponentów za Lata 2015–2020. National Agricultural Support Center. Data on the Production of Biocomponents for 2015–2020. 2020. Available online: https://bip.kowr.gov.pl/informacje-publiczne/odnawialne-zrodla-energii/informacje-dotyczace-rynku-biokomponentow (accessed on 4 October 2020).
29. Krajowy Ośrodek Wsparcia Rolnictwa. Rejestr wytwórców (Stan na 25.09.2020 r.). National Agricultural Support Center. Data on the Production of Biocomponents for 2015–2020. 2020. Available online: https://www.

kowr.gov.pl/uploads/pliki/DI/Biokomponenty_i_biopaliwa/W/Rejestr_wytworcow_2020.09.25.pdf (accessed on 4 October 2020).

30. Mączyńska, J.; Kupczyk, A.; Rutkowski, D.; Tucki, K. Aktualne problemy sektorów biokomponentów i biopaliw ciekłych w Polsce (Current problems of biocomponents and liquid biofuels sectors in Poland). *Gospod. Mater. Logistyka* **2019**, *2*, 23–27. [CrossRef]
31. Golisz, E.; Kupczyk, A.; Mączyńska, J. Biopaliwa ciekłe w świetle dyrektywy 2018/2001 w sprawie promowania stosowania energii ze źródeł odnawialnych (Liquid biofuels in the light of Directive (EU) 2018/2001 on the promotion of the use of energy from renewable sources). *Gospod. Mater. Logistyka* **2019**, *6*, 20–25.
32. GUS. *Energia ze Źródeł Odnawialnych W 2014 Roku*; Energy from Renewable Sources in 2014; GUS: Warsaw, Poland, 2015.
33. GUS. *Energia ze Źródeł Odnawialnych W 2016 Roku*; Energy from Renewable Sources in 2016; GUS: Warsaw, Poland, 2017.
34. GUS. *Energia ze Źródeł Odnawialnych W 2018 Roku*; Energy from Renewable Sources in 2018; GUS: Warsaw, Poland, 2019.
35. Eurostat. Share of Energy from Renewable Sources in Transport, 2004—2018. Available online: https://ec.europa.eu/eurostat/statistics-explained/index.php?title=File:Share_of_energy_from_renewable_sources_in_transport,_2004-2018_(%25_of_gross_final_energy_consumption).png (accessed on 3 October 2020).
36. Polska Izba Paliw Płynnych. Nowelizacja Ustawy o Biopaliwach (Polish Chamber of Liquid Fuels Amendment to the act on biofuels). Available online: http://paliwa.pl/strona-startowa/archiwum/nowelizacja-ustawy-o-biopaliwach (accessed on 4 October 2020).
37. Energy Regulatory Office Wykaz Podmiotów, które są Zobowiązane do Realizacji Narodowego Celu Wskaźnikowego (List of Entities that are Required to achieve the National Indicative Target). Available online: https://rejestry.ure.gov.pl/o/17 (accessed on 4 October 2020).
38. CMS. Nowelizacja Ustawy o Biokomponentach i Biopaliwach—Sposób na Kryzys Paliwowy Czy też Szansa na Nowe Inwestycje? Amendment to the act on Biocomponents and Biofuels—A Way to Deal with the Fuel Crisis or a Chance for New Investments? Available online: https://www.cire.pl/item,184194,14,0,0,0,0,0,nowelizacja-ustawy-o-biokomponentach-i-biopaliwach---sposob-na-kryzys-paliwowy-czy-tez-szansa-na-nowe-inwestycje.html (accessed on 4 October 2020).
39. Dawid, H.; Reimann, M. Evaluating Market Attractiveness: Individual Incentives Versus Industry Profitability. *Comput. Econ.* **2005**, *24*, 321–355. [CrossRef]
40. Borowski, P.F. New Technologies and Innovative Solutions in the Development Strategies of Energy Enterprises. *High Tech Innov. J.* **2020**, *1*, 39–58. [CrossRef]
41. Dzwigol, H.; Aleinikova, O.; Umanska, Y.; Shmygol, N.; Pushak, Y. An entrepreneurship model for assessing the investment attractiveness of regions. *J. Entrep. Educ.* **2019**, *22*, 1–7.
42. Zykiene, I.; Snieška, V.; Bruneckiene, J.; Burksaitiene, D. Assessment of regions using an index for a location's attractiveness for business development. *Econ. Res. Ekon. Istraz.* **2020**, 1–19. [CrossRef]
43. Golisz, E.; Kupczyk, A. Stan aktualny i perspektywy rozwoju sektora biopaliw transportowych w Polsce (Current state and development prospects for the transport biofuels sector in Poland). *Gospod. Mater. Logistyka* **2016**, *3*, 32–40.
44. Golisz, E.; Samson-Bręk, I.; Borowski, P.; Kupczyk, A. Konieczne zmiany w sektorze biopaliw transportowych w Polsce (Necessary Changes in the Market of Biofuels Transport in Poland). *Gospod. Mater. Logistyka* **2013**, *11*, 2–12.
45. Sikora, M. Aktualny stan i atrakcyjność sektorów biopaliw w Polsce. Cz. 2 (Current state and attractiveness of the biofuel sectors in Poland. Part 2). *Przemysł Ferment. I Owocowo Warzywny* **2016**, *1*, 27–29. [CrossRef]
46. Kupczyk, A.; Mączyńska, J.; Sikora, M.; Gawron, J. Identyfikacja obecnego stanu oraz atrakcyjność sektorów biopaliw transportowych w Polsce (Identification of the current state and the attractiveness of the transport biofuel sectors in Poland). *Rocz. Nauk. Stowarzyszenia Ekon. Rol. Agrobiz.* **2017**, *19*, 139–144.
47. Borowski, P.F. Zonal and Nodal Models of Energy Market in European Union. *Energies* **2020**, *13*, 4182. [CrossRef]
48. Gallo, M.; Marinelli, M. Sustainable Mobility: A Review of Possible Actions and Policies. *Sustainability* **2020**, *12*, 7499. [CrossRef]
49. Sierpiński, G.; Staniek, M.; Kłos, M.J. Decision Making Support for Local Authorities Choosing the Method for Siting of In-City EV Charging Stations. *Energies* **2020**, *13*, 4682. [CrossRef]

50. Kupczyk, A.; Mączyńska, J.; Redlarski, G.; Tucki, K.; Bączyk, A.; Rutkowski, D. Selected Aspects of Biofuels Market and the Electromobility Development in Poland: Current Trends and Forecasting Changes. *Appl. Sci.* **2019**, *9*, 254. [CrossRef]
51. Razmjoo, A.; Kaigutha, L.G.; Rad, M.V.; Marzband, M.; Davarpanah, A.; Denai, M. A Technical analysis investigating energy sustainability utilizing reliable renewable energy sources to reduce CO_2 emissions in a high potential area. *Renew. Energy* **2020**, *164*, 46–57. [CrossRef]
52. Tsilingiridis, G.; Sidiropoulos, C.; Pentaliotis, A. Reduction of air pollutant emissions using renewable energy sources for power generation in Cyprus. *Renew. Energy* **2011**, *36*, 3292–3296. [CrossRef]
53. Lorenzi, G.; Baptista, P. Promotion of renewable energy sources in the Portuguese transport sector: A scenario analysis. *J. Clean. Prod.* **2018**, *186*, 918–932. [CrossRef]
54. Borowski, P.F. Bamboo as an innovative material for many branches of world industry, Annals of Warsaw University of Life Sciences-SGGW. *For. Wood Technol.* **2019**, *107*, 13–18. [CrossRef]
55. Borowski, P.F. Nexus between water, energy, food and climate change as challenges facing the modern global, European and Polish economy. *AIMS Geosci.* **2020**, *6*, 397–421. [CrossRef]
56. Rządowy Projekt Ustawy o Zmianie Ustawy o Biokomponentach i Biopaliwach Ciekłych Oraz Niektórych Innych Ustaw (A Government Bill to Amend the Act on Biocomponents and Liquid Biofuels and Certain Other Acts). Available online: https://www.sejm.gov.pl/Sejm8.nsf/PrzebiegProc.xsp?nr=3626 (accessed on 4 October 2020).
57. WysokieNapiecie.pl. Zmiany w Biopaliwach Nie Wprowadzą Rewolucji na Stacjach (Changes in biofuels will not revolutionize the petrol stations). Available online: https://wysokienapiecie.pl/21296-zmiany-w-biopaliwach-nie-wprowadza-rewolucji-na-stacjach/ (accessed on 4 October 2020).
58. Sawicki, B. Etanol i współuwodornienie są szansą na zwiększenie udziału biopaliw w Polsce (Ethanol and co-hydrogenation are an opportunity to increase the share of biofuels in Poland). Available online: https://biznesalert.pl/bioetanol-wodor-biopaliwa-polska-zwiekszenie-energetyka/ (accessed on 4 October 2020).

3

Advanced Turbine Cycles with Organic Media

Marian Piwowarski * and Krzysztof Kosowski *

Faculty of Mechanical Engineering, Gdansk University of Technology, Gabriela Narutowicza Street 11/12, 80-233 Gdansk, Poland

* Correspondence: marian.piwowarski@pg.edu.pl (M.P.); kosowski@pg.gda.pl (K.K.);

Abstract: Organic Rankine Cycle (ORC) power plants have become very popular and have found their applications in systems with renewable sources of energy. So far their overall efficiencies are not very impressive and only for the upper temperature of about 300 °C do they exceed 20%. A drawback of these cycles is the limitation of the cycle upper temperature due to the heat exchanger technology and the materials used. However, it is possible to overcome these difficulties by certain modifications of the thermodynamic cycles, a proper choice of the working medium and the optimization of cycle parameters. In the paper the problems of choosing the working medium and the question of higher temperature at the turbine inlet have been discussed. Different modifications of the schemas of the thermodynamic cycles have also been taken into account. The variants of power plants with regenerators, reheaters and heat exchangers have been considered. The proposed increase in temperature (in some cases up to 600 °C or higher) and innovative modifications of the thermodynamic cycles allow to obtain the power plant efficiency of above 50%. The modified cycles have been described in detail in the paper. The proposed cycles equipped with regenerators and reheaters can have the efficiency even slightly higher than classical steam turbine plants with a reheater and regenerators. Appropriate cycle and turbine calculations have been performed for the micro power plants of turbine output in the range of 10 kW–300 kW (up to several MW in some cases). The best arrangements achieved very high values of the overall cycle efficiency.

Keywords: thermodynamic cycles; ORC; efficiency

---

1. Introduction

At present, due to coal-fired plants, the power systems in many countries, are characterized by a relatively low efficiency [1,2]. In the case of distributed energy systems, however, the efficiency of electricity production is even lower [3–5], and the Organic Rankine Cycle power plants achieve the efficiency which exceeds 10% [6–8] (seldom about 20%). Higher values of efficiency are obtained by ultra-supercritical steam turbine power plants [9,10], combined gas-steam thermal cycles [11] or modified ORC cycles [12]. In distributed energy, on the other hand, the technology of microturbines is developed and, although the efficiency is still relatively low, the research is carried out intensively [13–16]. For many years the installations of electric output of several hundred kW and higher have been used in power plants based on biomass combustion [17,18]. Biofuels have become very popular and have been treated as renewable sources of energy. They can significantly vary in quality, heating value and other parameters which can cause some serious problems when used as fuel for stroke engines and gas turbines. In the case of gas turbines it is possible to overcome these difficulties by some modifications of the gas turbine engine. Instead of a combustion chamber a high temperature heat exchanger and an external furnace must be applied. In the external chamber it is possible to combust any fuel and the hot exhaust gases are used in the heat exchanger to warm up the air from the compressor before it enters the turbine. Thus, a gas turbine engine is modified to work as an air turbine which consists, in the simplest arrangement, of a compressor, a heat exchanger and a turbine. Air is a working medium for both the compressor as well as the turbine [19,20]. This solution allows us to burn fuels of varying parameters in the external combustion chamber without any harmful effects on the operation of the air turbine set.

In the case of vapor turbine power plants low-boiling media (as a rule organic ones) can be applied as working fluids apart from the traditional medium, i.e., steam. Thus, instead of steam Rankine cycle we have Organic Rankine Cycle. In ORC power plants the most important property of the organic media is their lower boiling temperature which make them more appropriate for use in the range of lower temperatures. Owing to that, the ORC plants are very often used with sources of heat of rather low temperatures (heat emitted from engines, gas turbines and industrial processes or agricultural and domestic waste). It is a very popular opinion that above 350 °C steam power plants have a higher efficiency than power plants with organic media [21]. At lower temperatures thermodynamic cycles with organic fluids can achieve a better efficiency than steam plants [22]. Our analysis shows that this statement is not entirely correct and may refer to simple cycles and relatively low temperatures which characterize current practical applications. If we consider more complex cycles and higher temperatures we may draw different conclusions.

The schema of the simplest typical ORC cycle is presented in Figure 1. It consists of a turbine, an electric generator, a condenser (low temperature heat exchanger), a pump and a vapor generator (high temperature heat exchanger). If the expansion in the turbine ends in the zone of superheated vapor, an additional heat exchanger (so called "regenerator") can be applied in order to use the heat of superheating for warming up the working liquid between the condenser and the boiler (Figure 2).

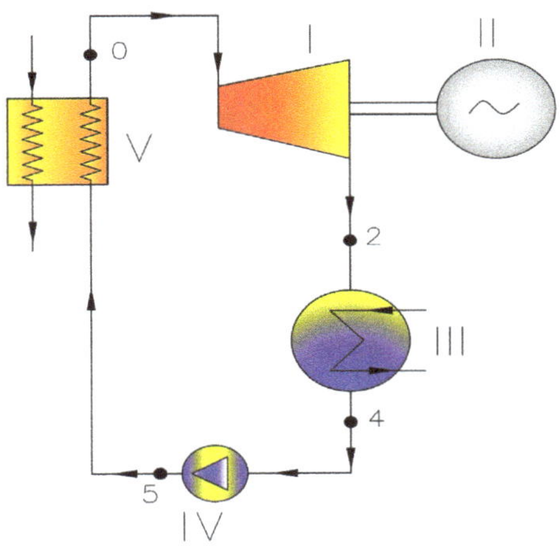

Figure 1. Standard Organic Rankine Cycle (ORC) cycle for wet fluids; where: I—turbine, II—generator, III—condenser, IV—main pump, V—vapor generator.

The analysis includes some other, more complex, schemas:

- Cycle with saturated live steam and a regenerator for warming up the working fluid (Variant 1—Figure 2);
- Cycle with superheated live steam and a regenerator for warming up the working fluid (Variant 2—Figure 2);
- Cycle with superheated live steam and a regenerator for warming up, vaporization and partial superheating of the working fluid (Variant 3—Figure 2);
- Cycle with supercritical live steam and a regenerator for warming up, vaporization and partial superheating of the working fluid (Variant 4—Figure 2);
- Cycle with superheated live steam, a regenerator for warming up, vaporization and partial superheating of the working fluid, and with a compressor (Variant 5—Figure 3);

- Cycle with superheated live steam, a regenerator for warming up, vaporization and partial superheating of the working fluid, with a compressor and an additional heat exchanger (Variant 6—Figure 4);
- Cycle with superheated live steam, a regenerator for warming up, vaporization and partial superheating of the working fluid, with a compressor, an additional heat exchanger and interstage superheating (Variant 7—Figure 5);
- Cycle with superheated live steam, a regenerator for warming up, vaporization and partial superheating of the working fluid, a compressor, an additional heat exchanger, interstage superheating and a vapor heater (Variant 8—Figure 6).

The interpretation of the above cycles in the temperature-entropy diagram is shown in Figure 7 with hexamethyldisiloxane (MM) as an example.

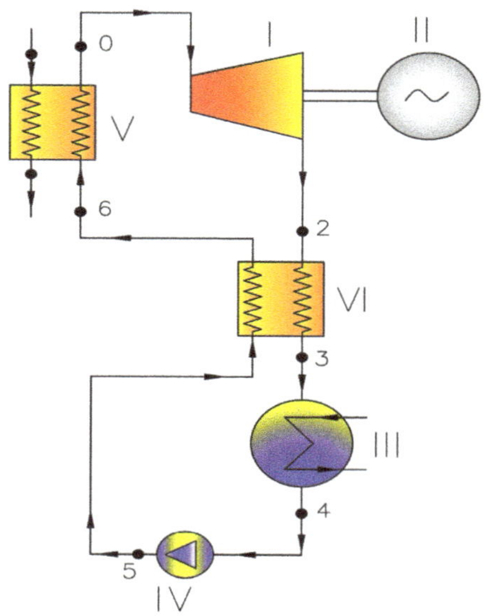

Figure 2. ORC cycle with regenerator; where: I—turbine, II—generator, III—condenser, IV—pump, V—vapor generator, VI—regenerator (Variant 1, 2, 3, and 4).

When choosing the working media, the most important features were taken into account, such as low toxicity, chemical stability, low potential for decomposition and low flammability [7]. The following media have been included in the analysis:

- Acetone (chemical formula—$C_{11}H_{24}$, commercial name—acetone, [23]);
- Octamethylcyclotetrasiloxane (chemical formula—$C_8H_{24}O_4Si_4$, commercial name—D4, [24]);
- Dodecamethylcyclohexasiloxane (chemical formula—$C_{12}H_{36}O_6Si_6$, commercial name—D6, [24]);
- Octamethyltrisiloxane (chemical formula—$C_8H_{24}O_2Si_3$, commercial name—MDM, [24]);
- Hexamethyldisiloxane (chemical formula—$C_6H_{18}OSi_2$, commercial name—MM, [24]);
- 1,1,1,3,3-pentafluoropropane (chemical formula—$C_3H_3F_5$, commercial name—R245fa, [24]);
- 1,1,1,3,3-pentafluorobutane (chemical formula—$C_4H_5F_5$, commercial name—R365mfc, [24]);
- Methylcyclohexane (chemical formula—C_7H_{14}, commercial name—c1cc6, [24]);
- N-propylcyclohexane (chemical formula—C_9H_{18}, commercial name—c3cc6, [24]);
- Undecane (chemical formula—$C_{11}H_{24}$, commercial name—C11, [24]);
- 1-chloro-3,3,3-trifluoroprop-1-ene (chemical formula—$C_3H_2ClF_3$, commercial name—R1233zd, [25]).

The calculations have been performed assuming the schema of the thermodynamic cycle, the working medium and its upper and lower parameters. Standard thermodynamics formulae have been used, occasionally applying the iteration method. The parameters of the working media have been determined using REFPROP [26] media library. Unfortunately, there is no sufficient information concerning the chemical stability of most of the organic media. Thus, the maximum temperature for particular fluids have been assumed according to the data provided by REFPROP media library. The thermodynamic calculations have been carried out for 11 various fluids and 8 different cycles. The values of initial pressure and temperature have been determined taking into account particular schema of the cycle, media stability limitation and the maximum value of the cycle efficiency. The pressure in the condenser has been estimated assuming the temperature of the cooling water equal to 15 °C. The efficiencies of particular elements (typical of micro power plants) are shown in Table 1 [3,4,6,7,27].

Table 1. Assumed values of the efficiencies of particular cycle elements.

Description	Symbol	Value	Unit
Turbine efficiency	η_T	0.85	[-]
Compressor efficiency	η_{com}	0.85	[-]
Pump efficiency	η_{PG}	0.80	[-]
Mechanical efficiency	η_m	0.98	[-]
Leakage efficiency	η_n	0.98	[-]
Generator efficiency	η_G	0.90	[-]
Regenerator efficiency	η_R	0.95	[-]
Pressure drop in heaters and regenerators	p_i/p_{i-1}	0.98	[-]

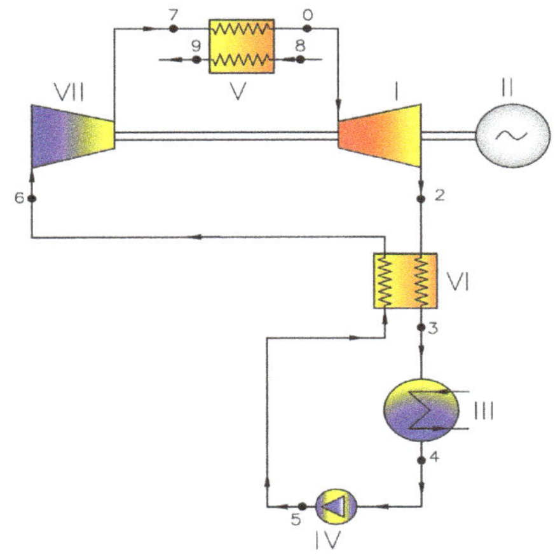

Figure 3. ORC cycle with regenerator and compressor; where: I—turbine, II—generator, III—condenser, IV—pump, V—vapor generator, VI—regenerator, VII—compressor (Variant 5).

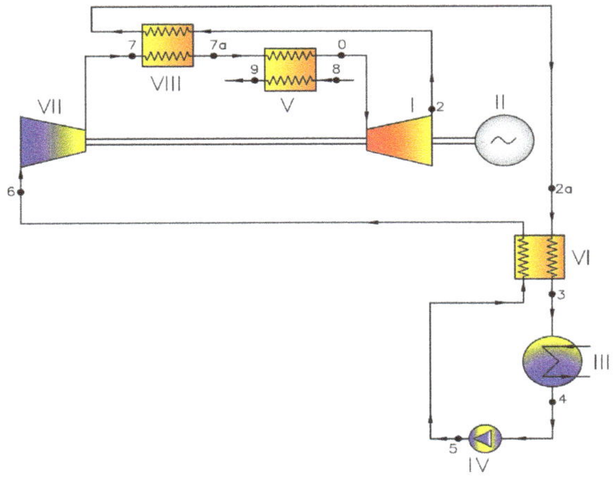

Figure 4. ORC cycle with regenerator, compressor and reheater; where: I—turbine, II—generator, III—condenser, IV—pump, V—vapor generator, VI—regenerator, VII—compressor, VIII—heat exchanger (Variant 6).

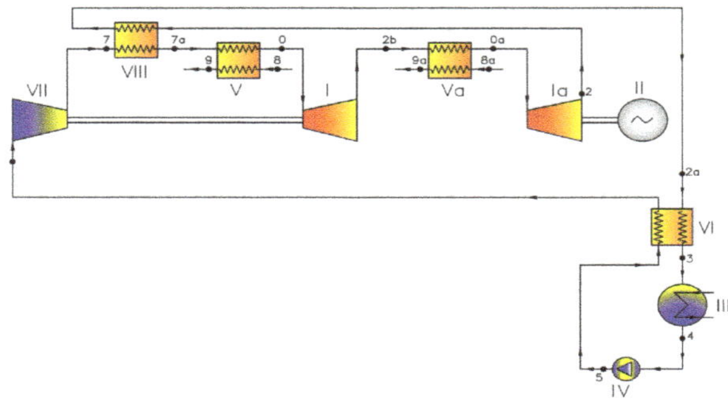

Figure 5. ORC cycle with regenerator, compressor, reheater and interstage superheater; where: Ia, Ib—turbines, II—generator, III—condenser, IV-pump, V—vapor generators, Va—interstage superheater, VI—regenerator, VII—compressor, VIII—heat exchanger (Variant 7).

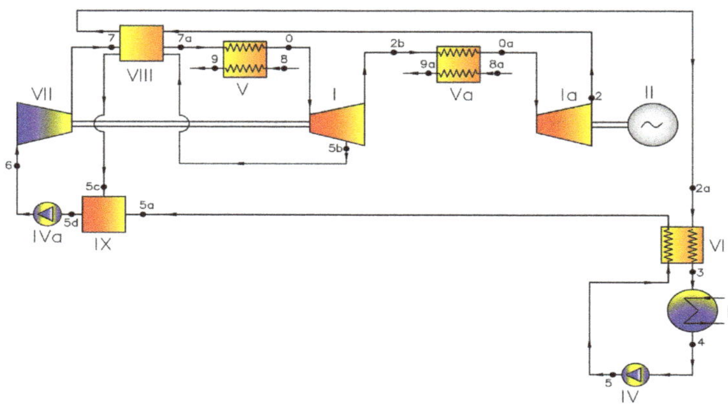

Figure 6. ORC cycle with regenerator, compressor, reheater, interstages superheater and heat exchanger; where: Ia, Ib—turbines, II—generator, III—condenser, IV, IVa—pumps, V—vapor generators, Va—interstage superheater, VI—regenerator, VII—compressor, VIII—heat exchanger, IX—vapor heater (Variant 8).

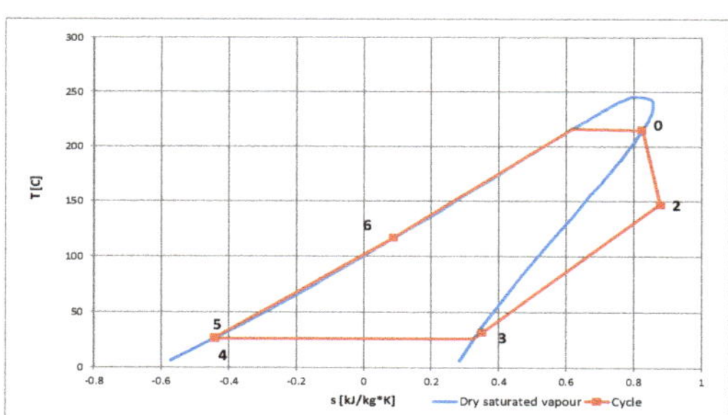

(**a**) Variant 1

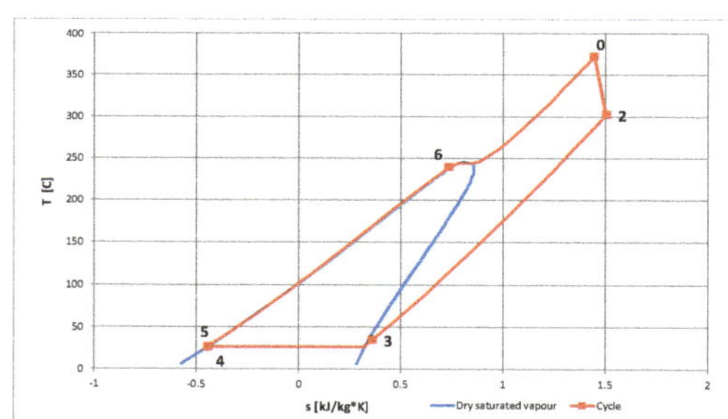

(**b**) Variant 2

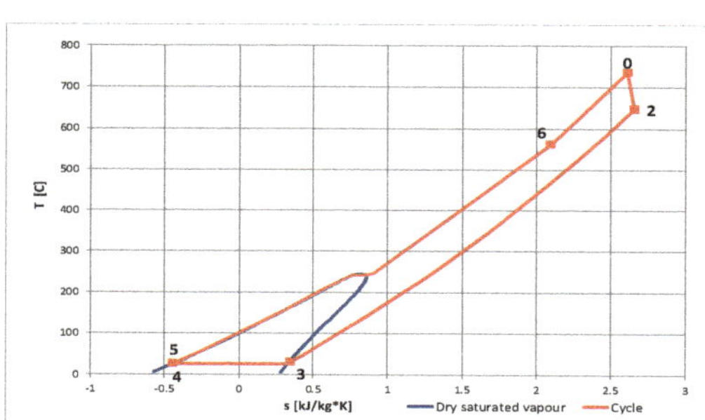

(**c**) Variant 3

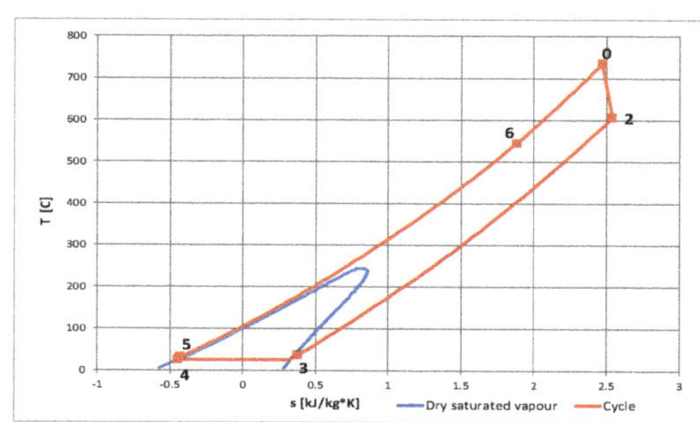

(**d**) Variant 4

Figure 7. *Cont.*

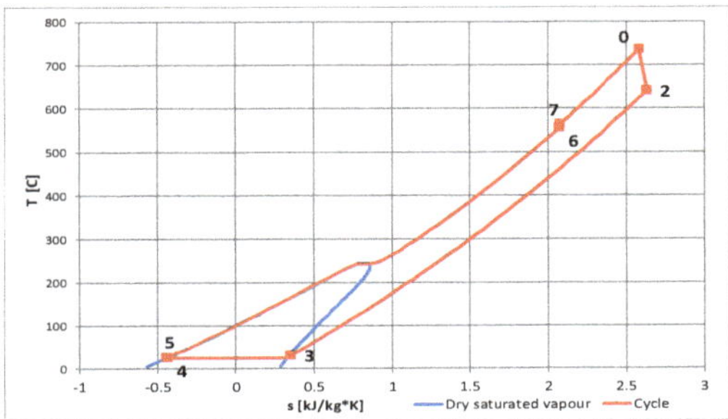

(**e**) Variant 5

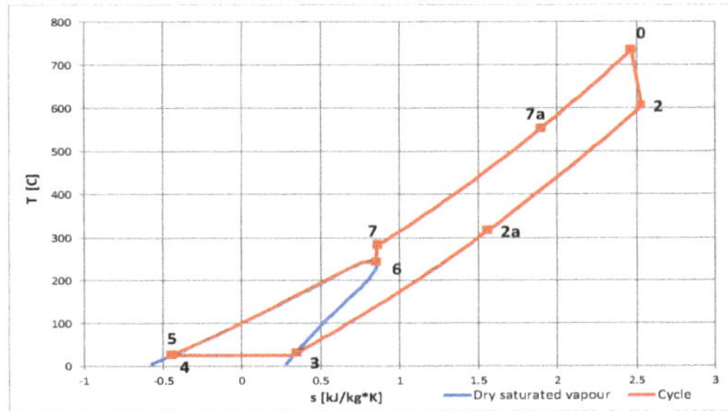

(**f**) Variant 6

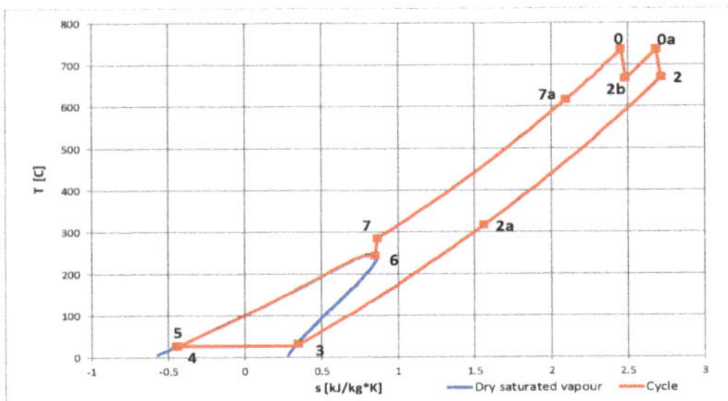

(**g**) Variant 7

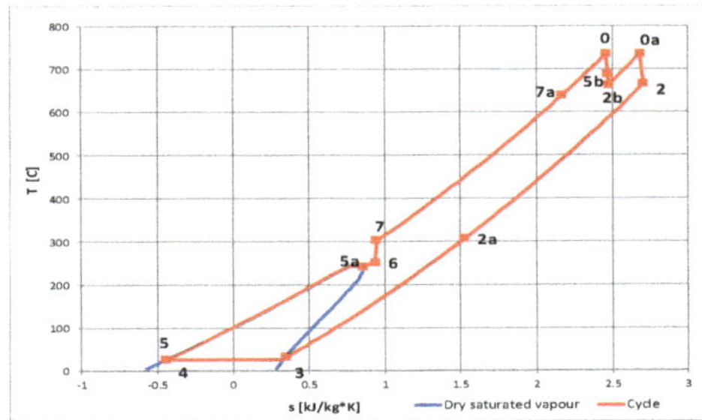

(**h**) Variant 8

Figure 7. Cycle interpretation in T-s diagram for all analyzed variants (hexamethyldisiloxane (MM) working fluid). The numbers in the graphs refer to the characteristic cycle points shown in Figures 2–6: (a) Variant 1-Figure 2; (b) Variant 2-Figure 2; (c) Variant 3-Figure 2; (d) Variant 4-Figure 2; (e) Variant 5-Figure 3; (f) Variant 6-Figure 4; (g) Variant 7-Figure 5; (h) Variant 8-Figure 6.

2. Results and Discussion

In the case of the simplest standard ORC cycle (Variant 1, Figure 2) the overall efficiency varies from 12% to 28% depending on the working medium and the parameters, Figure 8a.

For ORC cycle with superheated steam (Variant 2, Figure 2) higher values of efficiency can be obtained from 20% (R1233zd) to about 34% (C11), Figure 8b.

In Variant 3 (Figure 2)—the cycle with superheated live steam and the regenerator for warming up, vaporization and partial superheating of the working fluid the efficiencies vary from 28% (R245fa) to about 47% (C11), Figure 8c.

Variant 4 (Figure 2) is similar to Variant 3 except for the supercritical parameters at the turbine inlet allowing to obtain the efficiency from 30% (R245fa) to about 47% (D6)—Figure 8d.

In Variant 5 (Figure 3) the unusual element for ORC cycles appears: the compressor for vapor (Figure 3). The efficiencies of these plants vary from 30% (R245fa) to about 47% (C11)—Figure 8e.

In Variant 6 (Figure 4) the additional heat exchanger is applied and the efficiencies vary from 28% (R245fa) to above 47% (C11)—Figure 8f.

In Variant 7 (Figure 5) the interstage superheater is introduced and the efficiencies vary from 37% (R245fa) to nearly 50% (C11), Figure 8g.

Variant 8 (Figure 6) appears to be the most complex of the considered cycles (the heat exchanger with vapor bleeding from high pressure turbine (HP) is added). In this case the highest efficiencies ranging from 37% (R245fa) to nearly 52% (MM) may be obtained—Figure 8h.

The numbers in the Figure 7 refer to the characteristic cycle points shown in Figures 2–6, respectively.

In particular variants of the power plants and the applied working fluids the values of the upper temperature and pressure vary remarkably. For example, the temperature and pressure of live vapor in Variant 1 are presented in Figure 9. The temperature changes from 145 °C (R245fa) to 372 °C (D6), while the pressure from 0.95 MPa (D6) to about 3.5 MPa (acetone). In Variant 8 the temperatures and pressure are shown in Figure 10. The temperature changes from 380 °C (R245fa) to 770 °C (C11), while the pressure from 9.59 MPa (D4) to about 22.14 MPa (acetone).

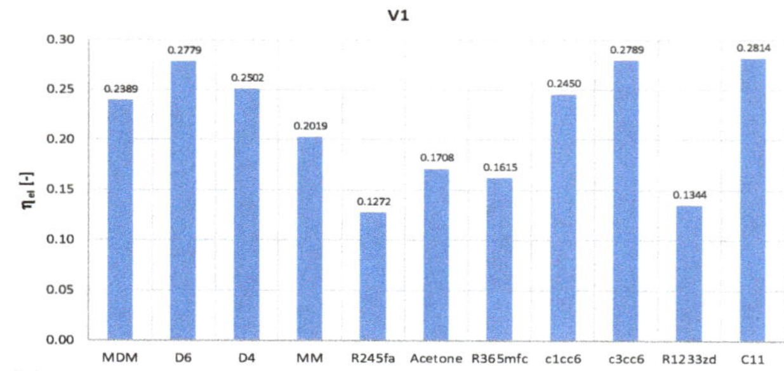

(a) Variant 1

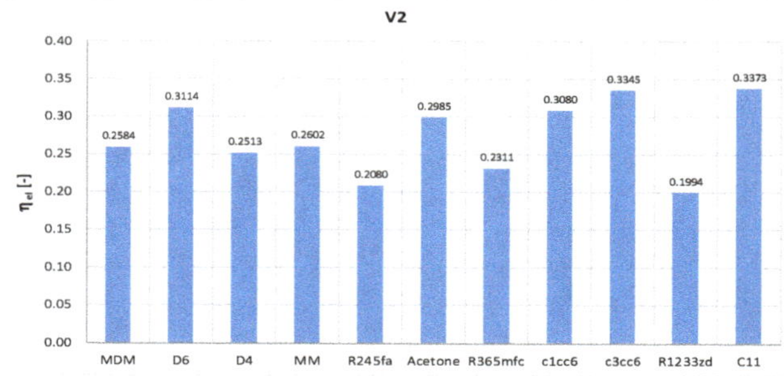

(b) Variant 2

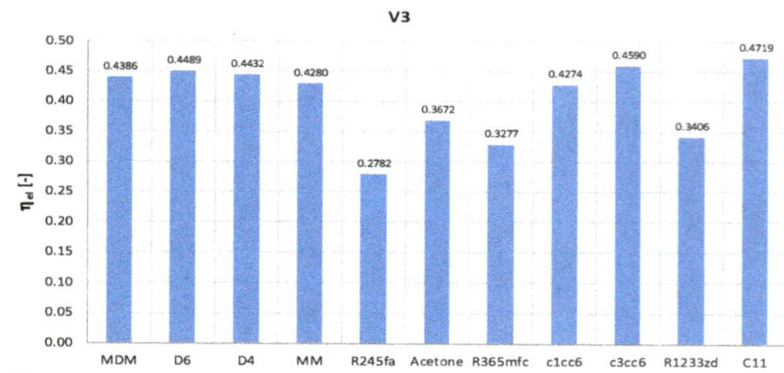

(c) Variant 3

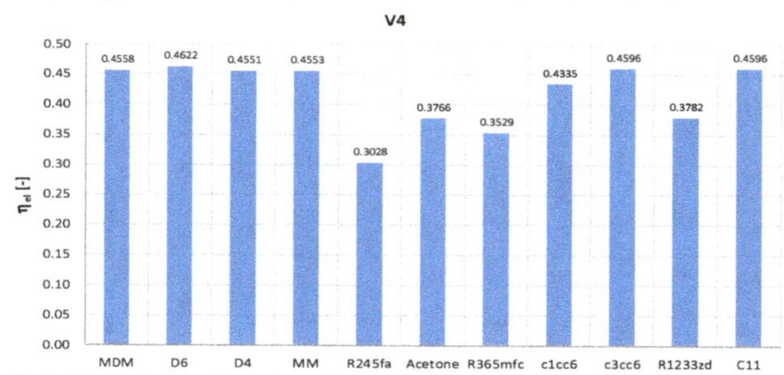

(d) Variant 4

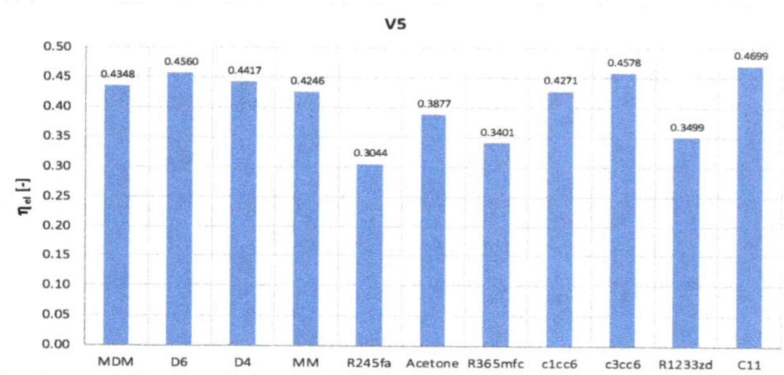

(e) Variant 5

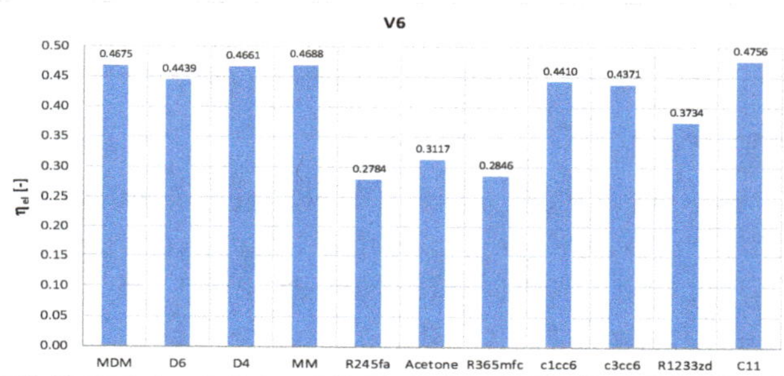

(f) Variant 6

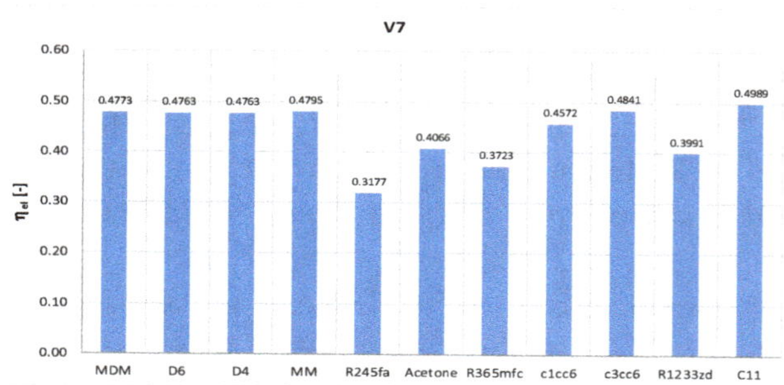

(g) Variant 7

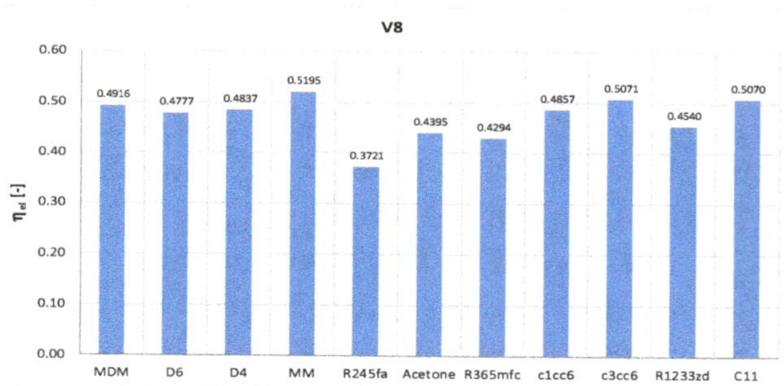

(h) Variant 8

Figure 8. Overall efficiency for different working media: (**a**) Variant 1; (**b**) Variant 2; (**c**) Variant 3; (**d**) Variant 4; (**e**) Variant 5; (**f**) Variant 6; (**g**) Variant 7; (**h**) Variant 8.

As far as the interstage superheater is considered, the pressure varies even more remarkably, Figure 11. In Variant 8, for example, it changes from 0.012 MPa (D6) to 1.55 MPa (R245fa).

Thus, comparing the efficiencies of various considered cycles the values of upper temperatures should be taken into account. Very high overall efficiency (52%) was achieved for the MM medium in Variant 8 (Figure 12).

The examples of turbine flow parts in Variants 1, 6 and 8 are presented in Figures 13–15. All the examples can be treated as micro turbines (turbines of very small output) with vapor of MM as the working medium. The variants have been designed as multistage turbines with disc rotors (impulse turbines). In Figure 13 a 4-stage turbine of 75 kW is shown and in Figure 14 a flow part of the 8-stage turbine of 240 kW is presented. While Figure 15 shows flow parts of 5-stage high pressure (HP) turbine and of 3-stage low pressure (LP) turbine of total output equal 220 kW.

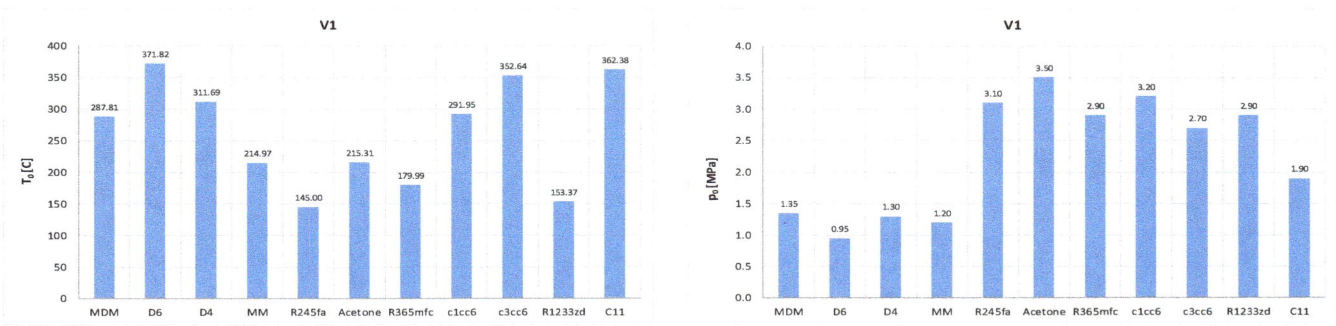

(**a**) temperature (**b**) pressure

Figure 9. Values of temperature (**a**) and pressure (**b**) of live vapor optimized for different working media (Variant 1).

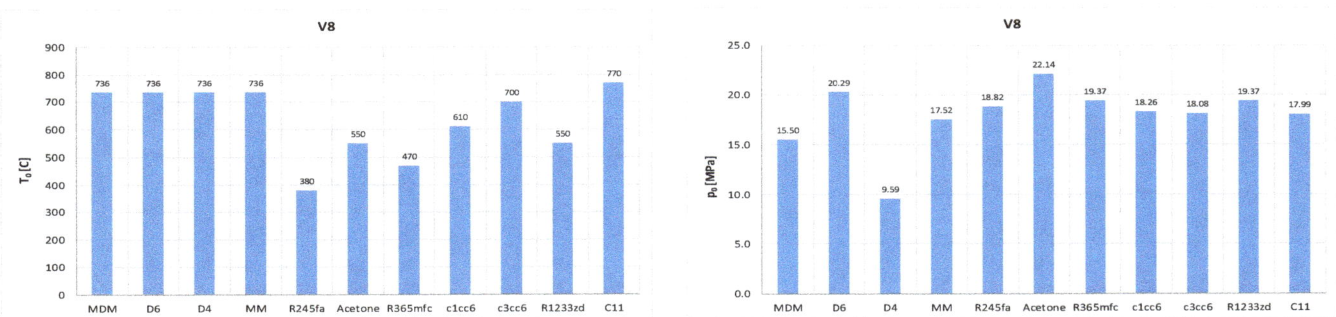

(**a**) (**b**)

Figure 10. Values of temperature (**a**) and pressure (**b**) of live vapor optimized for different working media (Variant 8).

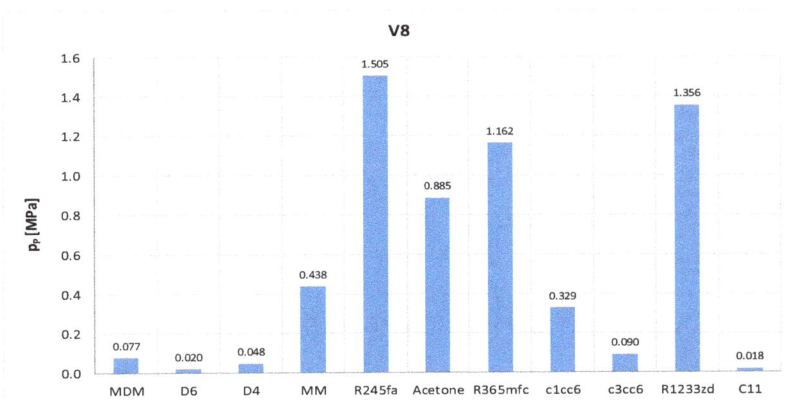

Figure 11. Variant 8—pressure of interstage superheater.

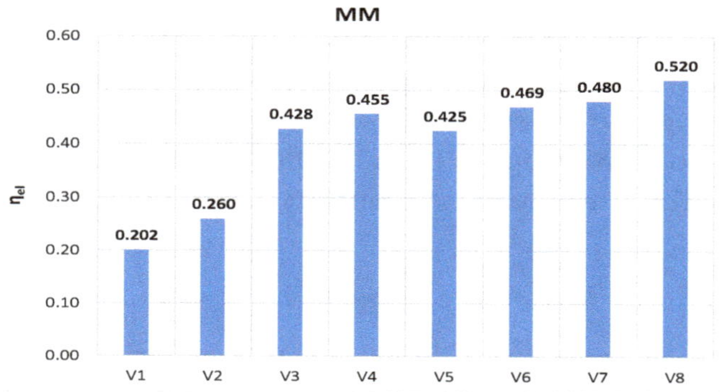

Figure 12. Maximum overall efficiency in particular variants of power plants (MM working fluid).

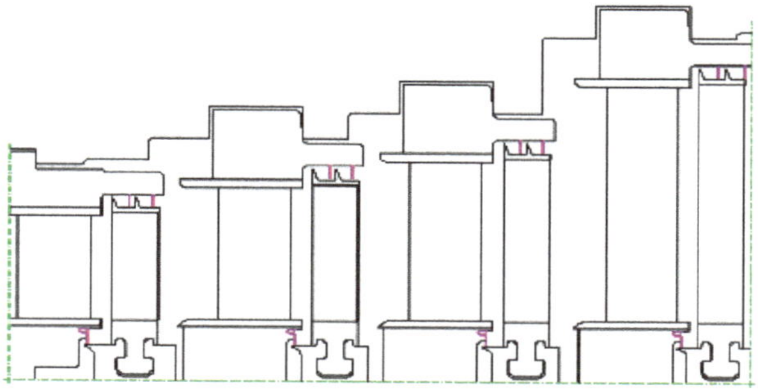

Figure 13. Turbine flow part for Variant 1 and MM working fluid.

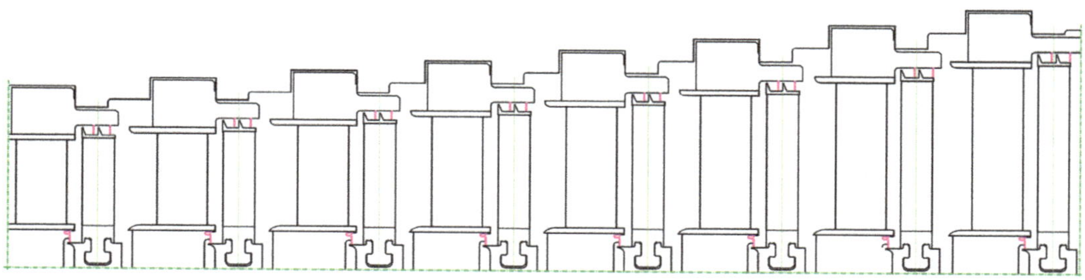

Figure 14. Turbine flow part for Variant 6 and MM working fluid.

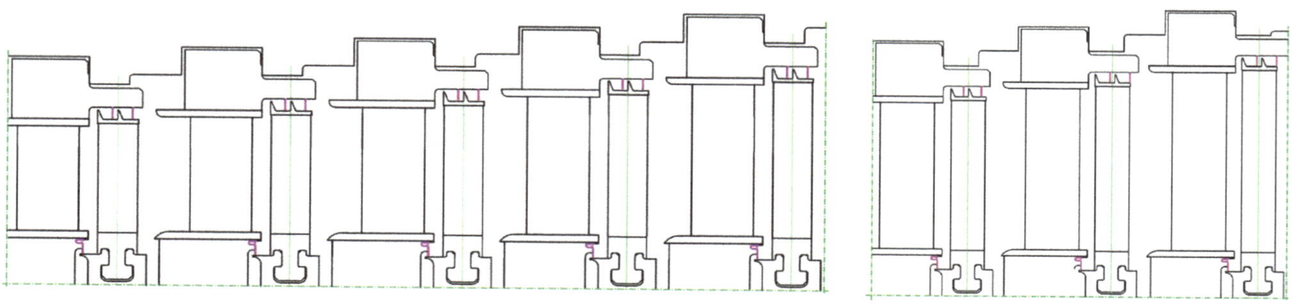

(**a**) low pressure turbine (**b**) high pressure turbine

Figure 15. Flow part of high (**a**) and low pressure (**b**) turbine for Variant 8 and MM working fluid.

3. Conclusions

Nowadays, only very simple ORC cycles are applied and described in bibliography (Figures 1 and 2). They are limited to 4 or 5 main elements: the turbine, the condenser, the pump, the vapor generator and the regenerator. The temperature of live vapor in these solutions is relatively low (lower than 300 °C). The presented analyses have been performed for far more complex cycles, for higher values of temperature (even 700 °C) and for higher values of the initial pressure. New elements, such as additional heat exchangers, interstage superheaters and compressors have been included in the cycles. These types of ORC power plants have not been discussed in bibliography so far. The obtained results of the presented analysis can be considered for future applications.

The performed analyses clearly prove that it is possible to build ORC power plants of relatively high efficiency even exceeding that of 50%, which is a very competitive value. In such cases ORC power plants can achieve an efficiency which is higher than that of modern steam turbine plants with supercritical parameters. The statement seems to stand in contradiction with the opinion that above 350 °C steam power plants have a higher efficiency than power plants with organic media. It is also very important to underline that even the most complex cycle (Variant 8 with MM working fluid) is cheaper and much simpler than modern supercritical steam power plants of large output with similar maximum efficiency.

Author Contributions: Conceptualization, K.K. and M.P.; methodology, K.K. and M.P.; formal analysis K.K. and M.P.; data curation, K.K. and M.P.; writing—original draft preparation, K.K. and M.P.; writing—review and editing, K.K. and M.P.; supervision, K.K. and M.P.; funding acquisition, K.K. and M.P. All authors have read and agreed to the published version of the manuscript.

References

1. Kamiński, J.; Kudełko, M. The prospects for hard coal as a fuel for the Polish power sector. *Energy Policy* **2010**, *38*, 7939–7950. [CrossRef]
2. Szczerbowski, R.; Ceran, B. Development perspectives of the Polish power generation sector according to the climate preservation conference COP21 policies. *E3S Web Conf.* **2017**, *14*, 1–10. [CrossRef]
3. Kosowski, K.; Tucki, K.; Piwowarski, M.; Stepien, R.; Orynycz, O.; Wlodarski, W. Thermodynamic Cycle Concepts for High-Efficiency Power Plants. Part B: Prosumer and Distributed Power Industry. *Sustainability* **2019**, *11*, 2647. [CrossRef]
4. Mikielewicz, D.; Kosowski, K.; Tucki, K.; Piwowarski, M.; Stępień, R.; Orynycz, O.; Włodarski, W. Influence of Different Biofuels on the Efficiency of Gas Turbine Cycles for Prosumer and Distributed Energy Power Plants. *Energies* **2019**, *12*, 3173. [CrossRef]
5. Kosowski, K.; Piwowarski, M.; Stepien, R.; Włodarski, W. Performance characteristics of a micro-turbine. *J. Vib. Eng. Technol.* **2014**, *2*, 341–350.
6. Kosowski, K.; Piwowarski, M.; Stepien, R.; Włodarski, W. Design and investigations of the ethanol microturbine. *Arch. Thermodyn.* **2018**, *39*, 41–54. [CrossRef]
7. Landelle, A.; Tauveron, N.; Haberschill, P.; Revellin, R.; Colasson, S. Organic Rankine cycle design and performance comparison based on experimental database. *Appl. Energy* **2017**, *204*, 1172–1187. [CrossRef]
8. Quoilin, S.; Lemort, V. Technological and Economical Survey of Organic Rankine Cycle Systems. In Proceedings of the 5-th European Conference Economics and Management of Energy in Industry, Villamura, Portugal, 14–17 April 2009.
9. Piwowarski, M. Optimization of steam cycles with respect to supercritical parameters. *Pol. Marit. Res.* **2009**, *S1*, 45–51. [CrossRef]
10. Tumanovskii, A.G.; Shvarts, A.L.; Somova, E.V.; Verbovetskii, E.Kh.; Avrutskii, G.D.; Ermakova, S.V.; Kalugin, R.N.; Lazarev, M.V. Review of the Coal-Fired, Over-Supercritical and Ultra-Supercritical Steam Power Plants, Pleiades Publishing. *Therm. Eng.* **2017**, *64*, 83–96. [CrossRef]
11. Piwowarski, M.; Kosowski, K. Design analysis of combined gas-vapour micro power plant with 30 kW air turbine. *Pol. J. Environ. Stud.* **2014**, *23*, 1397–1401.
12. Włodarski, W. Experimental investigations and simulations of the microturbine unit with permanent magnet generator. *Energy* **2018**, *158*, 59–71. [CrossRef]
13. Włodarski, W. Control of a vapour microturbine set in cogeneration applications. *ISA Trans.* **2019**. [CrossRef [PubMed]

14. Włodarski, W. A model development and experimental verification for a vapour microturbine with a permanent magnet synchronous generator. *Appl. Energy* **2019**, *252*. [CrossRef]
15. Kosowski, K.; Stępień, R.; Włodarski, W.; Piwowarski, M.; Hirt, Ł. Partial admission stages of high efficiency for a microturbine. *J. Vib. Eng. Technol.* **2014**, *2*, 441–448.
16. Braimakis, K.; Karellas, S. Energetic optimization of regenerative Organic Rankine Cycle (ORC) configurations. *Energy Convers. Manag.* **2018**, *159*, 353–370. [CrossRef]
17. Duvia, A.; Gaia, M. ORC Plants for Power Production from Biomas from 0.4MWe to 1.5MWe: Technology, Efficiency, Practical Experiences and Economy. In Proceedings of the 7th Holzenergie Symopsium, Zurich, Switzerland, 18 October 2002.
18. Duvia, A.; Tavolo, S. Application of ORC Units in the Pellet Production, Field: Technical-Economic Considerations and Overview of the Operational Results of an ORC Plant in the Industry Installed in Mudau (Germany). In Proceedings of the 7th Holzenergie Symopsium, Zurich, Switzerland, 18 October 2002.
19. Kosowski, K.; Tucki, K.; Piwowarski, M.; Stępień, R.; Orynycz, O.; Włodarski, W.; Bączyk, A. Thermodynamic Cycle Concepts for High-Efficiency Power Plans. Part A: Public Power Plants 60+. *Sustainability* **2019**, *11*, 554. [CrossRef]
20. Mikielewicz, D.; Kosowski, K.; Tucki, K.; Piwowarski, M.; Stępień, R.; Orynycz, O.; Włodarski, W. Gas Turbine Cycle with External Combustion Chamber for Prosumer and Distributed Energy Systems. *Energies* **2019**, *12*, 3501. [CrossRef]
21. Mikielewicz, D.; Mikielewicz, J. Cogenerative micro power plants—A new direction for development of power engineering? *Arch. Thermodyn.* **2008**, *29*, 109–132.
22. Mikielewicz, J.; Mikielewicz, D. Comparative study of selected fluids for use in supercritical Organic Rankine Cycles. *Arch. Thermodyn.* **2009**, *30*, 3–15.
23. Set of Product's Card of Chempur Poland Ltd. Available online: http://en.chempur.pl (accessed on 25 February 2020).
24. Set of Product's Card of Sigma-Aldrich Ltd. Available online: https://www.sigmaaldrich.com (accessed on 25 February 2020).
25. Set of Product's Card of SCHIESSL Poland Ltd. Available online: https://www.schiessl.pl/en (accessed on 25 February 2020).
26. Lemmon, E.; Huber, M.; McLinden, M. *NIST Standard Reference Database 23: Reference Fluid Thermodynamic and Transport Properties-REFPROP*, 9th ed.; National Institute of Standards and Technology, Standard Reference Data Program: Gaithersburg, MD, USA, 2010.
27. Wajs, J.; Mikielewicz, D.; Jakubowska, B. Performance of the domestic micro ORC equipped with the shell-and-tube condenser with minichannels. *Energy* **2018**, *157*, 853–861. [CrossRef]

4

Design and Optimization of an Integrated Turbo-Generator and Thermoelectric Generator for Vehicle Exhaust Electrical Energy Recovery

Prasert Nonthakarn [1,*], Mongkol Ekpanyapong [1], Udomkiat Nontakaew [2] and Erik Bohez [1]

1 School of Engineering and Technology, Asian Institute of Technology, Pathum Thani 12120, Thailand
2 Faculty of Engineering, King Mongkut's University of Technology North Bangkok, Bangkok 10800, Thailand
* Correspondence: prasertnonthakarn@gmail.com

Abstract: The performance of turbo-generators significantly depends on the design of the power turbine. In addition, the thermoelectric generator can convert waste heat into another source of energy. This research aims to design and optimize an integrated turbo-generator and thermoelectric generator for diesel engines. The goal is to generate electricity from the vehicle exhaust gas. Electrical energy is derived from generators using the flow, pressure, and temperature of exhaust gases from combustion engines and heat-waste. In the case of turbo-generators and thermoelectric generators, the system automatically adjusts the power provided by an inverter. Typically, vehicle exhausts are discarded to the environment. Hence, the proposed conversion to electrical energy will reduce the alternator charging system. This work focuses on design optimization of a turbo-generator and thermoelectric generator for 2500 cc. diesel engines, due to their widespread usage. The concept, however, can also be applied to gasoline engines. Moreover, this model is designed for a hybrid vehicle. Charging during running will save time at the charging station. The optimization by variable van angles of 40°, 50°, 62°, 70°, and 80° shows that the best output power is 62°, which is identical to that calculated. The maximum power outputted from the designed prototype was 1262 watts when operating with an exhaust mass flow rate of 0.1024 kg/s at 3400 rpm (high performance of the engine). This research aims to reduce fuel consumption and reduce pollution from the exhaust, especially for hybrid vehicles.

Keywords: thermoelectric generation; turbo-generator; exhaust heat recovery

1. Introduction

Currently world energy consumption is continuously increasing, and this is associated with a high rate of environmental damage due to the use of fossil fuels. This has been a concern in light of a shortage of energy resources and environmental protection. In the past few years, a pressing need has arisen to develop green energy systems and breakthrough technologies to solve these problems. Since transportation is one of the main sectors responsible for the utilization of energy, transmissions, and new vehicle components are being developed in order to achieve maximum energy savings, and as a consequence less fuel consumption.

The prime source for powering vehicles is the internal combustion engine, or ICE. The principle approach for releasing energy is the conversion of fuel in the combustion chamber into usable energy. Unfortunately, the ICE has poor energy conversion efficiency. In the case of gasoline, diesel, and hybrid electric vehicles, the percentage of utilized fuel during running is around 25%, with the remaining 75% lost in the form of exhaust gas enthalpy. This lost energy is crucial, and is largely due to exhaust gas from waste heat (40%), the cooling system (including heat radiation, 30%), and friction (5%) [1]. Consequently, the amount of energy lost from the exhaust and cooling systems is twice the amount of energy than is actually utilized for mechanical energy. It can be estimated that 20% of the lost energy can be converted to only 10% of the total electricity [2]. In this case, the exhaust gas or cooling system can be used to generate energy, and thus fuel efficiency is increased.

In order to increase fuel efficiency, various energy saving technologies have been developed. Many researchers have focused on the utilization of the exhaust gas of vehicles for power generation. Engine exhaust gas energy recovery is one of the more promising technologies, through simultaneous energy saving and emission reduction. Engine exhaust gas has a certain pressure and heat due to its high pressure and high temperature, which can be used to generate electricity for power storage, thereby reducing fuel consumption and increasing energy recovery efficiency. Moreover, the high-grade mechanical energy of the remaining pressure can be recycled through the exhaust gas directly expanding, which is known as the direct recovery method, and through low-grade thermodynamic energy used in heat transmission, known as the indirect recovery method. Each method has different measurements and solutions for energy recovery [3].

This study focused on designing and optimizing turbo-generator coupling with a thermoelectric generator for 2500 cc. diesel engines. The work is composed of two main categories: Electrical energy recovery, and mechanic energy recovery units. The design proposes to convert exhaust gas energy to electrical power by mechanical exhaust energy, which is then converted to electricity through a turbine, and further proposes to convert waste heat directly to electric energy. When using diesel engines without turbos, there is no obstacle that will reduce the pressure and flow rate of the exhaust stream from the combustion unit. Electrical energy recovery is a way to utilize recovered energy to supply electrical power to the automobile's interior equipment, or to bring this energy to charge the battery for further use. Mechanical energy recovery, on the other hand, is a way to recover energy from the exhaust gas, and is directly attached to the inverter in order to make the power available to the vehicle. Power can be generated by the different turbine geometry components; for example, the rotor, nozzle and volute. Electrical energy is collected during driving time.

The structure of this paper is as follows. First, the turbo-generator model and thermoelectric generator are presented. Next, the simulation results are illustrated. Then, the experimental results, which validate the idea, are discussed. Finally, we conclude the paper and make recommendations for future work.

2. Exhaust Gas Heat Recovery Technology

The energy emitted by the engine exhaust gas is pressure and heat. Exhaust gas recovery transforms the high pressure and heat back to reusable energy in the form of electrical power. The high speed of hot waste gas produces a great amount of mechanical energy that can be utilized for power generation; this is known as the direct method. At present, waste heat recovery technology from the engine by the direct method is available, and is known as mechanical turbo-compounding and electric turbo-compounding. Mechanical turbo-compounding using a conventional turbocharger depends on the turbine to generate energy from the exhaust gas flow expelled from an internal combustion engine. Alternatively, thermal energy derived from the exhaust gas is indirectly converted to additional electricity using the Rankine cycle (organic or steam), and is directly recovered through heat transmission of the thermoelectric generator.

Energy recovery using the direct technology method is easier to assemble than the indirect method. Electrical turbo-compounding can be easily assembled, and can be directly connected to the turbocharge axle or directly attached to the secondary turbocharge shaft [4–6]. Melanie Michon [7] reported the results of study on the comparison of an automotive turbo-generator exhaust gas energy recovery system using low voltage and high voltage machines. Aman M.I. Mamat [8] developed a highly efficient turbine suitable for a low pressure turbine, or LPT, using turbo-compounding in small and large segment passenger cars that use gasoline engines with a capacity of 1.0 L. The turbine was designed for two purposes, to recover the energy of exhaust gas at low pressure with ratios of 1.05–1.3, and to drive a small generator with fixed operation conditions at 50,000 rpm. The resulting power output is up to 1.0 kW. Commercial Computational Fluid Dynamics (CFD) software (ANSYS Unveils Release 16.0) was used in simulation to determine the low-pressure turbine performance. Hence, the performance was verified with comprehensive turbine testing derived from the Imperial College turbine test rig (cold-flow test facility). The caterpillar concept has been considered in electrical turbo-compounding research [9–11]. These researchers have shown that using a high-performance turbocharger reduces fuel consumption by 5% in basic engine cycles (brake specific fuel consumption). Besides that, it can reduce fuel consumption overall with by up to approximately 9%–10%. Electrical turbo-compounding is designed as part of the turbo-compound and is directly attached to the exhaust pipe, to obtain exhaust gas heat flowing out with a generator running at a very high rotational speed. The turbine design in this case will produce a large amount of power when compared with using only a turbine for the compressor driver. The energy in the exhaust residual during usual use is taken to drive the high-speed generators that are included in the same unit, which will transform the remaining energy to electric power. Energy recovery technology can achieve high power from exhaust gas, and work as a power plant. The maximum energy recovery from the exhaust is approximately 1.8 kW at the initial stage. Bowman Power Group Ltd. presents turbo-generators for power generation (power plants) [12]. Bowman designed and applied a turbo-generator for installation in a power plant generator, which brings the electrical power produced to the electric grid parallel to the original alternator. The turbo-generator engine is mounted with a 250 kW to 1 MW engine that is equipped with a turbocharger, and mounted after the turbocharger. The installation can produce around 10%–20% more power. Controlled Power Technologies Ltd. (Basildon, UK). Federal-Mogul Controlled Power Ltd. (Michigan, USA).

Unit 4, Westmayne Industrial Park, Bramston Way, Laindon designed and built a turbo-generator integrated gas energy recovery system (TIGERS) that has the same functionality as the turbo-generator [13]. TIGERS are water-cooled generators coupled to an exhaust driven turbine. The small turbo-generator is installed at the exhaust pipe. This pipe is another pipe created for flow control to the generator. The turbo-generator uses the valve to increase or decrease the amount of exhaust passing through this series of turbo-generator. The opening of the more or less valve is controlled by the control system relative to the velocity of the exhaust gas engine. The exhaust gas flows through the turbine, causing the power transmission to be powered by the generator. Power output depends on the engine load. The system produces up to 600 watts of electricity power.

Other technologies, such as the Rankine cycle (organic or steam indirect), produce power indirectly. Beside these, thermoelectric generator technologies obtain energy directly through heat transmission. The Rankine cycle uses the principle of changing the status of liquids. At any time, the liquid may be pressurized and heated, and it transforms into pushed vapor due to expansion. The waste heat recovery (WHR) evaporator transforms the energy in the exhaust by installing it in a position after the manifold exhaust in the production phase, by which the working fluid expands in the turbine. The efficiency of energy recovery is based on the efficiency of the heat exchanger. Furthermore, using the organic Rankine cycle to recover energy requires a large space to install the system, and it is difficult to assemble the parts [14,15].

A thermoelectric generator is another way of bringing the energy from the exhaust gas back to a usable state, and is convenient and easy to install in an engine. The Peltier–Seebeck effect describes how a thermoelectric generator transforms the heat in the exhaust gas into electric power. The current arises from the difference of temperature between one side of the thermoelectric generator and the muffler. It is a device with no moving parts (a solid-state device) that directly transforms the heat to electricity. The advantage of a thermoelectric module without moving parts is that the power-generating device is simple, robust, reliable, modular, and maintenance free. Moreover, its production of electricity is environmentally friendly.

X. Liu [16] proposed simulation and experimental results of positioning of the thermoelectric generator installation. In this work, the thermoelectric generator is placed between the catalytic converter and muffler, where there is no system parts to block the flow of exhaust. As a result, it does not cause the exhaust pressure to reverse inside the exhaust pipe, and there will be a large area for heat transfer, which improves the performance of the heat transmission. Tae Young Kim [17] studied the efficiency of the heat released with recycled exhaust gas. The experiment used 40 thermoelectric generators equipped with a large diesel engine, and operated with the engine at a rotation cycle of 1000, 1500, and 2000 rpm. The thermoelectric generator was installed on the top and bottom of the exhaust gas channel series. The experimental results illustrated that the energy released directly varies based on the engine load and speed, and the maximum energy value at 2000 rpm is 119 watts. Xiuxiu Sun [18] compared two characteristics of thermoelectric generators: Series installation and parallel installation. The results of the experiments show that parallel installation provides a better performance than series installation. Hua Tian [19] presented the effect of the energy that has been released, and the efficiency of thermoelectric generators. The results conclude that energy and efficiency can be increased by increasing the heat source temperature, and reducing the heat in the desired cooling areas. The performance is based on the ratio of two types of materials—Skutterudite and bismuth telluride—to make thermoelectric generators reliable and easy to use. Ilker Temizer [20] developed a prototype for recycling heat. In this work, the thermoelectric generator is mounted with a diesel engine. Experiments were carried out looking at different engine speeds and different engine loads. The speed of the engine rotation was set at five different levels, and the engine load was set at two different levels. The experimental results show the maximum output was 156.7 watts at the maximum speed and maximum engine load. Changxin Liu [21] studied prototype generation by installation of a thermoelectric generator with a car exhaust muffler. The experimental results found that, when operating the engine with a heat pipe at a temperature of 473 K, the thermoelectric generator can produce an energy output of to 250 watts, with a heat efficiency of 5.35%.

Based on the hybrid electric charge for a hybrid car, it was found that the average electric charge would be 1–5 h per 100 km driving distance. For charging, the car will be parked at a power station, which wastes time. The average speed of a used car driving on the highway is 100 km/h. At that speed, the engine speed is 3500 rpm.

Contemporary's cars have a higher level of power consumption than ever before, because of the equipment used to drive the engine and because of electrical demands such as the air conditioning system (using an electric compressor), electric power steering, electronic throttle control, electric windows, and power seats. In addition, the equipment included for convenience when driving such as phone chargers, refrigerators, heaters, and car entertainment equipment also require electricity.

In this research, the recovery of energy in the exhaust is different from previous works. It can summarize with the major differences as shown in Table 1. The main difference is this research will design the turbo-generator and thermoelectric for 2500 cc. diesel engines of light-truck.

Table 1. Summary of parameters for turbo-generators and the thermoelectric generator model.

Engine	Mobile	Type	Energy Receive	Speed	Weigh Extra	Reference
n/a	Yes (Heavy-Duty)	TG	600 W	45,000 rpm	11 kg	[11]
n/a	n/a	TG	1.8 kW	80,000 rpm	n/a	[12]
Diesel	No (Power Plant)	TG	30 kW	60,000 rpm	50 kg	[13]
Gasoline 1.0 L	Yes (Passenger)	TG	1.0 kW	50,000 rpm	n/a	[14]
Diesel	Yes (Heavy-Duty)	TEG	119 W	n/a	n/a	[18]
Diesel	Yes	TEG	156.7 W	n/a	n/a	[21]
Diesel	No (Power Plant)	TEG	250W	n/a	n/a	[22]
Diesel 2.5 L	Yes (Light-Truck)	TG + TEG	1262 W	50,000 rpm	15 kg	Author work

TG (turbo-generator), TEG (thermoelectric-generator), and n/a (not available).

3. Turbo-Generators and Thermoelectric Generator Modeling

The turbo-generators and thermoelectric generator system developed by the designing and installing of turbo-generators after exhaust manifold and installing the thermoelectric generator in the next position. Therefore, this section outlines the process of design for turbo-generators, followed by turbo-generator development.

3.1. Exhaust Energy

External heat balance is based on measurements of mechanical energy, and energy taken outside by the cooling liquid and exhaust gases. The external balance equation takes the following form [2]:

$$Q = Q_e + Q_{ch} + Q_w + Q_n + Q_r, \tag{1}$$

where Q is the total heat inputted to the engine (J/h), Q_e is heat that can be exploited, Q_{ch} is cooling loss, Q_w is the loss of exhaust, Q_n is an incomplete loss, and Q_r is the residual of the balance/remaining. The loss of exhaust is the largest amount of energy derived from heat and pressure, which can be calculated from the following equation:

$$Q_w = \dot{m} \times c_p \times \Delta T, \tag{2}$$

where $\dot{m}$ is the exhaust gas mass flow rate, c_p is the specific heat of exhaust gas, and ΔT is the exhaust gas temperature difference.

To understand an overview of all the energy generated by the internal combustion engine using the functional principle of the Otto cycle, and to be able to see and understand the energy mix in the exhaust gas released along with the exhaust muffler, Figure 1 shows the relationship of pressure and volume in the cylinder box of the combustion engine within four strokes. The bottom part of the figure shows the piston and the cylinder that move up and down during operation of a four-stroke engine. The piston continuously moves from the bottom death center to the top death center. The movement causes various strokes, including the intake, compression, combustion, and exhaust strokes, each of which work together with the displacement of the piston and closing/opening of valve. Briefly, the energy in the exhaust gas in the figure is the area of the exhaust energy (area F-G-H-I) that is released from the discharge of the exhaust stroke in the spitting. The piston moves up from the bottom death center to the top death center at the same time that the exhaust valve opens. This makes the head of the piston press the exhaust gas. Then, the exhaust gas with heat and high pressure flows out of the cylinder through the manifold exhaust to the exhaust pipe and to the pipe end, respectively.

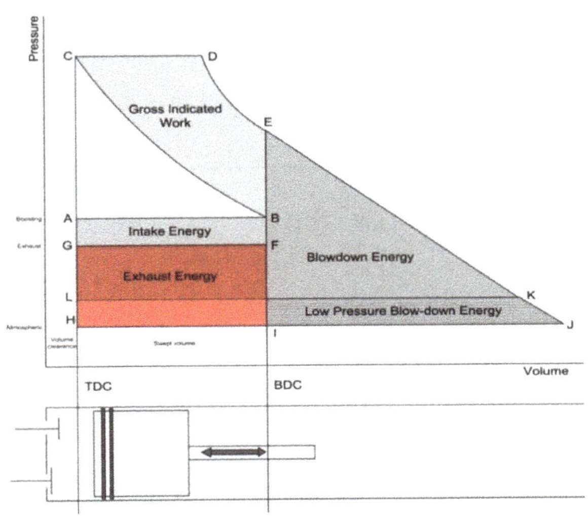

Figure 1. Exhaust energy

3.2. Flow Chart for Calculating the Exhaust Mass Flow of an Internal Combustion Engine

This flow chart (Figure 2) shows the process for calculating the mass flow rate of the exhaust gas, which will be used to design the rotor blades in the next stage.

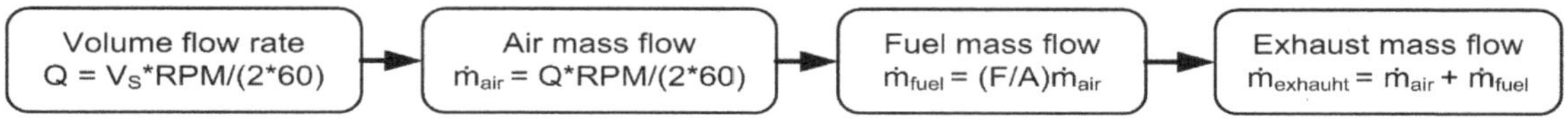

Figure 2. Exhaust gas.

The operating exhaust conditions are setup for a 2500 cubic centimeter diesel engine with the speed of the engine 3400 rpm (normal high use). Table 2 shows the flow characteristics of the exhaust gas that changed when the engine speed changes. To summarize, the increased engine rotation speed affects the mass flow rate and flow velocity, where the speed of the engine varies from 1000 to 5000 rpm. The exhaust mass flow rate calculated as the following:

$$\dot{m}_{exhaust} = \dot{m}_{air} + \dot{m}_{fuel} = 0.1024 \text{ kg/s}.$$

Table 2. Mass and velocity flow of exhaust gas on difference engine rotation speed.

Engine Rotation Speed (rpm)	Exhaust Gas Mass Flow Rate (kg/sec)	Exhaust Gas Flow Velocity (m/min)
1000	0.0320	900
2000	0.0640	1799
3000	0.0960	2699
3200	0.1024	3598
4000	0.1280	3778
5000	0.1600	4498

3.3. Specific Design

In the first step, the dimension design of the parts of the rotor used the details following the instructions as shown in Table 3 and Figure 3.

Table 3. Design parameters of the rotor.

Parameter	Recommended Range	Source
α_2	68–75	Dixon, Rohlik
β_3	50–70	Whitfiel and Baines
D_{3H}/D_{3S}	<0.4	Dixon, Rohlik
D_{3S}/D_2	<0.7	Dixon, Rohlik
D_3/D_2	0.53–0.66	Whitfiel and Baines
b_2/D_2	0.05–0.15	Whitfiel and Baines, Dixon, Rohlik
U_2/C_O	0.55–0.80	Balje diagram
W_3/W_2	2–2.5	Ribaud and Mischell
V_3/V_2	0.15–0.5	Whitfiel and Baines

The equation below describes the relationships of the theoretical work for turbines:

$$L_O = \frac{k_1}{k_1 - 1} \times R_{sp} \times T_s \times \left[1 - \left(\frac{P_s'}{P_s} \right)^{\frac{k_1 - 1}{k_1}} \right], \tag{3}$$

$$\frac{T_s'}{T_s} = \left(\frac{P_s'}{P_s} \right)^{\frac{K_1 - 1}{K_1}}, \tag{4}$$

$$L_o = \frac{k_1}{k_1 - 1} \times R_{sp} \times (T_s' - T_s), \tag{5}$$

where L_0 is the theoretical work by energy per mass of exhaust gas (kJ/kg), k_1 is the value of adiabatic exhaust gases, R_{sp} is the constant value of the exhaust gases relative to the gas constant to the molecular weight of the gas, P_s is the pressure of exhaust gases behind the turbine, P_s' is the pressure of exhaust gases in front of the turbine, T_s is the temperature of exhaust gases flowing into the turbine, and T_s' is the temperature of exhaust gases flowing out from the turbine.

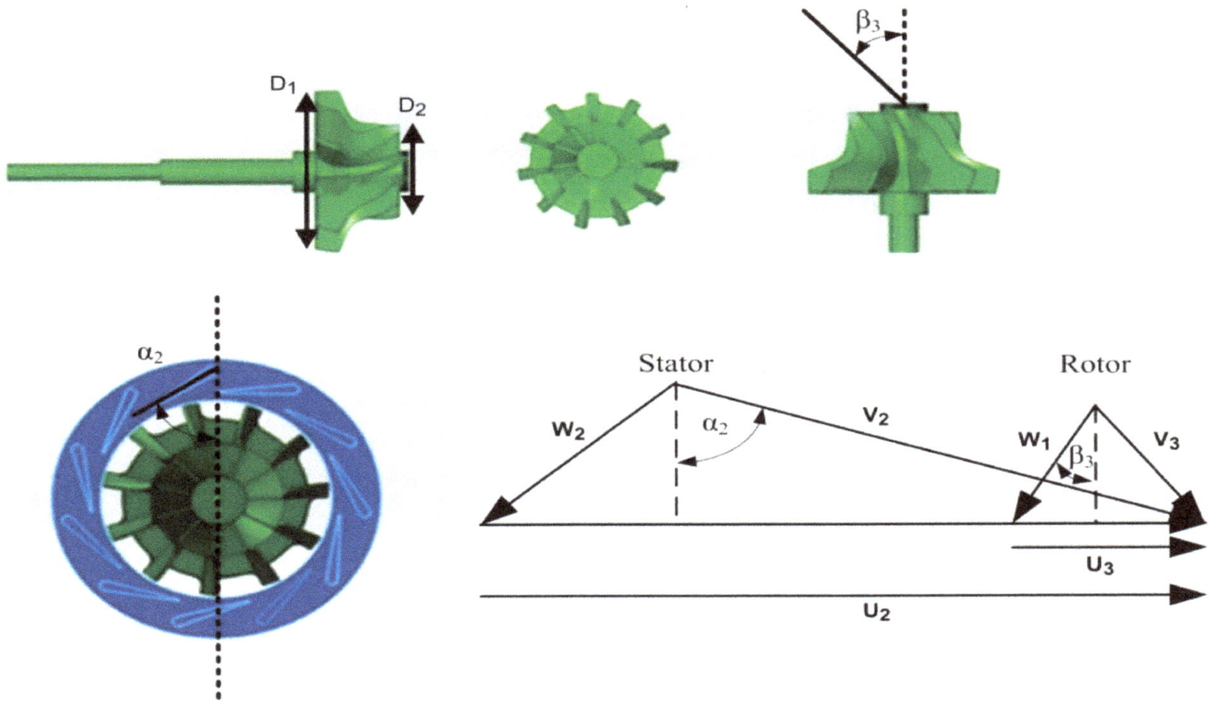

Figure 3. Velocity triangles for the rotor and stator.

The following equation explains the size of the work occurring in the turbine, which is related to the decrease of the temperature after the exhaust gas flows through the turbine.

$$N_t = G_s \times l_t = G_s \times l_0 \times \eta_t, \tag{6}$$

where N_t is the theoretical power (W), G_s is the intensity of exhaust gas flow (kg/s), and η_t is the efficiency of the turbine in isentropic.

The Euler equation shows the energy transfer in the rotor, which can be represented as a product of the torque by the angular velocity:

$$\dot{W} = \overline{\omega} TQ = \dot{m}(u_1 c_{01} - u_2 c_{02}) = \dot{m}.c_p(T_{10} - T_{20}), \tag{7}$$

where $\dot{W}$ is the energy transfer, Q is the heat transfer per unit mass flow, $\overline{\omega}$ is the angular velocity, $\dot{m}$ is the mass rate of flow, $(u_1 c_{01} - u_2 c_{02})$ is the rate of change of angular momentum, c_p is the specific heat at a constant pressure, and $(T_{10} - T_{20})$ 10-T20 is the temperature difference of the inlet and outlet.

3.4. *Formula for One Dimensional Calculation*

Once the energy transfer has been determined, the mass flow of exhaust gas required to achieve the specified power can be calculated from:

$$W = \frac{Power}{\dot{m}_{exhaust}}. \tag{8}$$

The blade speed at the inlet can be calculated from the velocity triangles, in which the relative work output is:

$$u_2 = \sqrt{w}. \tag{9}$$

The absolute velocity at the inlet:

$$v_2 = \frac{u_2}{\sin \alpha_2}. \tag{10}$$

where α_2 is absolute gas angle at radius.

The radius at the inlet of the rotor has the value of:

$$r_2 = \frac{60u_2}{2\pi \times rpm}. \tag{11}$$

The relative velocity at the inlet:

$$w_2 = \sqrt{v_2^2 - u_2^2}. \tag{12}$$

The rotor shroud radius and hub radius are:

$$r_{3s} = 0.75r_2. \tag{13}$$

The rotor hub radius:

$$r_{3h} = kr_{3s}. \tag{14}$$

where k is the hub-tip ratio at the inlet impeller.

The hub blade speed:

$$u_{3h} = \frac{2\pi r_{3h} rpm}{60}. \tag{15}$$

The shroud blade speed:

$$u_{3s} = \frac{2\pi r_{3s} rpm}{60}. \tag{16}$$

The mean of the exit radius is equal to its square value using the Balje diagram:

$$r_3 = \sqrt{0.5\left(r_{3s}^2 + r_{3h}^2\right)}. \tag{17}$$

Figure 4 shows the Balje diagram used to calculate the value of the exit radius. The value of Ω_s efficiency is in range 0.2–0.8 and the recommended value of $\frac{r_3}{r_2}$ is in range 0.53–0.66.

Consequently, the blade speed at the mean exit radius is:

$$u_3 = \frac{2\pi r_3 rpm}{60}. \tag{18}$$

The blade width at the outlet:

$$b_3 = r_{3s} - r_{3h}. \tag{19}$$

Using mass balance, the inlet blade height can be determined as:

$$b_2 = \frac{\dot{m}_{exhaust}}{2\pi r_2 v_2 \cos \alpha_2}. \tag{20}$$

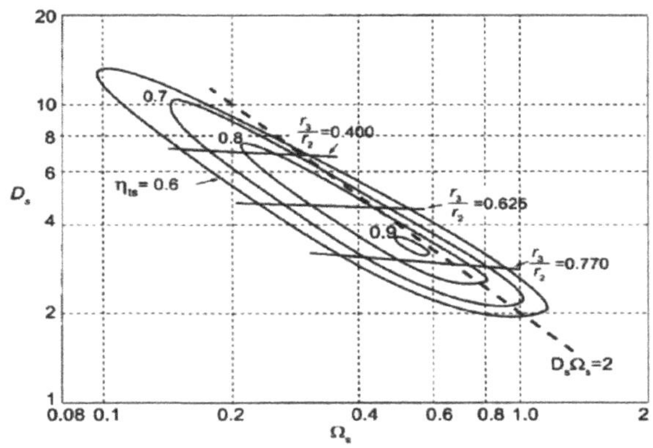

Figure 4. Balje diagram for a radial turbine [Reproduced with permission from [22], Seppo, A. Principles of Turbomachinery; John Wiley & Sons: New York, NY, USA, 2011.].

Using the velocity diagram for calculating the flow angle of relative velocity at the exit:

$$\beta_3 = \tan^{-1}\left(\frac{u_3}{v_3}\right). \tag{21}$$

The flow angle of relative velocity at the hub and shroud at the outlet are:

$$\beta_{3h} = \tan^{-1}\left(\frac{u_{3h}}{v_3}\right), \beta_{3s} = \tan^{-1}\left(\frac{u_{3s}}{v_3}\right). \tag{22}$$

Figure 5 shows an overview of the equations used for calculating the dimensions of the turbine, respectively, where the process for calculating values is continuously correlated using the MATLAB program.

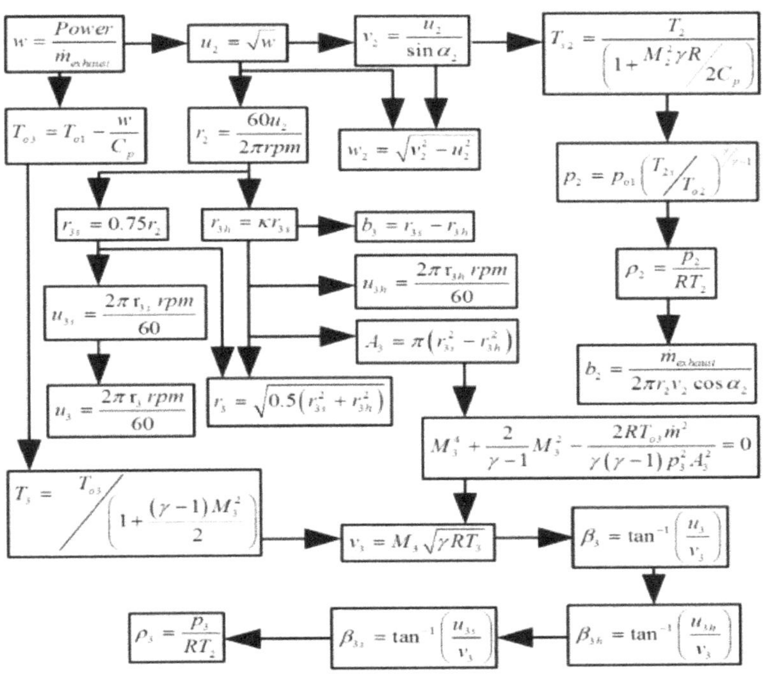

Figure 5. The dimension design of turbine.

3.5. The Dimension Design

The values of the calculated components from the previous sections are summarized and shown in Table 4 in order to create the turbine model as well as applying the values for the turbo-generator drawing.

Table 4. The results from the one-dimensional calculation of the radial turbine.

No	Description	Abbreviation	Unit	Value
1	The flow angle at inlet of impeller	α_2	degree	62
2	The radius at inlet	r_2	m	0.0298
3	The inlet blade height	b_2	m	0.0076
4	The rotor shroud radius	r_{3s}	m	0.0175
5	The rotor hub radius	r_{3h}	m	0.0075
6	The mean exit radius	r_3	m	0.0134
7	The flow angle of relative velocity at outlet	β_3	degree	17.4907
8	The flow angle of relative velocity at shroud	β_{3s}	degree	22.2716
9	The flow angle of relative velocity at hub	β_{3h}	degree	9.9650
10	The blade width at outlet	b_3	m	0.01

3.6. Drawing the Turbo-Generator

The turbo-generator parts use calculated values from Table 3 for drawing in Computer Aided Design (CAD) models of the rotor and volute with the size and shape shown in Figure 6. The overview of all parts is shown in Figure 7.

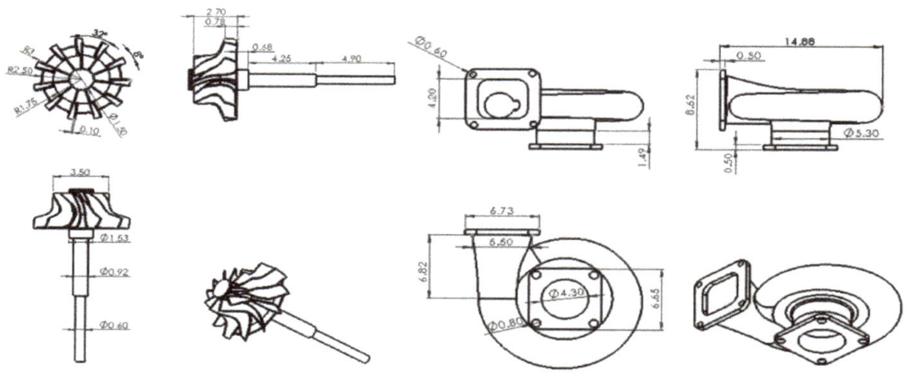

Figure 6. Rotor and volute design.

Part of the turbo-generator model:

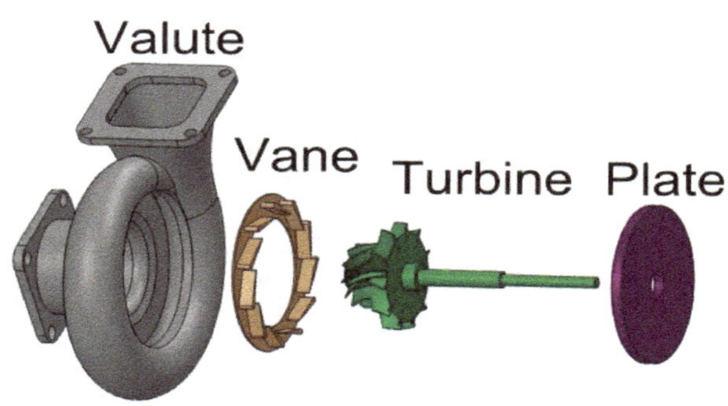

Figure 7. The turbo-generator components.

The size of the new turbine design is shown in Table 5, and compared with the turbine of current vehicles. The differences of dimension occur when the inlet blade angle increases and the outlet blade angle decreases.

Table 5. The geometric features of the turbine compared with turbines in current commercial cars.

Turbine	New Design Rotor	Commercial Rotor1/2 *
Number of blades	11	11
Inlet blade angle	62°	20°/30°
Outlet blade angle	17°	50°/60°
Turbine diameter (mm)	60	60
Vane angle	Varied 62°	No vane
Rotation speed (rpm)	52,000	
Mass flow rate (kg/s)	0.1024	
Temperature inlet/outlet (°C)	400/80	
Pressure outlet (bar)	1.1	

* Toyota Hilux model.

3.7. Drawing and Construction of a Thermoelectric Generator

The output voltage is directly proportional to the temperature change, which is the principle of a thermoelectric generator using the phenomenon characteristics known as the Seebeck effect [16], and displayed as the following equation:

$$V = \alpha \Delta T, \tag{23}$$

where α is the Seebeck coefficient $\left(VK^{-1}\right)$ and ΔT is the temperature difference of two sides of the surface in K.

The Reynolds number derived from the value of the heat source from the engine exhaust gas is:

$$R_e = \rho v D / \mu, \tag{24}$$

where ρ is the density, v is viscosity, D is the equivalent diameter, and μ is the viscosity of the fluid flowing through the tube.

The heat transfer coefficient of the hot side is determined by:

$$h_e = N_u k_e / D_h. \tag{25}$$

The Nusselt number is defined as the ratio of convection heat transfer to fluid conduction heat.

$$N_u = \frac{h_e D_h}{k_e}, \tag{26}$$

where N_u is Nusselt number, k_e is thermo conductivity of exhaust, and D_h is the hydraulic diameter.

The following equation shows convection on the plate:

$$N_u = 0.664\left(\mathrm{Re}^{0.5} \times \mathrm{Pr}^{0.33}\right), \tag{27}$$

where Re is the Reynolds number and Pr is the Prandtl number.

The heat conversion efficiency of waste heat recovery is calculated using the energy from the thermoelectric generator divided by the heat inserted into the thermoelectric generator.

$$\eta = \frac{P_{output}}{\dot{m} C_p (T_{in} - T_{out})}, \tag{28}$$

where η is the conversion efficiency, P_{output} is the thermoelectric generator power output, $\dot{m}$ is the exhaust gas mass flow rate, C_p is the exhaust gas specific heat, T_{in} is the exhaust gas system inlet temperature, and T_{out} is the exhaust gas system outlet temperature.

The equation to calculating the power generated by the thermoelectric generator is:

$$P = N\alpha_{pn} I \Delta T_{leg} - I^2 N R_{pn}, \tag{29}$$

where N is the number of thermoelectric couples employed, α_{pn} is the Seebeck coefficient, I is, respectively, the electric current, ΔT_{leg} is the thermoelectric leg temperature difference, and R_{pn} is the value of the thermoelectric resistivity couple.

The equation for the system efficiency, η, can be calculated by:

$$\eta = \frac{p_{TEG}}{P_{engine}} \times 100\%, \tag{30}$$

where p_{TEG} is the thermoelectric generator maximum output power and P_{engine} is the power of the engine.

The efficiency of a TE module used as a generator can be approximated by the following relationship, where Z is a material property, T_c is cold temperature, T_h is hot temperature, and T is $(T_h + T_c)/2$.

$$\eta_{TE\max} = \frac{W_{elec}}{Q_h} = \frac{\Delta T}{T_h} \cdot \frac{\sqrt{1+ZT-1}}{\sqrt{1+ZT+\frac{T_c}{T_h}}}, \tag{31}$$

$$Z = \frac{\left(\alpha_p - \alpha_n\right)^2}{\left(\left(\lambda_p \rho_p\right)^{1/2} + \left(\lambda_n \rho_n\right)^{1/2}\right)^2}, \tag{32}$$

where α_p is the Seebeck coefficient corresponding to p, α_n is the Seebeck coefficient corresponding to n, λ_p is thermal conductivity corresponding to p, λ_n is thermal conductivity corresponding to n,ρ_p is electrical resistivity the corresponding to p, and ρ_n is electrical resistivity corresponding to n.

The power output is:

$$P = Q_h - Q_c = I^2 R_L, \tag{33}$$

where I is the electrical current in the generator circuit, R_L is the electric resistance of semiconductor couple, Q_h is the heat absorbed from heat source, and Q_c is the heat absorbed from the cold source.

Power outputs and conversion efficiencies were calculated applying the numerical results, the electric current and absorbed heat shown in the equations below:

$$P = N\left[\alpha_{pn}(T_h - T_c)I - I^2 R\right], \tag{34}$$

$$R = \frac{L}{A}\left(\rho_n + \rho_p\right), \tag{35}$$

$$\eta = \frac{P}{Q_h}, \tag{36}$$

where P is the output power, N is the number of thermoelectric elements in the module, and α_{pn} is the Seebeck coefficient. L is the length of the legs and A is the cross-sectional area.

The equation for the thermocouple conversion efficiency is:

$$\eta = P/Q_h, \tag{37}$$

where Q_h is the absorbed and P is the output power.

Another important part of the process for designing a thermo-electric generator is finding the heat transfer correlations, which is calculated from the heat of the engine exhaust flowing through the exhaust pipe. The design will focus on the suitability and ease of installation with real engines in a limited space to make the best heat transfer efficiency and to not extremely affect the pressure of the exhaust. Figure 8 illustrates the heat transfer correlations of the thermo-electric generator.

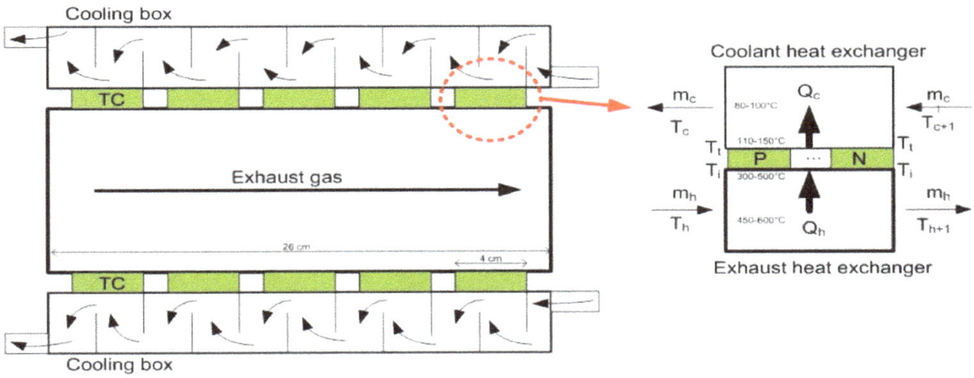

Figure 8. Heat transfer correlation of the thermoelectric module.

Energy balance of cooling heat exchanger calculated by:

$$Q_c = m_c c_p (T_{c+1} - T_c), \tag{38}$$

$$Q_c = h_c A_c (T_c - (T_c + T_{c+1})/2), \tag{39}$$

where h_c is the heat transfer coefficient for the coolant heat exchanger, A_c is the heat transfer area for the coolant side, m_c is the mass flow rate, T_c is the coolant water temperature, and T_t is the temperature of the cold side of the thermoelectric module.

Energy balance of exhaust heat exchanger calculated by:

$$Q_h = m_h c_p (T_h - T_{h+1}), \tag{40}$$

$$Q_h = h_h A_h (T_h - (T_h + T_{h+1})/2 - T_i), \tag{41}$$

where h_h is the heat transfer coefficient for the exhaust heat exchanger, A_h is the heat transfer area for the exhaust side, m_h is the mass flow rate, c_p is the specific heat of the exhaust, T_h is the temperature of the exhaust, and T_i is the temperature of the hot side of the thermoelectric module.

Heat and cooling transfer of the system calculated by:

$$Q_c = SIT_c + K(T_h - T_c) - 0.5I^2 R, \tag{42}$$

$$Q_h = SIT_h + K(T_h - T_c) - 0.5I^2 R, \tag{43}$$

where S is the Seebeck coefficient, R is the internal resistance, K is the thermal conductance of the module, and I is the total current of the generator.

Calculations for efficiency of a thermo-electric generator:

The exhaust gas mass flow rate was 0.1024 kg/s, the exhaust gas specific heat c_{pg} was $1kJ/kg.k$, and the supplied heat was: $the\ exhaust\ gas = m_g c_{pg} \Delta T = 0.1024 \times 1.148 \times (110 - 115) = 587.776$ W, $electrical\ power\ output = VI = 16 \times 5 = 80$ W

$$efficiency\ of\ thermoelectric\ \text{generator} = \frac{electrical\ power\ output}{heat\ supplied\ by\ the\ exhaust\ gas} = 13.61 \tag{44}$$

The thermoelectric module in Figure 9 composed of two units with dimensions of 240 mm × 100 mm × 300 mm. It is a combination of three main parts: The water cooling box, heat exchange box, and thermoelectric plate ($TECI - 12706$). The water cooling box has eight parts with dimensions of 70 mm × 40 mm × 300 mm. It was used as a thermo-electric plate to reduce the heat by removing the heat from the water using the radiator. The heat exchange box has dimensions of 100 mm × 100 mm × 300 mm to transfer the heat from hot water to the 80 thermoelectric plates. The design will focus on the suitability and ease of installation with real engines in limited space to make the best heat transfer efficiency and to not extremely affect the pressure of exhaust.

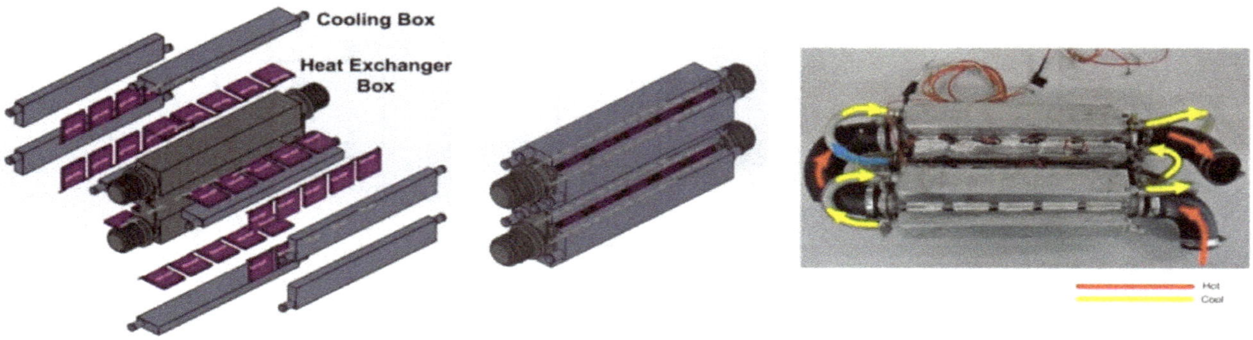

Figure 9. Thermoelectric generator module.

4. Simulation Results and Optimization

4.1. Turbine Performance

The steps of the CFD simulation started from design and drawing, which was shown in Figure 10 by modeling the adjustment of the vane angle at 40°, 50°, 62°, 80°, and 85° respectively. Then, the step was the model preparation for using in the simulation process. The step started from the CFX-Design Modeler procedure which is an important step of ANSYS simulation. The total number of faces that resulted from this procedure was 358 faces, separated into turbine 282 faces, exhaust gas 28 faces, and vane 48 faces. After that was the surface preparation procedure using the section of the CFX-Meshing. The total value of the model was 914,018 nodes and 3,461,517 elements, divided into turbine 509,734 nodes and 2,545,575 elements, exhaust gas 132,988 nodes and 688,147 elements and vane 271,296 nodes and 227,768 elements.

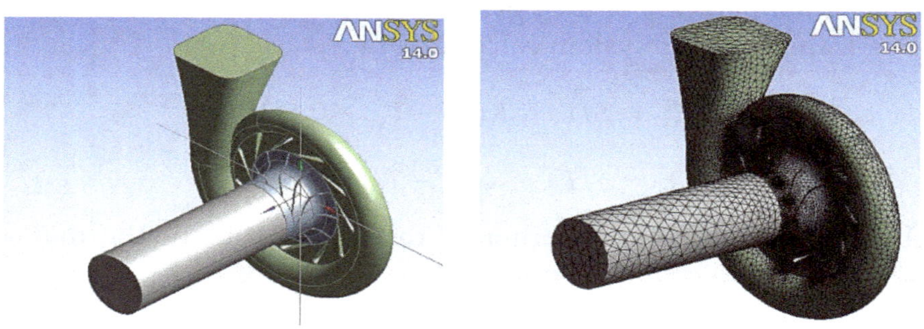

(A) **(B)**

Figure 10. Model preparation. (**A**) CFX-design modeler and (**B**) CFX-meshing.

The next step was the process of simulating the flow condition in the model. The simulation simulated the exhaust flow that flowing into the inlet of the turbine case and simulated the flow out of the front. There are various configurations in this step including the exhaust flow rate, pressure temperature, and flow characteristics details as follows: The CFX-Pre procedure had three domains and three interfaces, the exhaust domain had a static inlet, an inlet mass flow rate 0.1024 kg/s, and outlet pressure at 101.325 kPa. Turbine domain configured a rotation speed between 15,000 and 35,000 rpm and the vane domain was defined as the static flow at 400 °C as shown in Figure 11.

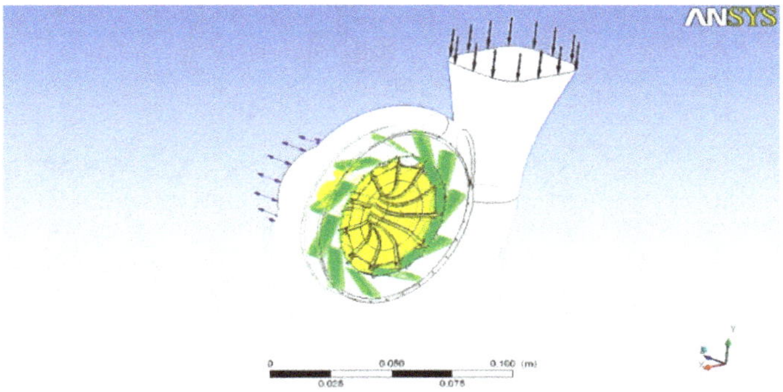

Figure 11. CFX-pre simulation.

The results of the simulation shown in Figure 12. In the simulation, there were eight different vane angle simulations including not vane, vane angle 40°, 50°, 62°, 80°, and 85°, and commercial 1, 2. The simulation was set with a rotation speed from 15,000 to 35,000 rpm.

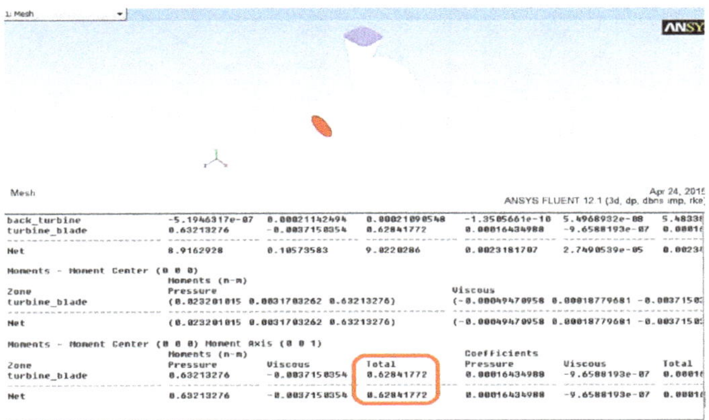

Figure 12. Moment on the turbine.

The torque on the turbo blade was 0.628 N − m. This was used to calculate the power, using this equation:

$$P = T \times 2 \times \pi \times \frac{N}{60} = 0.628 \times 2 \times \pi \times \frac{52000}{60} = 3401 \text{ watt}. \quad (45)$$

Table 6 shows the comparison of torque and power for adjusting the inlet blade at various angles without the vane and commercial turbines.

Table 6. Turbine performance.

	None	40°	50°	62°	70°	80°	Commercial 1	Commercial 2
Torque	0.138	0.355	0.525	0.624	0.578	0.502	0.259	0.289
Power	752	1937	2858	3401	3147	2734	1411	1573

The results of the simulation for the input parameter exhaust inlet was 0.1024 kg/s (3500 rpm; engine rotation speed), for the pressure outlet was 1.1 bars, and for the temperature was 400 °C. The inlet angle set to five variables with a vane angle of 40°, 50°, 62°, 70°, and 80° as shown in Figure 13. The results shown in Table 5 and Figure 14A suggest torque and power increased if the rotation speed increased. The turbine could generate from 1937 watts of power with an engine speed of 4200 rpm, when the vane angle varied from 40° (1937 watts) to 62° (3401 watts). After the vane angle increased to more than 62°, a reduction of power occurred.

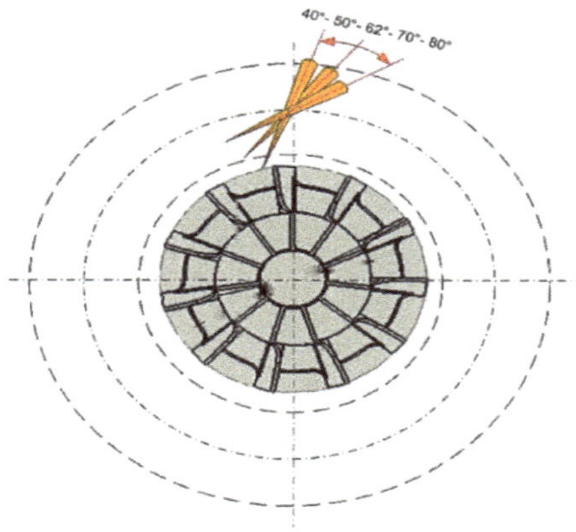

Figure 13. The variable vane angled; 40°, 50°, 62°, 80°, and 85°.

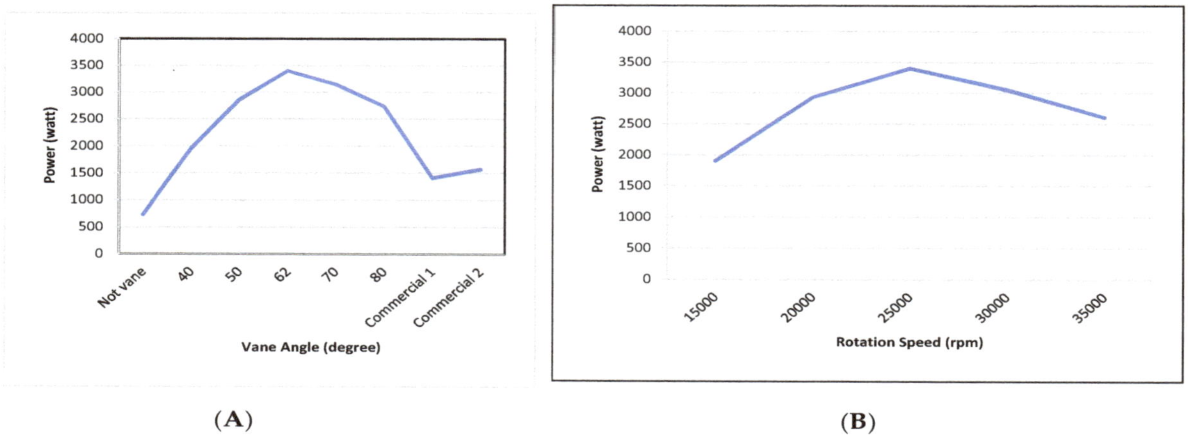

Figure 14. Generated power results. (**A**) Results showing the vane angle and power and (**B**) results showing the rotation speed and power.

The results from the simulation of the relationship between the generated power and rotation speed are shown in Figure 14B, the power output increased to 3401 watts at 25,000 rpm.

4.2. Pressure and Path Line of Flowing

The value of the pressure is shown in Figure 15. Pressure reached the maximum at the inlet and was lower at the outlet. The value at the inlet was 1.67 atm, at the volute was 1.46 atm, at the turbine blade was 1.20 atm, and at the outlet was 1 atm.

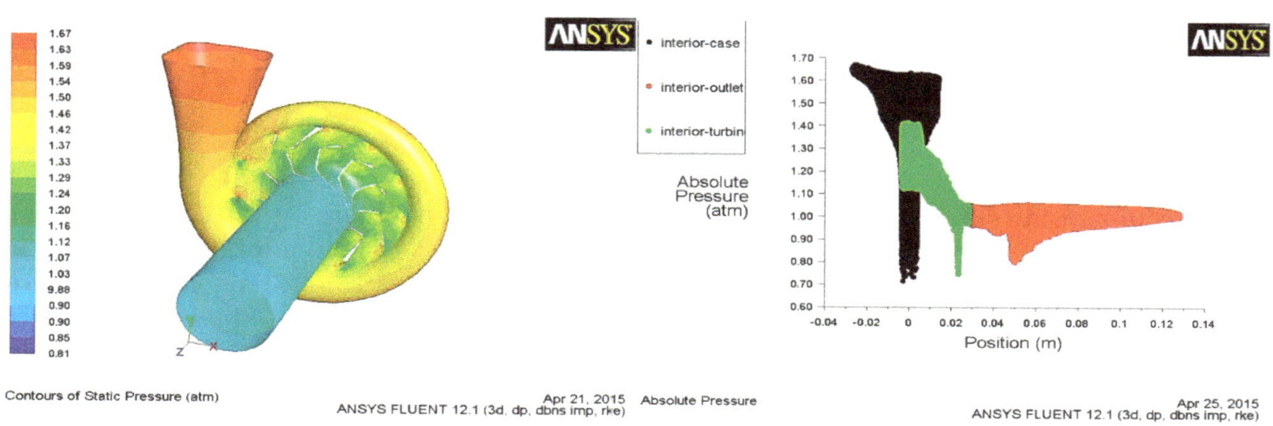

Figure 15. Pressure at all turbine walls.

The value of the pressure was varied as shown in Figure 16. The highest pressure was 1.5 atm and the lowest at 1 atm. Pressure in the turbine blade in Figure 17 shows the characteristic of the exhaust gas flow in the gap between the rotor and volute, expressed as a flowing line from the entrance through the rotor blade and flowed out in front of the rotor.

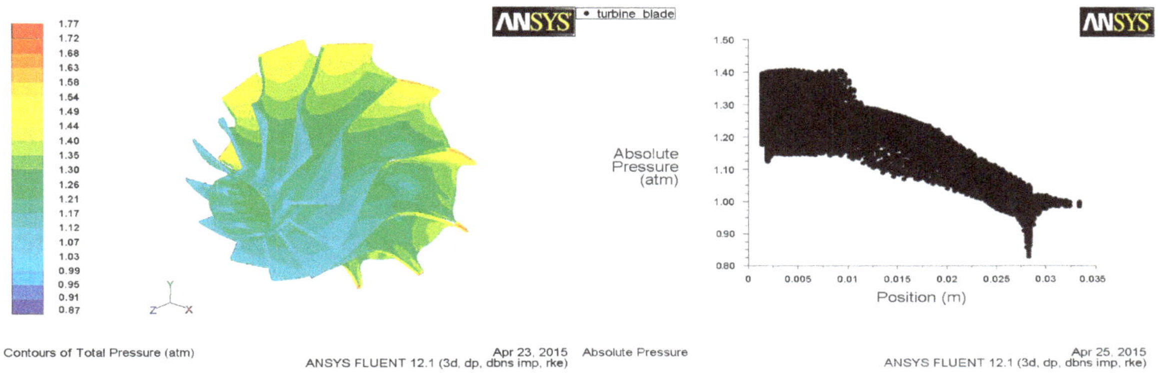

Figure 16. Pressure at the turbine blade.

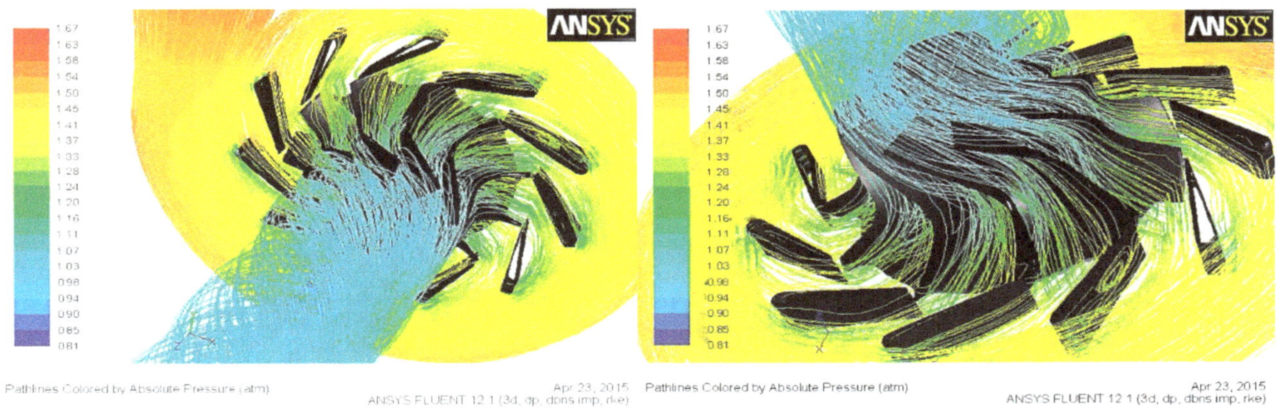

Figure 17. Path line of flowing at the turbine blade.

The sound speed at the turbine blade was 490–530 m/s. This sound speed was an acceptable value for the shock wave design as shown in Figure 18.

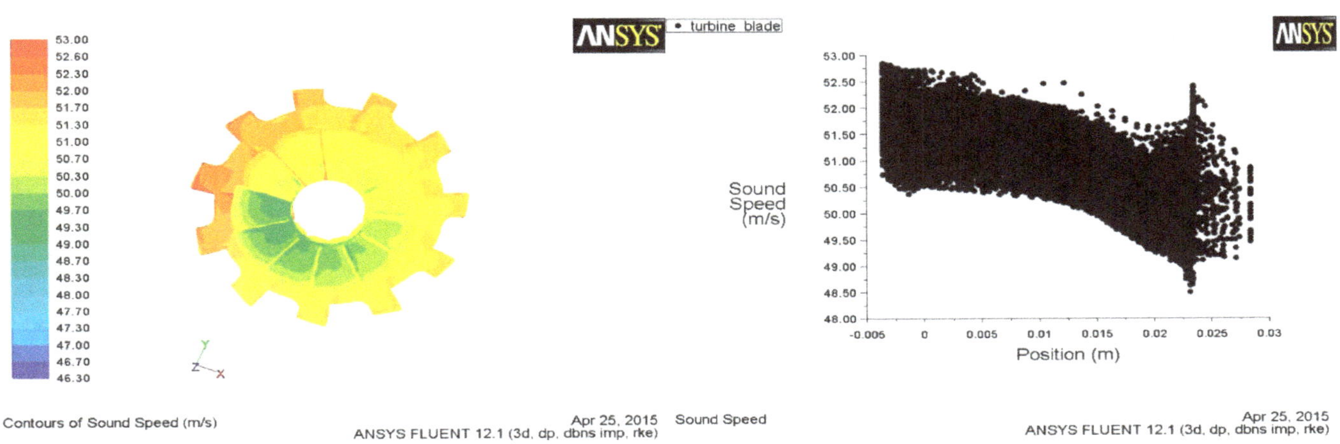

Figure 18. Sound speed at the turbine blade.

5. Experimental Results

5.1. Experimental Setup

The experimental facility shown in Figure 19 used in this study was based on a real internal combustion engine, and experiment testing works without load. The experimental design for the experiment consists of four main parts: The engine building, turbo-generator unit, thermoelectric-generator unit, and parameter output measurement. The engine was based on a 2500 cc. diesel engine (Toyota 2LII model). The engine consisted of a water-cooling system, oil lubricating system, and engine electric system. The turbo-generator consisted of a turbine, volute, vane, and generator. The connection of the turbine to the generator used coupling and the gear reducer rotation speed. The thermoelectric-generator consisted of a water-cooling box, heat exchange box, and thermoelectric plate. Finally, the performance measurement aimed to find the engine rotating speed, turbo-generator rotating speed, temperature and pressure measurement, exhaust gas measurement, and finally the power measurement of the turbo-generator and thermoelectric-generator.

Each experiment performed 10 times with the same condition. The values from results recorded after the engine runs 20 min so that the engine was working in the heat condition. The speed of the engine was adjusted with step 100 rpm started from 1000 to 3600 rpm. The values recorded consisted of the engine rotation speed, inlet and outlet temperature of the turbo-generator, inlet and outlet temperature of the thermoelectric-generator, the power value from the turbo-generator, and the power value from the thermoelectric-generator.

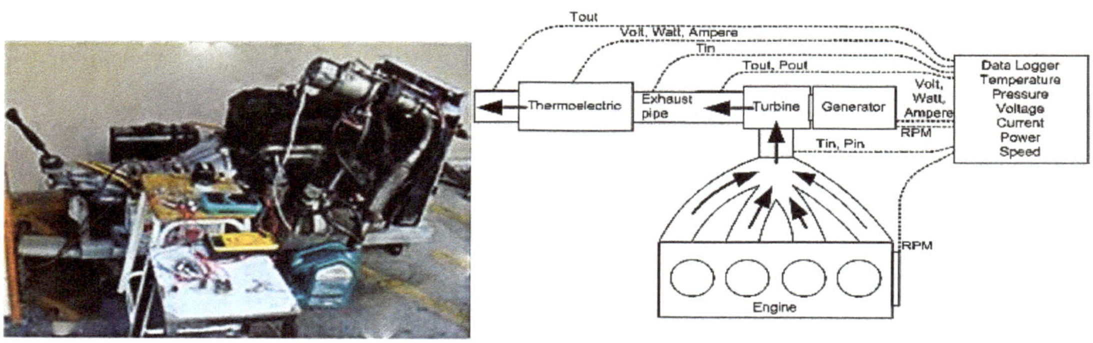

Figure 19. The real internal combustion engine test.

5.2. The Measurement Intrument

The set of instruments for measuring values from the experiments were created specifically. The main components were divided into three parts: The processing and interpret sections, data recording section, and the measuring sensors detailed as follows:

The first section was the processing section using the STM32F407 microcontrollers for receiving the values from the sensor and displaying the values with the LCD monitor.

Since you have not mentioned about the details of this sensors/apparatus, so that in order to understand the reported results the measurement accuracy of the discussed

The second part was the section for receiving values from the microcontroller to store in the SD card. The SD card module recorded the values in two ways: When the value changed and recorded the values every second.

The third part was the sensor that sent various values in both the digital and analog signal to the microcontroller composing of the following values: Engine rotation speed, turbo generator rotation speed, inlet temperature of the turbo generator, outlet temperature of the turbo generator, inlet temperature of thermoelectric, outlet temperature of thermoelectric, inlet pressure of the turbo generator, outlet pressure of the turbo generator, output power turbo generator, and output power of thermoelectric. The details of the sensor used as following: Engine rotation speed and turbo generator rotation speed used the hall effect proximity switch (NJK-5002C), inlet temperature of the turbo generator, outlet temperature of the turbo generator, inlet temperature of thermoelectric, and outlet temperature of the thermoelectric used resistance temperature detector (PT100 RTD), inlet pressure of the turbo generator, and outlet pressure of the turbo generator used MEMS pressure sensor (XGZP701DB1R), output power turbo generator, and output power of thermoelectric used current detection sensor module (WCS1800).

These instruments are comparable to the standard instruments and adjusted to the precise value to solve the problem of variance of experimental results, as the results of each trial result, the value of the measurement instrument is different, but there will be a tendency or appearance of the same rise or decline. This causes different values due to a number of factors such as how long it takes to start an extended engine to heat up the inside of the engine, and the heat in the exhaust is more as well. The acceleration of the engine at each time of the trial is not fixed. Sometimes the speed of the cycle may be accelerated or sometimes slow acceleration. This affects the exhaust pressure and the heat of the exhaust, and the installation in experiments does not start at the same heat, can affect the various values that come out. Therefore, it uses multiple experiments to find the average value that makes the measured value more reliable. Before the actual experiments, the trial operator was experimented 20 times to make the actual experiment the most accurate. In conclusion, each test was made when the engine was at a normal temperature (before starting the engine). The values recorded starting from the operating temperature of the engine that was 80 degrees Celsius. The engine acceleration gradually accelerated from the light to the peak. In each trial, there was a time equivalent of 1 h. The experiment would repeat 20 times in each experiment and then select 10 times of experimental results with the lowest variance and maximum reliability. The selected variance with uncertainty between ± 0.1%–3.0%.

5.3. Temperature of the Exhaust Turbo-Generator

The temperature experiment on the turbo-generator was undertaken by running the engine and adjusting the engine idle speed to the maximum speed. The temperature sensor measurement was installed at the inlet and outlet of the turbo-generator as shown in Figure 20. The graph shows the correlation value between the rotation speed engine and exhaust gas temperature. The trial operation repeated 10 times for calculating the average value, as shown in the picture. The experiments consisted of measurements inlet and outlet exhaust gas temperature in the turbo-generator and recorded the values from the starting point of the engine (600 rpm) and accelerated the speed of the engine to the maximum speed (4700 rpm). The X-axis displays the engine rotation speed in rounds per minute (rpm). The Y-axis displays the exhaust gas temperature in degrees Celsius. The red line shows the outlet exhaust gas temperature and blue line shows the inlet exhaust gas temperature.

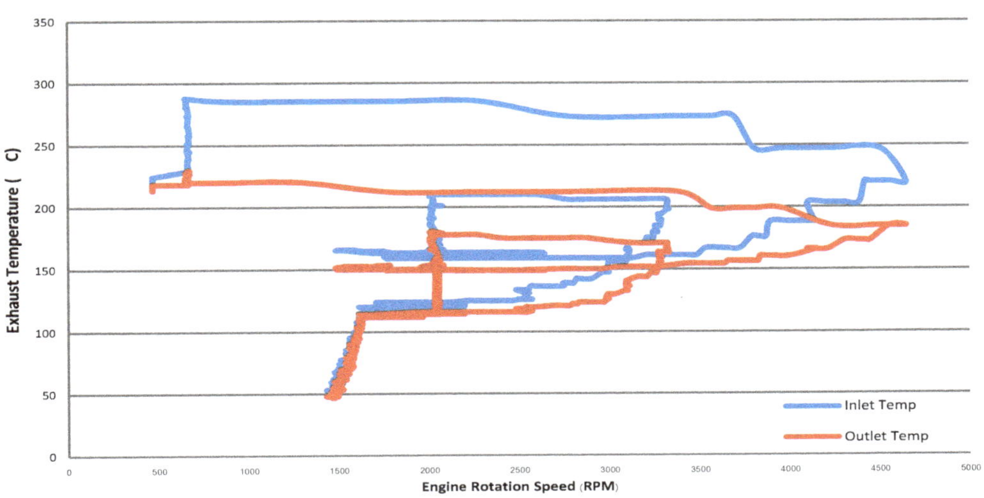

Figure 20. The inlet and outlet exhaust temperature.

The temperature of the exhaust gas depends on the rotation speed of the engine and the engine running time. At the beginning of the rotation, the exhaust gas temperature was lower. It will increase as the speed of the engine increases, as a result of the heat combustion of exhaust gas. The inlet temperature was higher than the outlet temperature since the direct proportion characteristics.

5.4. Pressure of the Exhaust Turbo-Generator

Pressure measurements of the turbo-generator were taken by running the engine and adjusting the speed of the accelerator from idle speed to the maximum speed. The measurements were taken at the inlet and outlet of the turbo-generator as illustrated in Figure 21. The graph shows the correlation value between the rotation speed engine and exhaust gas pressure. The trial was operated 10 times for calculating the average value as shown in the picture. The experiments consisted of the measurements inlet and outlet exhaust gas pressure in the turbo-generator and recording the values from the starting point of the engine (600 rpm) and increase the speed of the engine to the maximum speed (4700 rpm). The X-axis displays the engine rotation speed in rounds per minute (rpm). The Y-axis displays the exhaust gas pressure in kilopascals. The red line shows the inlet exhaust gas pressure and the blue line shows the outlet exhaust gas pressure.

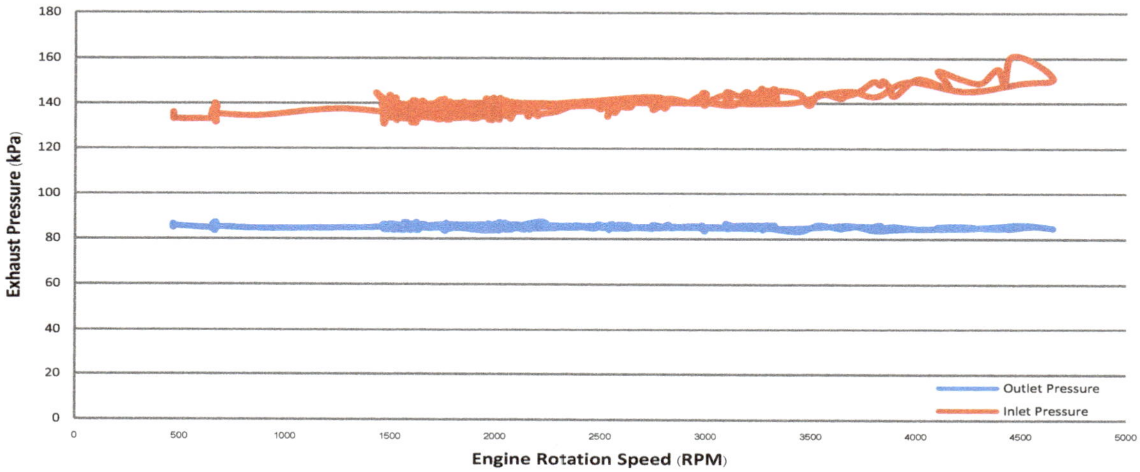

Figure 21. Inlet and outlet exhaust pressure.

The rotation speed of the engine and runtime engine defines the pressure of the exhaust gases. During the initial stage, the rotation speed of the pressure was low. It will increase as the speed increases, as a result of the exhaust gas flow rate. The outlet pressure was lower than the inlet pressure.

5.5. Rotation Speed of the Turbo-Generator Test Based on a Variable Vane Angle

The rotational test of the rotation speed of the turbo-generator was undertaken by running the engine and adjusting the accelerator speed from idle to maximum speed, along with an adjustment of the vane angle testing set. The results are shown in Figure 18. The graphs show the correlation value between the rotation speed engine and the turbo-generator rotation speed with 10 times the repeat experiments for calculating the average value, as shown in the graph. The experiments measured the engine rotation speed in the pulley position of the engine, measured the turbo-generator rotation speed in the axis position of the turbo-generator and recorded the value from the starting point of the engine at a lighter speed (700 rpm), and then accelerated the engine speed to the maximum speed (3600 rpm). The X-axis shows the engine rotation speed in rpm. The Y-axis shows the turbo-generator rotation speed in the rpm. The red line shows a 52° vane angle turbo-generator rotation speed, the green line showing a 62° vane angle turbo-generator rotation speed, and the blue line displaying a 72° vane angle turbo-generator rotation speed.

Figure 22 shows three different conditions of the engine speed and vane angle. When the vane angle was 62°, speed increased at a stable rate and had a tendency to continuously increase. When the vane angle was 52°, the turbo-generator started to rotate at a high engine speed. When the vane angle was 72° the turbo-generator speed increased as the engine speed increased, and it tended to decrease at a higher engine speed. Increasing the number of vanes decreased the axial flow velocity and increased the turbulence energy [23].

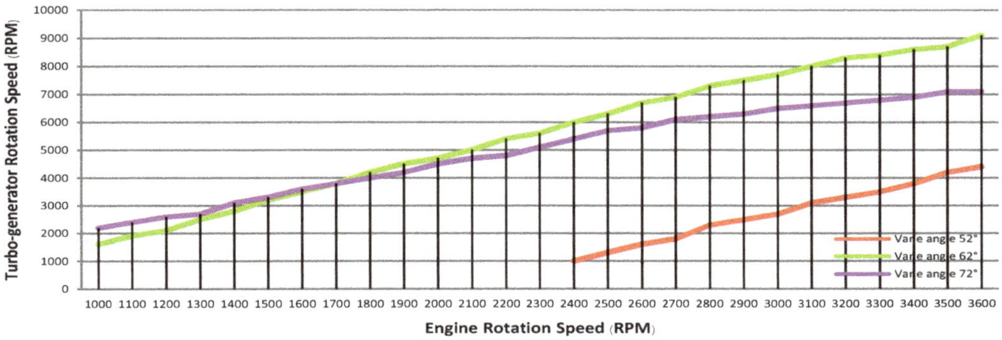

Figure 22. Rotation speed of the turbo-generator test.

5.6. Power of the Turbo-Generator Test Based on Adjusting the Vane Angle

The power test of the rotation speed of the turbo-generator was undertaken by running the engine and by adjusting the accelerator speed from idle speed to maximum speed. The vane angle was adjusted for each testing set as shown in Figure 19. The graphs display the correlation value between the rotation speed engine and the turbo-generator power by performing 10 iterations of the experiment and calculating the average value. The experiment consisted of measuring the engine rotation speed in the pulley position of the engine and the turbo-generator power in the axis of output power of turbo-generator. Then, the values recorded from the starting point of the engine at 700 rpm and accelerated the speed of the engine to the maximum speed (3600 rpm). The X-axis is a rotation speed engine in rpm. The Y-axis is the turbo-generator power in watts. The red line represents the 62° vane angle turbo-generator rotation power, the green line represents the 72° vane angle turbo-generator rotation power, and the blue line represents the 52° vane angle turbo-generator rotation power.

Figure 23 illustrates the correlation value of the engine rotation cycle with the capacity outputted from the turbo-generator with the angle adjustment of three vane levels. The total output of energy was increased when the cycle speed of the turbo-generator engine was increased with adjusting the angle of the vane at 62° and 72°. The engine could produce energy from the beginning at low speed up to the maximum speed while the turbo-generator adjusted the angle of the vane at 52° must be added to the engine rotation cycle to 2400 rpm to start producing power output. As shown in the graph, the turbo-generator with the vane at 62° had a capacity of producing the maximum energy starting from 150 watts at the engine rotation around 900 rpm and increased by rotation cycle to 900 watt at 3600 rpm.

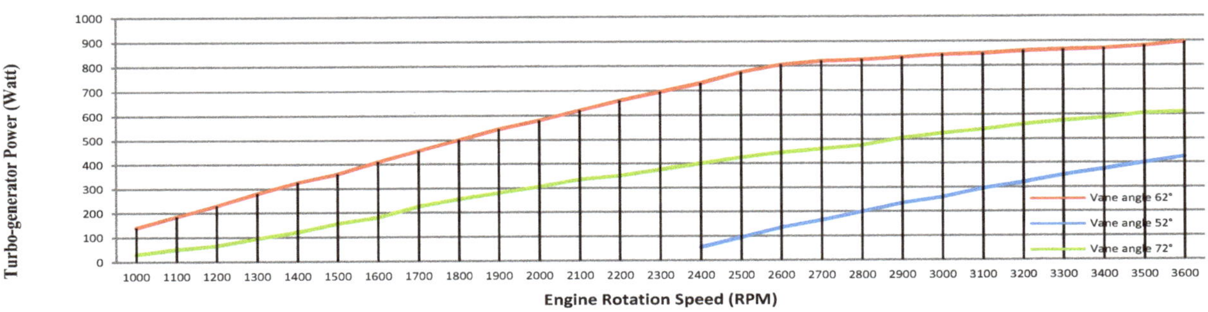

Figure 23. Power of the turbo-generator test.

Turbine efficiency:

$$\eta_t = \frac{T_{00} - T_{20}}{T_{00} - T_{2s}} = 51.1\%. \tag{46}$$

Total Shaft Output Power:

The reduction of turbo-generator energy can be divided into three speed of different power: Idle speed (145 watts), running speed (870 watts), and high speed (890 watts).

5.7. Power of the Thermoelectric-Generator Test

The power output from the thermoelectric generator shown in Figure 24. The graph shows the relationship value between the engine running time and the thermoelectric-generator output power, with 10 repeat experiments for calculating the average value. The experiments consist of measuring the engine running time when starting the engine and measuring the thermoelectric-generator output power in the power cord position. The values were recorded from the starting point of the engine at 700 rpm, and accelerated the speed of the engine to the maximum speed (3600 rpm). The X-axis is the engine running time, showing the value in seconds. The Y-axis is the thermoelectric-generator output power display value in watts. The red line is the inlet exhaust gas temperature, the yellow line is the outlet exhaust gas temperature, the green line is the output voltage, the purple line is the output current, and the blue line is the output power. The line-graph illustrates the exhaust temperature and power when running the engine. Overall, the power output is related to the temperature difference. To begin with, the power output started at 271 watts, before it increased to just under 392 watts at a higher temperature. This fell slightly when the engine was stopped, before dropping to its lowest point of 288 watts at the end of testing.

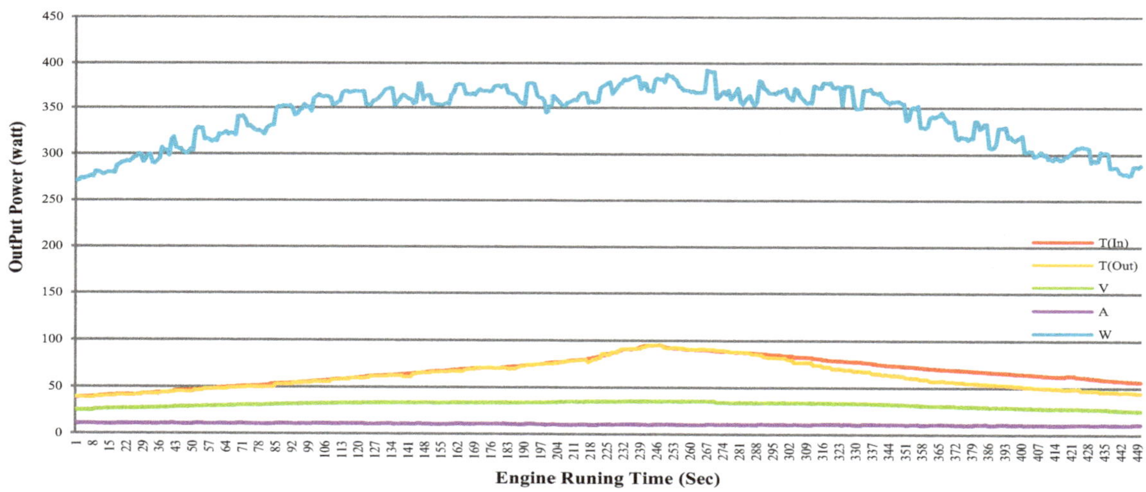

Figure 24. Power of the thermoelectric-generator test.

Figure 25 shows the correlation between the rotation speed of the engine and output power of the whole system. The trial operation was repeated 10 times for calculating the average value. The X-axis represents the engine rotation speed. The Y-axis represents the output power. The red line represents the energy from the turbo-generator. The green line represents the energy from the thermo-electric generator, and the blue line represents the total power from the turbo-generator and thermo-electric generator. The total power saving when combining both the turbo-generator and thermo-electric generator units was 1262 watts.

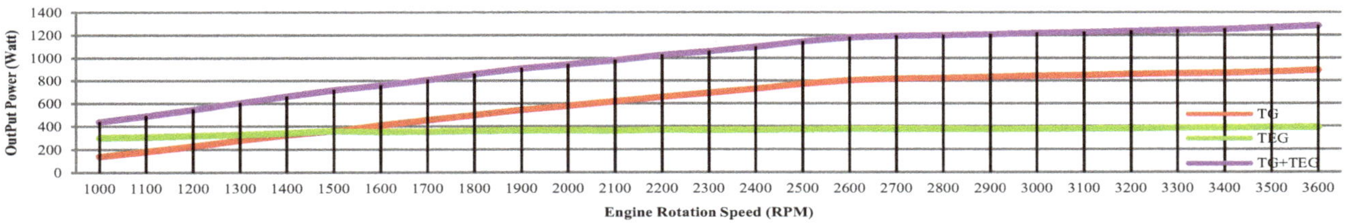

Figure 25. Total output power.

6. Conclusions

Turbo-generators and thermoelectric generators are ideal for the recovery of waste energy in exhaust gas. The temperature and flow pressure of the combustion engine is transformed into electric power. That power can be converted into electrical energy to support the electrical supply in a vehicle. Normally, exhaust gas is discarded to the environment as waste gas. The main difference between this research and previous works is the type of energy from exhaust gases for recycling. Most of previous works used either pressure or heat whilst this research focus on the discharge of energy from the exhaust with both pressure and heat recovery to the form of electrical energy. In addition, this research designed the prototype for a 2500 cubic centimeters diesel engine, which is widely used. The research aimed to design a turbo-generator and thermoelectric generator that could convert exhaust gas energy to electrical energy. The turbo-generator model could generate up to 870 watts of power at 3400 rpm (top speed of the engine), and the thermoelectric generator could generate up to 392 watts of power. Both systems combined could generate up to 1262 watts of power. The electric power can be used in electric charging; generally, the alternator will produce 12 volts, or 35 amperes (420 watts), to support in-vehicle usage. The experiments were a comparison of the results from the simulation from the design and optimization of the vane angle of 40°, 50°, 62°, 70°, and 80° with the results of the installation and the actual test. The result shows that everything was consistent in the design with the angle vane 62°. Besides that, in the simulation process, the result shows that the angle vane gave the maximum torque and power, which corresponded to the same direction as the outcome of the actual trial. The integrated turbo-generator and thermoelectric generator could potentially be used as part of the charging system. Energy recovery during driving can be used for battery charging and as an energy storage device. In this model, the turbo-generator and thermoelectric generator had the capacity to generate up to 1262 watts that could be utilized in hybrid vehicles. The power from the recovery of waste energy could support the electrical needs of other parts. At present, the device converts the power into electricity, and for the convenience of the driver, it could be utilized for features such as electric air conditioning, electric power steering, refrigerators, and mobile phone chargers.

Author Contributions: P.N. contributed to the conceptualization, methodology, validation, and data collection, and wrote the paper; M.E., U.N. and E.B. made corrections to the paper and gave some useful recommendations.

Acknowledgments: The authors would like to thank the Rajamangala University of Technology Srivijaya for scholarship in doctoral study.

References

1. Yang, J. Potential applications of thermoelectric waste heat recovery in the automotive industry. In Proceedings of the International Conference on Thermoelectrics, Clemson, SC, USA, 19–23 June 2005; pp. 155–159.
2. Xiong, Y.; Xu, L.; Zhong, Y. *The Theory and Application of Energy-Saving Technology on Vehicle*; China Petrochemical Press: Beijing, China, 2006; pp. 9–13.
3. Fu, J.; Liu, J.; Yang, Y.; Yang, H. A study on the prospect of engine exhaust gas energy recovery. In Proceedings of the International Conference on Electric Information and Control Engineering, Wuhan, China, 15–17 April 2011.

4. Jorge, V. State of the Art of Thermoelectric Generators Based on Heat Recovered from the Exhaust Gases of Automobiles. In Proceedings of the 7th European Workshop on Thermoelectric, Pamplona, Spain, 3–4 October 2002.
5. La Grandeur, J.; Doug Crane, D.; Eder, A. Vehicle Fuel Economy Improvement through Thermoelectric Waste Heat Recovery. In Proceedings of the 11th Diesel Engine Emissions Reduction (DEER) Conference, Chicago, IL, USA, 21–25 August 2005.
6. Fairbanks, J. Overview of High Efficiency Thermoelectric and Potential Applications. In Proceedings of the High Efficiency Thermoelectric Workshop, San Diego, CA, USA, 17–20 February 2004.
7. Michon, M. Switched Reluctance Turbo-Generator for Exhaust Gas Energy Recovery. In Proceedings of the International Conference on Electric Machines & Drives Conference (IEMDC), Niagara Falls, ON, Canada, 15–18 May 2011; pp. 1609–1614.
8. Aman Mamat, M.I. Design and Development of a Low Pressure Turbine for Turbocompounding Applications. *Int. J. Gas Turbine Propuls. Power Syst.* **2012**, *4*, 1–8.
9. Hopmann, U.; Algrain, M.C. Diesel Engine Electric Turbo Compound Technology. In *Proceedings of the Future Transportation Technology Conference & Exhibition*; SAE International: Warrendale, PA, USA, 2003. [CrossRef]
10. Hopmann, U. Diesel engine waste heat recovery utilizing electric turbocompound technology. In Proceedings of the Catterpillar, DEER Conference, San Diego, CA, USA, 30 August–2 September 2004.
11. Hopmann, U.; Algrain, M. Diesel engine waste heat recovery utilizing electric turbocompound technology. Presented at the 2003 DEER Conference, Newport, Rhode Island, 25–28 August 2003; Caterpillar Inc.: Deerfield, IL, USA, 2003.
12. Bowman Power Group Ltd. Available online: http://www.bowmanpower.com/power_generators.php (accessed on 20 August 2017).
13. Controlled Power Technologies Limited. Available online: http://www.cpowert.com/products/tigers/ (accessed on 20 August 2017).
14. Srinivasan, K.K.; Mago, P.J.; Krishnan, S.R. Analysis of exhaust waste heat recovery from a dual fuel low temperature combustion engine using an Organic Rankine Cycle. *Energy* **2010**, *35*, 2387–2399. [CrossRef]
15. Yamada, N.; Minami, T.; Mohamad, M.N. Fundamental experiment of pumpless Rankine-type cycle for low-temperature heat recovery. *Energy* **2011**, *36*, 1010–1017. [CrossRef]
16. Liu, X.; Deng, Y.D.; Chen, S.; Wang, W.S.; Xu, Y.; Su, C.Q. A case study on compatibility of automotive exhaust thermoelectric generation system, catalytic converter and muffler. *Case Stud. Therm. Eng.* **2014**, *2*, 62–66. [CrossRef]
17. Kim, T.Y.; Negash, A.A.; Cho, G. Waste heat recovery of a diesel engine using a thermoelectric generator equipped with customized thermoelectric modules. *Energy Convers. Manag.* **2016**, *124*, 280–286. [CrossRef]
18. Liang, X.; Sun, X.; Tian, H.; Shu, G.; Wang, Y.; Wang, X. Comparison and parameter optimization of a two-stage thermoelectric generator using high temperature exhaust of internal combustion engine. *Appl. Energy* **2014**, *130*, 190–199. [CrossRef]
19. Tian, H.; Sun, X.; Jia, Q.; Liang, X.; Shu, G.; Wang, X. Comparison and parameter optimization of a segmented thermoelectric generator by using the high temperature exhaust of a diesel engine. *Energy* **2015**, *84*, 121–130. [CrossRef]
20. Temizer, İ.; İlkılı, C. The performance and analysis of the thermoelectric generator system used in diesel engines. *Renew. Sustain. Energy Rev.* **2016**, *63*, 141–151. [CrossRef]
21. Liu, C.; Pan, X.; Zheng, X.; Yan, Y.; Li, W. An experimental study of a novel prototype for two-stage thermoelectric generator from vehicle exhaust. *J. Energy Inst.* **2016**, *89*, 271–281. [CrossRef]
22. Seppo, A. *Principles of Turbomachinery*; John Wiley & Sons: New York, NY, USA, 2011.
23. Luft, M.; Olszowiec, P. Losses of IC engine—A chance for electrical energy recuperation. *Zesz. Nauk. Akad. Morska Szczec.* **2012**, *30*, 78–82.

5

Evaluation of Waste Plastic Oil-Biodiesel Blends as Alternative Fuels for Diesel Engines

Chalita Kaewbuddee, Ekarong Sukjit *, Jiraphon Srisertpol, Somkiat Maithomklang, Khatha Wathakit, Niti Klinkaew, Pansa Liplap and Weerachai Arjharn

Suranaree University of Technology, 111 University Avenue, Suranaree Sub-District, Muang District, Nakhon Ratchasima 30000, Thailand; annen_ch@hotmail.com (C.K.); jiraphon@sut.ac.th (J.S.); somkiat_mai@outlook.co.th (S.M.); vkata@sut.ac.th (K.W.); niti_nick@hotmail.com (N.K.); pansa@sut.ac.th (P.L.); arjharh@g.sut.ac.th (W.A.)

* Correspondence: ekarong@sut.ac.th

Abstract: This study examined the use of waste plastic oil (WPO) combined with biodiesel as an alternative fuel for diesel engines, also commonly known as compression ignition engines, and focused on comparison of the basic physical and chemical properties of fuels, engine performance, combustion characteristics, and exhaust emissions. A preliminary study was conducted to determine the suitable ratio for the fuel blends in consideration of fuel lubricity and viscosity, and these results indicated that 10% biodiesel—derived from either palm oil or castor oil—in waste plastic oil was optimal. In addition, characterization of the basic properties of these fuel blends revealed that they had higher density and specific gravity and a lower flash point than diesel fuel, while the fuel heating value, viscosity, and cetane index were similar. The fuel blends, comprised of waste plastic oil with either 10% palm oil biodiesel (WPOP10) or 10% castor oil biodiesel (WPOC10), were selected for further investigation in engine tests in which diesel fuel and waste plastic oil were also included as baseline fuels. The experimental results of the performance of the engine showed that the combustion of WPO was similar to diesel fuel for all the tested engine loads and the addition of castor oil as compared to palm oil biodiesel caused a delay in the start of the combustion. Both biodiesel blends slightly improved brake thermal efficiency and smoke emissions with respect to diesel fuel. The addition of biodiesel to WPO tended to reduce the levels of hydrocarbon- and oxide-containing nitrogen emissions. One drawback of adding biodiesel to WPO was increased carbon monoxide and smoke. Comparing the two biodiesels used in the study, the presence of castor oil in waste plastic oil showed lower carbon monoxide and smoke emissions without penalty in terms of increased levels of hydrocarbon- and oxide-containing nitrogen emissions when the engine was operated at high load.

Keywords: waste plastic oil; biodiesel; castor oil; emission; diesel engine

1. Introduction

The demand for and consumption of energy is expected to increase, especially for fossil fuels. In Thailand, fossil fuels, also known as conventional energy, are widely used in various forms of transportation and industrial plants because of their convenience and ability to provide a high heating value. Fossil fuels are a nonrenewable resource which continues to be used by humans, with demand steadily increasing. Thus, these fossil fuels will soon be entirely consumed. For this reason, many countries are beginning to rely more on alternative energy or renewable energy sources.

Thailand still lacks any significant alternative sources of energy. Moreover, it produces insufficient energy to meet the demand, resulting in the import of over 49% of its consumed energy in 2017 [1]. The value of crude oil and imported petroleum products in Thailand increased by 39.8% and 23.0%, in 2016 and 2017, respectively, due to the higher oil demand. Thailand has mainly imported crude oil from Middle Eastern countries. In Thailand, the share of energy from renewable sources is expected to increase steadily. To increase this share and reduce primary energy consumption, waste plastic oil has been proposed as a new option for use in transportation. While there is less demand for transportation energy, this initiative represents a move toward the direction of diversification of fuels through energy conversion technologies. It also focuses on using oil from plastic waste in diesel engines. Plastic waste is a petroleum waste that comes from both household and industrial sectors, leading to a large amount of plastic waste. These wastes require hundreds of years for decomposition and are a burden to manage. Most plastic is recycled using mechanical recycling, while only 2% of chemicals are recycled [2]. Generally, the waste management process that is currently popular is the landfill method, which normally requires a lot of landfill space and has an impact on the environment, resulting in soil pollution.

Plastic waste is composed of hydrocarbons, which are the main component of conventional fuels. This raises the possibility of recycling these plastic wastes through their conversion into fuel. Products can also be obtained from the production process, in addition to being used as an energy source similar to conventional fuels. It is also able to provide environmental benefits in terms of waste management for maximum benefits and reduction in the amount of plastic waste, reduced plastic waste disposal, and also minimizing the problem of finding places for garbage landfills. The use of plastic waste as a renewable energy feedstock also helps in mitigating the energy crisis.

Several studies have investigated the use of waste plastic oil in diesel engines as an alternative fuel. Waste plastic pyrolysis oil has properties that are similar to diesel fuel, including the heating value, density, and cetane index, and can be used as a substitute for diesel fuel [3]. The literature also shows that diesel engines use waste plastic oil to provide stability in performance and a similar efficiency [4]. The different types of plastics are also basically impacted by their different compositions. Recent studies have shown that the oil product of HDPE (high-density polyethylene), mixed with LDPE (low-density polyethylene), has a higher heating value than LDPE, PP (polypropylene), and HDPE alone. It was revealed that LDPE produces the highest yields [5]. Waste plastic oil has also been studied with regard to engine power, and it was found that there was no significant difference from diesel fuel [6]. The thermal efficiency of waste plastic oil was higher when compared to diesel fuel [7,8]. However, one study examined the exhaust emissions of a four-cylinder, direct-injection diesel engine running on diesel blended with different ratios of waste plastic oil and found that the amount of nitrogen oxides increased because of the longer ignition delay [9] and that there was greater hydrocarbon emission in comparison to diesel fuel [10].

In addition, it is expected that biodiesel will be used as a renewable energy source in the energy transportation sector. A great deal of research supports the use of biodiesel as a suitable alternative in replacing diesel fuel. The presence of oxygen in fuel molecules is expected to result in cleaner biodiesel combustion, leading to improvements when considering emission. However, there are only a few reports on the use of biodiesel mixed with waste plastic oil. For example, Ramesha et al. [11] reported that B20 algae biodiesel blended with waste plastic oil can be a suitable fuel for diesel engines. The waste plastic oil-biodiesel blend showed an increase of 16% in brake thermal efficiency with respect to diesel engines. Additionally, the carbonaceous gas emissions, including hydrocarbons and carbon monoxide, were decreased, but nitrogen oxides slightly increased, as compared to diesel fuel. In the study by Senthilkumar et al. [12], waste plastic oil was mixed with Jatropha biodiesel for diesel engines. The brake thermal efficiency and brake specific fuel consumption of the waste plastic oil-biodiesel blend were higher than the waste plastic oil. The hydrocarbon and carbon monoxide emissions decreased when waste plastic oil was blended with Jatropha biodiesel.

In the present work, waste plastic oil-biodiesel blends were used as an alternative fuel in a diesel engine without any engine modifications. The selected biodiesels were produced from castor oil and palm oil through a transesterification process and were then blended with waste plastic oil. Palm is an important economic crop and main feedstock for biodiesel production in Thailand. To avoid the use of edible feedstock, castor oil was considered because of its benefits of high oxygen content in fuel molecules and excellent fuel lubricity. These properties are attributed to the presence of ricinoleic acid, which is the main component of castor oil [13]. The oxygen in the fuel molecules contributes to better combustion processes in terms of emissions. In this study, we evaluated the effect of biodiesel addition to waste plastic oil in terms of basic physical and chemical fuel properties of the resulting fuel mixture, mainly focusing on fuel lubricity and viscosity, engine performance, combustion characteristics, and exhaust gas emissions of a single-cylinder diesel engine. In the section of combustion characteristics, basic parameters comprised of in-cylinder pressure and crank angle were recorded during the engine test. After that, heat release rate of test fuels was calculated on the basic principles of the first law of thermodynamics, which the specific heat ratio was calculated based on the in-cylinder pressure and combustion chamber volume through the assumption of polytropic process.

2. Materials and Methods

As part of this research, some physical and chemical properties of test fuels were determined. Experimental tests were carried out on a single-cylinder diesel engine (model Kirloskar TV1 with product code 240PE). The engine was connected to an eddy current dynamometer to simulate the load and was tested at a constant speed of 1500 rpm under variable load conditions, i.e., low, medium, and high load conditions (25%, 50%, and 75% of the maximum torque). The gas detector was used to determine the levels of carbon monoxide (CO), nitrogen oxides (NO_X), hydrocarbon (HC), and smoke. The fuel consumption of the engine was tested by adjusting the load of the diesel engine. Therefore, the scope of this research project was limited to the following:

- To find a suitable ratio of the fuel blend by mainly considering fuel lubricity and viscosity; the fuel blends were prepared at various ratios, i.e., 0%, 5%, 10%, and 15% (by volume) of the biodiesel to the waste plastic oil.
- To test the suitable fuel blends with a single-cylinder diesel engine without any engine modifications. The objective of this test was to evaluate engine performance, combustion characteristics, and exhaust emissions.

2.1. *Materials*

In this experimental investigation, waste plastic oil (WPO) was used as the main fuel, whereas the other fuels included castor oil methyl ester (COME) and palm oil methyl ester (POME) as components for blending with WPO. The diesel fuel was commercial diesel fuel (B7) containing 7% biodiesel, according to the department of energy business in Thailand. The waste plastic oil used in this research project was produced from waste plastic by the pyrolysis method. The pyrolysis process is a chemical process of heating that decomposes plastics in the absence of oxygen. The master plant that processes waste plastic to oil is located at Suranaree University of Technology.

2.1.1. Waste Plastic Oil

The raw materials used in this study were from plastic waste, such as plastic waste bags collected from waste in Suranaree Subdistrict, Nakhon Ratchasima, Thailand. The composition of these plastics includes polyethylene (PE) and polystyrene (PS) and about 70% was contaminated organic matter. The waste plastics obtained from mechanical biological treatment (MBT) were processed into raw materials using an agglomerator, which processed the plastic into small pieces that could be continuously fed into the oil processing plant. The waste plastic oil was recycled using pyrolysis and did not undergo distillation.

The pyrolysis process involves the breakdown of large molecules into smaller molecules by chemically decomposing organic matter through heating in an oxygen-free environment. Waste plastic is processed to maintain a temperature of 300–350 °C inside the reactor, where the waste plastic is then vaporized and the outlet gas condensed through the condenser unit at this high temperature. The obtained liquid was taken as fuel, and this process happened constantly in converting the waste plastic back into usable oil. All gases from this process were treated before being released into the atmosphere. The exhaust gas was treated through scrubbers and chemical treatment for neutralization. From the pyrolysis process, the following output products were collected: Waste plastic oil (70%), gas (10%), and solid (20%), with values based on the weight of the input. The plastics yielded approximately 600 L per ton.

2.1.2. Production of Castor Oil Biodiesel and Palm Oil Biodiesel

The experimental work was carried out in a laboratory at Suranaree University of Technology. Castor oil was used in the transesterification process to convert castor oil into castor oil methyl ester. Methanol and potassium hydroxide (KOH) catalyst were used for the reaction. The reaction was carried out using methanol and castor oil in a 9:1 molar ratio with 0.5% KOH (by weight of oil). The KOH was first dissolved in methanol and was then mixed with the castor oil. This mixture was heated and stirred using an electric heater and a magnetic stirrer. The reaction was carried out at a constant temperature of 50 °C for about 120 min. Then, the mixture was poured into a separating funnel to separate the methyl ester of castor oil and glycerol. The layers were separated and were allowed to settle for a minimum period of 8 h, with glycerol at the bottom layer and the ester at the top layer. The castor oil methyl ester was then washed with water to remove any traces of methanol or potassium hydroxide that was not reacted. The castor oil methyl ester was heated to 120 °C for moisture removal.

However, palm oil was also used with methanol and potassium hydroxide (KOH) in the reaction. The reaction was carried out by taking methanol and palm oil in a 12:1 molar ratio and 2% KOH (according to the weight of the oil). KOH was dissolved in methanol and this mixture was then mixed with palm oil. This mixture was heated and stirred using an electric heater and a magnetic stirrer. The reaction was carried out at a constant temperature of 60 °C for about 30 min. Then, the mixture was poured into a separating funnel to separate the methyl ester of the palm oil and glycerol. The layers were separated and allowed to settle for 24 h, with glycerol at the bottom layer and the ester at the top layer. The palm oil methyl ester was then washed with water to remove any traces of methanol or potassium hydroxide that was not reacted. The palm oil methyl ester was heated to 120 °C for moisture removal.

2.2. Gas Chromatography Analysis

The column for GC–MS analysis was a DB-wax capillary column (60 m length × 0.25 mm inner diameter, 0.25 μm film thickness). Helium was used as a carrier gas with a constant flow rate of 1.0 mL/min. The oven temperature was programmed to operate from 70 °C to 250 °C, with the initial temperature of 70 °C that was held for 3 min, followed by a rate of heating of 3 °C/min to a temperature of 180 °C and then a rate of 10 °C/min to a final temperature at 250 °C, which was held for 25 min. The inlet was held at 250 °C with a split ratio of 20:1. The injection volume was 1 μL per sample. The mass spectrometer was scanned from mass to charge ratio (m/z) of 35 to 550 with the source at 250 °C.

2.3. Experimental Setup

An experimental investigation was tested to evaluate and compare the results obtained for the use of different types of test fuels. This study aimed to investigate the effect of waste plastic oil blended with biodiesel on engine performance and the emission of a single-cylinder diesel engine. WPOC10 and WPOP10 were selected for experimental comparison based on initial experiment data regarding the lubrication and viscosity of the blended fuels, whereby 10% biodiesel and 90% waste plastic oil was determined to be the optimal ratio for further testing in the engine. The total number of samples was four test fuels (using either WPOC10, WPOP10, WPO, diesel), which were prepared for testing with the equipment and measuring tools used in the laboratory as follows: A four-stroke, single-cylinder diesel engine (Kirloskar TV1) with a water cooler system, direct injection, and a rated output power of 3.5 kW at 1500 rpm, unmodified and under different loading conditions. The engine was mounted on a fixed bed floor in the laboratory room and the load was applied on the engine. A picture of the experimental setup is shown in Figures 1 and 2, and the engine specifications are given in Table 1.

Figure 1. Experimental setup for engine testing.

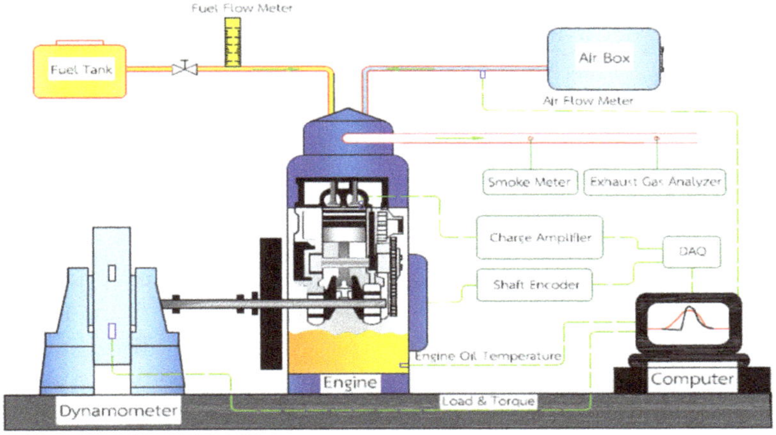

Figure 2. A schematic diagram of the experimental installation.

Table 1. Test engine specifications.

Engine	Specification
Model	Kirloskar TV1
Engine type	Single cylinder, 4 stroke, Water cooler, Direct injection
Bore	87.5 mm
Stroke	110 mm
Connecting rod length	234 mm
Swept volume	661 cc
Rate output	3.5 kW at 1500 rpm
Compression ratio	12–18
Dynamometer	Eddy current, Water cooler

The observation results during testing in the engine were used to evaluate the use of different fuel blends in comparison to the commercial diesel fuel and waste plastic oil as a reference. Three repetitions were carried out in each test to obtain the average values for analysis. The used confidence intervals corresponded to a 95% confidence level with respect to statistical significance of the result trends.

2.3.1. Testing by Adjusting Engine Load

Engine tests were done by loading the engine to a level that simulated the workload condition of the engine. Three engine loading conditions (25%, 50%, and 75% of the maximum engine torque) were chosen for engine experiments, and the engine loads were adjusted by the eddy current dynamometer, which acted as a brake and an enabling measurement of the engine's torque. The knob on the dynamometer loading unit was rotated to gradually adjust the engine load to ensure that the load on the load indicator gradually changed for each load value before the test. The experiment was performed using various blends at different loads, from low to high loads, at a constant speed. Furthermore, before testing the engine output with different fuels, the engine was always cleaned prior to installing the fresh oil to ensure the new oils were tested according to the intended compositions.

The fuel consumption was measured using a burette and stopwatch to determine the change in the fuel volume divided by the testing time. The burette with a unit of volume in mL was used to determine the volume of spent fuel. The tests were performed by counting the time needed using a stopwatch to determine the length of time before the fuel ran out. This required a fixed test fuel volume in a burette of 10 mL, and this measurement was repeated three times. The determination of each output value, such as the brake-specific fuel consumption, thermal efficiency, combustion, and amount of gas emission, was based on recordings at each engine load change using the four tested fuels.

2.3.2. Emission Testing

In each engine test, the engine load was gradually adjusted using the knob, and the actual engine speed was measured through encoder wheel monitoring, to measure the tested value of the exhaust gas emission of fuel blends.

The data corresponding to the parameters of engine combustion characteristics were collected through the signals from an in-cylinder pressure sensor and shaft encoder, both of which were analyzed using IC Engine Soft of Apex Innovations' software. The amount of nitrogen oxides (NO_X), hydrocarbon (HC), carbon monoxide (CO), and smoke in emissions was measured by the engine exhaust emission analyzer using a Testo 350 Gas analyzer for CO, NO_X, and HC, and a Testo 308 for smoke by installing the equipment for the experiment, as shown in Figure 2.

The Testo 350 and Testo 308 exhaust gas analyzers were used to measure the exhaust gas. The exhaust emission probe was placed in the tailpipe of the engine and the exhaust gas emissions were measured. The test engine was run to idle for approximately 10 min in order to ensure the stability of the engine before measuring the exhaust gas emissions. The Testo 350 was analyzed by using a nondispersive infrared and electronic chemical method. Furthermore, the Testo 308 was analyzed using the principles of absorption photometry.

3. Results and Discussions

3.1. Test Fuels

In order to improve the properties of waste plastic oil by combination with biodiesel, such as through increasing the oxygen content in the waste plastic oil for better combustion and to improve the viscosity and lubricity of the waste plastic oil, COME and POME were blended with waste plastic oil at different volumetric ratios, ranging from 0% to 15%, and the basic physical and chemical properties of the blended fuels were investigated, which mainly focused on fuel lubrication and viscosity. It can be concluded that the presence of 10% biodiesel in waste plastic oil is the optimal ratio because the smallest scar diameter was obtained after lubrication testing and the viscosity was within the acceptable criteria prescribed by the standard specification for diesel fuel, as shown in Figure 3. The lubricity testing was evaluated by a high-frequency reciprocating rig (HFRR) and was conducted according to EN ISO

12156 [14]. From the preliminary experiment, 10% biodiesel was enough to maintain the lubrication of the blended fuel and there was no significant improvement in the lubrication of the blend when exceeding this percentage of biodiesel in waste plastic oil. Therefore, a combination of either 10% castor oil biodiesel or 10% palm oil biodiesel with 90% waste plastic oil (WPOC10 and WPOP10, respectively) was selected for further investigation in the engine test to study the effect of biodiesel addition to waste plastic oil on engine performance, combustion characteristics, and exhaust emissions.

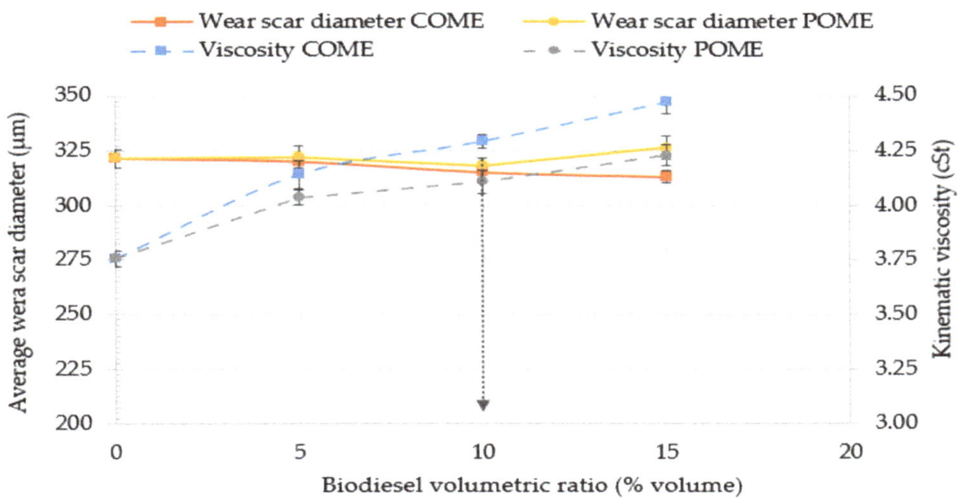

Figure 3. Lubricity and viscosity of biodiesel blending.

The fatty acid profile of castor oil and palm oil are summarized in Table 2, which confirms that the most abundant fatty acid in castor oil is ricinoleic acid, comprising about 85% by weight of the total fatty acid content, while the major constituents of palm oil are palmitic and oleic acid, at about 46% and 37% of the weight. Palmitic acid is a fatty acid that naturally occurs in vegetable and animals, and it is the main component of human milk fat. Furthermore, oleic acid is also the major component of many oils and fats.

Table 2. Fatty acid composition of palm oil and castor oil.

Fatty Acid		% wt.	
		POME	COME
Lauric	C12:0	0.92	0.02
Myristic	C14:0	1.28	0.06
Palmitic	C16:0	46.29	1.63
Stearic	C18:0	4.63	1.66
Oleic	C18:1	37.07	3.85
Ricinoleic	C18:1 OH	-	85.6
Linoleic	C18:2	8.67	6.04
Linolenic	C18:3	0.03	0.43
Arachidic	C20:0	0.36	0.08
Other		0.75	0.63

The physicochemical properties of fuels were carried out, based on the ASTM standards, and the properties of the test fuels are given in Table 3.

Table 3. Properties of the test fuels.

Properties	Test Method	Diesel	WPO	POME	COME	WPOP10	WPOC10
Kinematic viscosity at 40 °C (cSt)	ASTM D445	4.19	3.76	6.46	18.61	4.11	4.29
Specific gravity at 15.6 °C	ASTM D1298	0.828	0.825	0.875	0.910	0.830	0.845
Density at 15.6 °C (kg/m^3)	ASTM D1298	827	824	874	909	829	844
Flash point (°C)	ASTM D93	63	41	96	108	45	48
Gross calorific value (MJ/kg)	ASTM D240	42.45	40.58	36.79	37.95	39.18	39.64
Cetane index	ASTM D976	60.2	60.0	48.7	39.4	59.6	54.5

Characterization of Waste Plastic Oil

The waste plastic oil or pyrolysis oil used in this study was extracted from mixed plastic wastes. The chemical compounds contained in the waste plastic oil were analyzed by gas chromatography–mass spectrometry (GC–MS), using a gas chromatograph Agilent 7890A coupled to a mass spectrometer Agilent 7000B. The results of the GC–MS analysis of waste plastic oil and diesel are presented in Figure 4, and it is an important chemical compound contained in plastic oil and the percentage of the area is shown in Table 4.

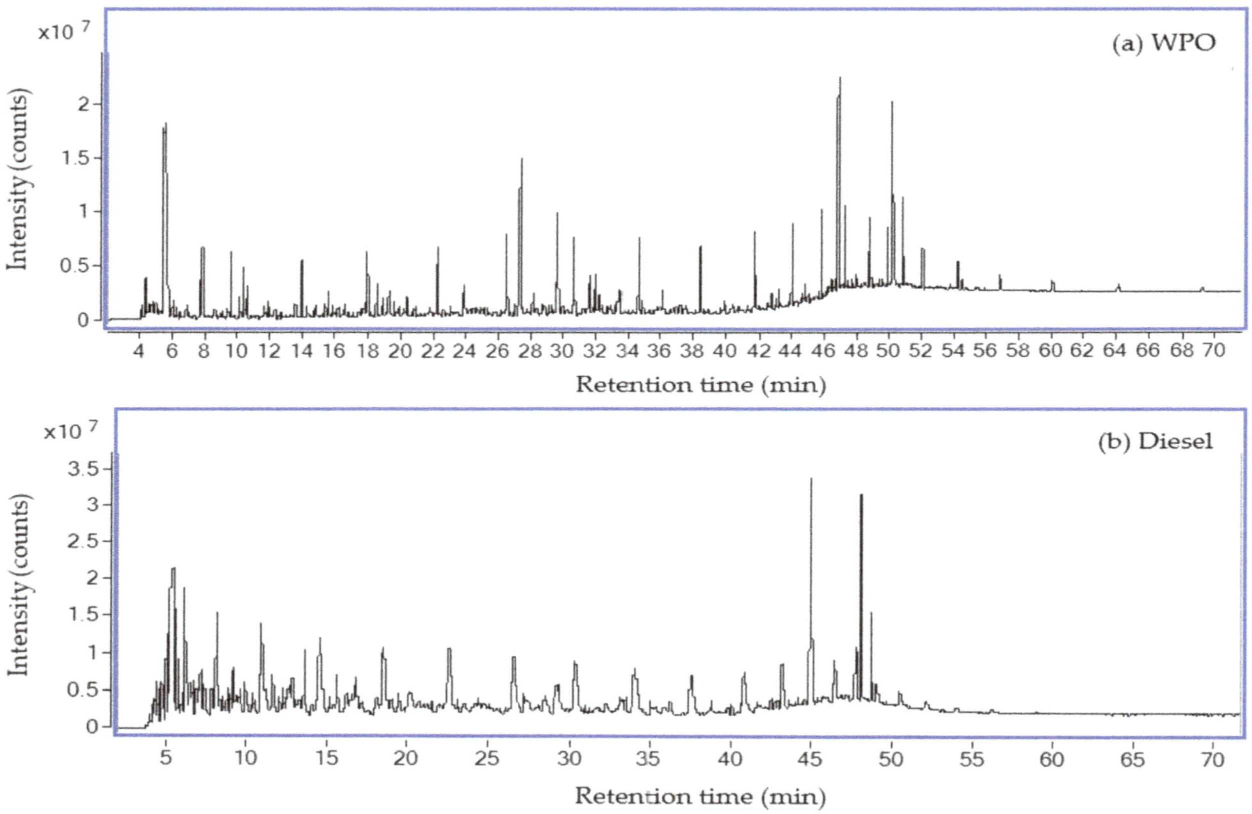

Figure 4. Total ion current chromatogram for: (**a**) Waste plastic oil and (**b**) diesel.

Table 4. Components identified from waste plastic oil and diesel by GC–MS analysis.

Carbon Content	% Area	
	WPO	Diesel
C_4–C_{11}	12.85	17.56
C_{12}–C_{20}	74.39	73.28
>C_{20}	12.76	9.16

The waste plastic oil consisted of different hydrocarbons contents, which separate according to the light and heavy fractions, from the lowest carbon atom (C_4) to the highest carbon atoms (>C_{20}) and can be divided into three groups. The C_4–C_{11} group represented the light fraction or gasoline, and typical gasoline consists of hydrocarbons between five and nine carbon atoms. The C_{12}–C_{20} group represented the middle fraction or diesel. Diesel has a high percentage of carbon atoms of C_{16}–C_{20} [15]. Table 4 presents the results of the comparison between fuels, and a similar trend was observed between waste plastic oil and diesel. The waste plastic oil and diesel produced the highest C_{12}–C_{20} fraction.

3.2. Engine Performance

Figure 5 illustrates the results of the brake specific fuel consumption (BSFC) of the engine for four kinds of test fuels according to three engine-operating loads. The results showed that the BSFC increased at the low loading (25% of the maximum torque) rather than at medium and high loading, respectively (50% and 75% of the maximum torque). The results also showed that the increment in the engine load seems to result in less specific fuel consumption for all the fuels [16,17]. The increase in engine loading resulted in an increase of fuel flow rate, brake thermal efficiency, and exhaust gas temperature while also decreasing the brake specific fuel consumption at the same time.

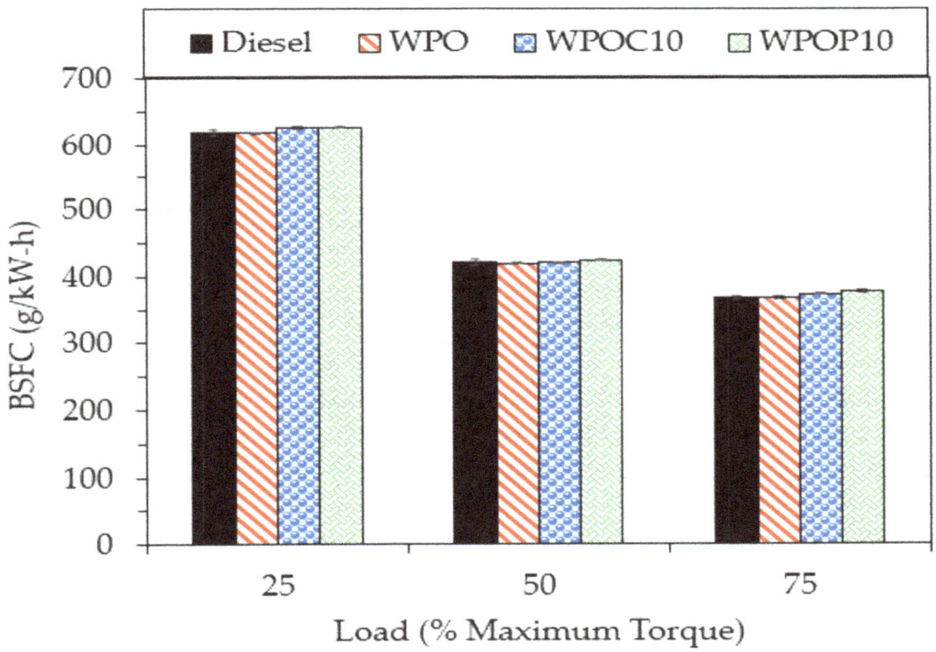

Figure 5. Variation of brake specific fuel consumption.

The brake specific fuel consumption was slightly increased for biodiesel-waste plastic oil. In general, brake specific fuel consumption was found to increase when the biodiesel quantity of the blends was increased, which was due to its lower heating value [16,18–20].

Figure 6 shows the variation of brake thermal efficiency (BTE) and engine load. Higher BTE was obtained with the use of waste plastic oil and its blends. The addition of biodiesel tended to improve the combustion of waste plastic oil. This may be attributed to the increase in oxygen content, due to oxygen in the fuel molecule of the fatty acid in biodiesel, resulting in more effective combustion [20–25]. In addition, the proper lubricating properties of biodiesel may play a role in reducing the friction to the level that the brake efficiency was enhanced from pure waste plastic oil [26,27].

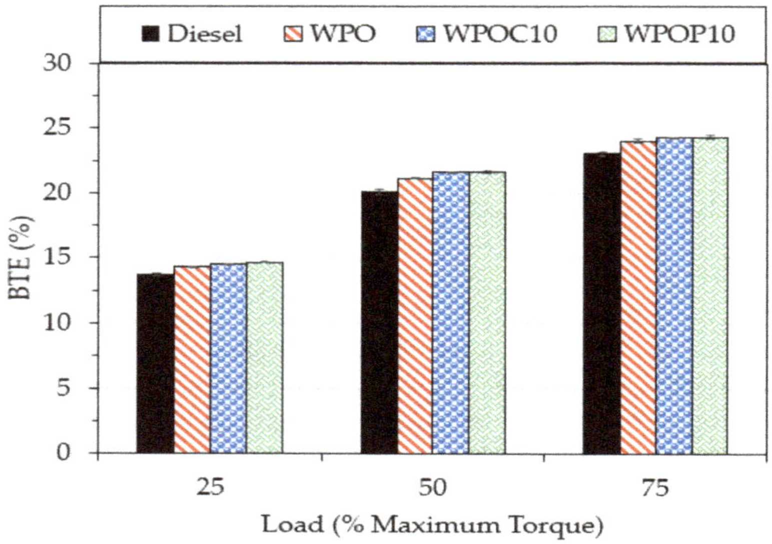

Figure 6. Variation of the brake thermal efficiency.

3.3. Combustion Characteristics

The combustion characteristics were examined as in-cylinder pressure and the rate of heat release based on the basic principles of the first law of thermodynamics, as shown in Figure 7. It was found that the combustion of waste plastic oil was similar to diesel fuel for all tested engine loads. The addition of castor oil biodiesel to waste plastic oil caused a delay in the start of the combustion rather than the addition of palm oil biodiesel. This was explained by the lower cetane number in castor oil. Additionally, the higher viscosity of castor oil biodiesel can be used to justify the delay in the combustion process due to the difficulty of fuel injection and the quality of fuel spray [27,28]. When considering peak of heat release rate, it was found that a higher peak was obtained for fuel blends with castor oil as biodiesel compared to palm oil. The accumulation of fuel volume during the longer ignition delay, which impacted the higher peak of premixed combustion, was used to justify the higher peak of the heat release rate obtained by the combustion of WPOC10 with respect to WPOP10 [29,30].

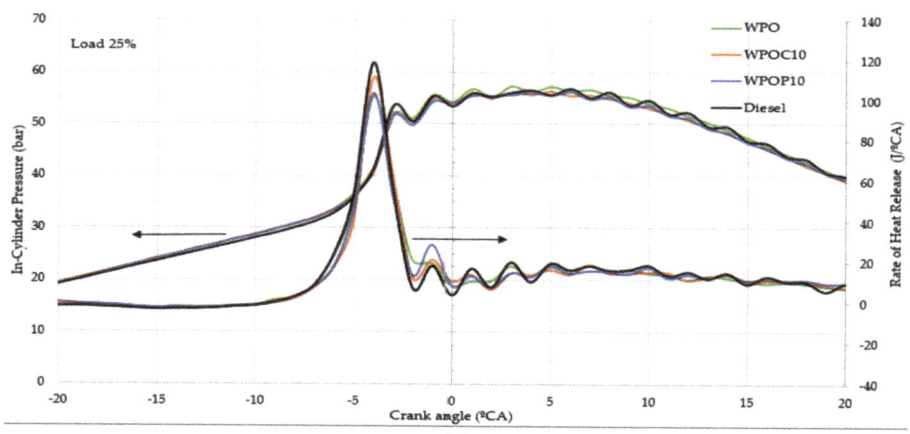

(a)

Figure 7. *Cont.*

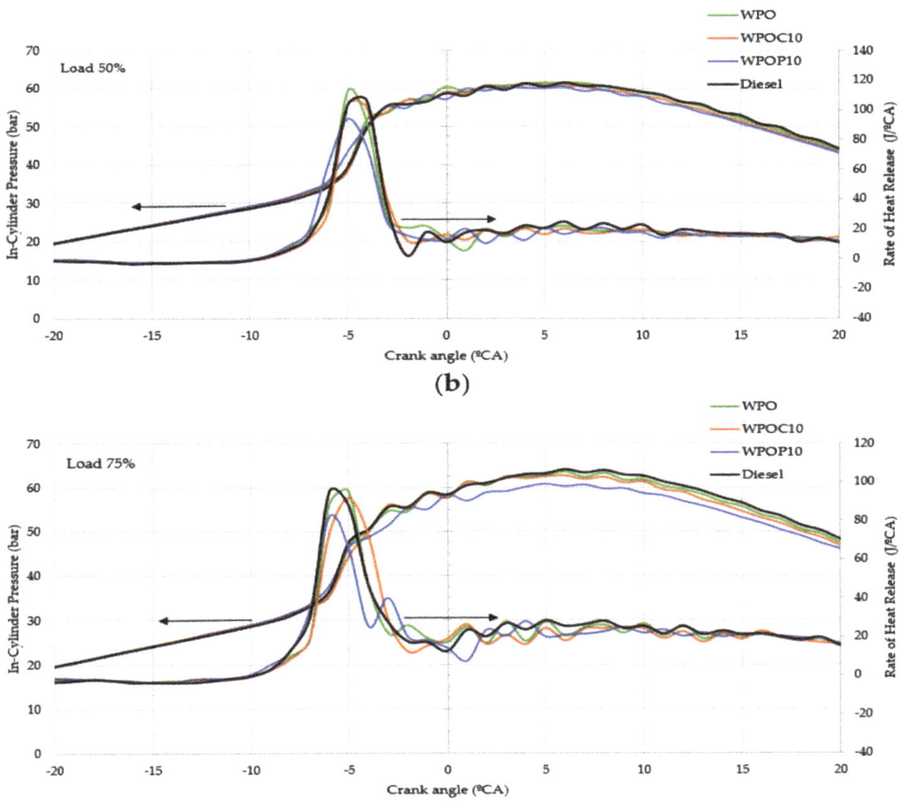

Figure 7. In-cylinder pressure and rate of heat release at different levels of maximum torque: (**a**) 25%; (**b**) 50%; (**c**) 75%.

3.4. Emissions

From Figure 8, the blends of the biodiesel-waste plastic oil showed a disadvantage in carbon monoxide emissions because of the higher viscosity and lower calorific value. Therefore, the combustion temperature was lowered due to ineffective atomization of the fuel blends, leading to an increase in carbon monoxide emissions. Comparing the two biodiesels used in this study, the presence of castor oil in fuel blends showed lower CO emissions. Although, castor oil possesses higher viscosity than palm oil, which can generate poor fuel atomization that results in more incomplete combustion. The higher oxygen content of castor oil may improve the quality of the combustion and can be compensated for by the effect of higher viscosity, leading to lower CO emissions by the addition of castor oil biodiesel to waste plastic oil as compared to the addition of palm oil biodiesel.

From Figure 9, it can be seen that the amount of nitrogen oxide emissions increased with increasing engine load, and the nitrogen oxide emission levels of plastic waste oil were higher compared to diesel fuel. Comparing waste plastic oil and its blends, lower NOx emissions were found when biodiesel was added. The reduction in the peak of the heat release in the premixed combustion phase was caused by the combustion of biodiesel blends, which tended to attenuate the increase in combustion temperature and did not favor NOx formation. The results of biodiesel addition were similar to diesel fuel blends and resulted in NO_X reduction, which was also observed in another study by Pumpuang et al. [22] using blends of diesel with castor oil ethyl ester biodiesel. Considering the addition of castor oil and palm oil biodiesel, the castor oil biodiesel blends showed higher NOx emissions than those of palm oil biodiesel. The longer ignition delay due to the lower cetane value, caused by the addition of castor oil biodiesel, led to a higher combustion temperature, and this can explain the higher NOx emissions observed with the combustion of waste plastic oil blended with castor oil biodiesel.

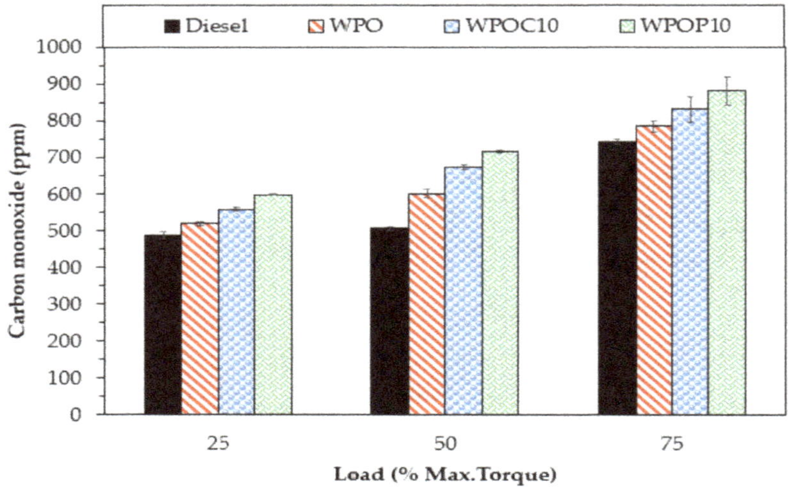

Figure 8. Carbon monoxide emissions.

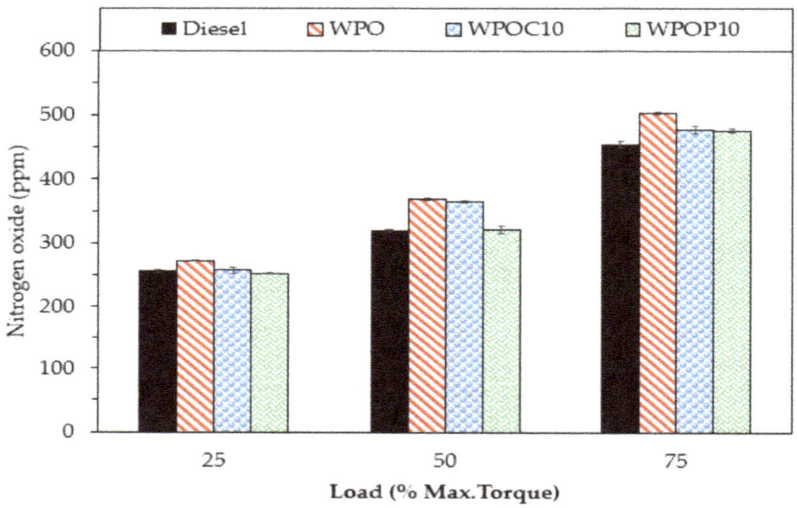

Figure 9. Nitrogen oxide emissions.

The variation of hydrocarbons with engine loads is shown in Figure 10. Higher levels of hydrocarbon emissions were found with the combustion of waste plastic oil with respect to diesel fuel. However, the addition of biodiesels may have contributed to reducing hydrocarbon emissions [31]. In the case of palm oil biodiesel, the shorter ignition delay due to the lower cetane index, in comparison to castor oil biodiesel, can improve (reduce) hydrocarbon emissions by allowing more time for the combustion process, resulting in lower levels of hydrocarbon emission. However, the blend with castor oil biodiesel, containing a higher oxygen content for the same volumetric percentage when comparing the two biodiesels, was blended with the waste plastic oil. This is expected to promote lower levels of hydrocarbon emission compared to the blend of palm oil biodiesel. The effect of the extremely high viscosity of castor oil, however, tended to increase the emission of hydrocarbons and could counteract the beneficial reductions in hydrocarbon emission due to the higher oxygen content [32]. These effects were more obviously seen at low engine operating loads where the temperature in the combustion chamber was not high enough to vaporize all the injected fuels. Consequently, WPOC10 produced higher hydrocarbons with respect to WPOP10.

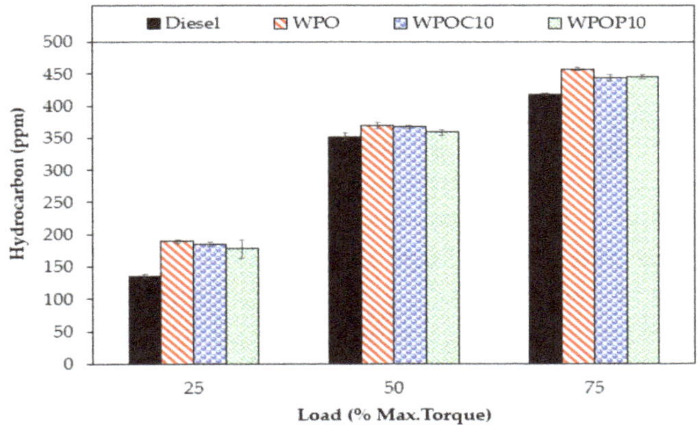

Figure 10. Hydrocarbon emissions.

Figure 11 shows the smoke emissions obtained by the combustion of the tested fuels. The combustion of the waste plastic oil was evidently associated with less smoke emission than diesel fuel. The addition of biodiesels to waste plastic oil tended to result in increased smoke emissions of these fuel blends. This increase in smoke emissions was apparent when the engine was operated at higher load conditions. Comparing castor oil and palm oil biodiesel, it was found that lower smoke emissions were obtained with the use of castor oil biodiesel. The two main factors had opposing effects on particulate matter emission, which was directly related to the smoke emissions. First, the higher viscosity of castor oil biodiesel caused more difficulty in fuel injection. The ineffective fuel atomization resulted in more incomplete combustion, which was related to a higher amount of unburnt fuel, resulting in higher smoke emissions. Second, the higher oxygen content present in the castor oil biodiesel and hydroxyl group belonged to ricinoleic acid as the primary fatty acid of castor oil can contribute to the enhancement of combustion quality, leading to less smoke emission [33]. It was notable that the effect of higher oxygen content may be more likely to reduce smoke emissions in this study as a result of lower smoke emissions associated with the combustion of WPOC10 compared to WPOP10.

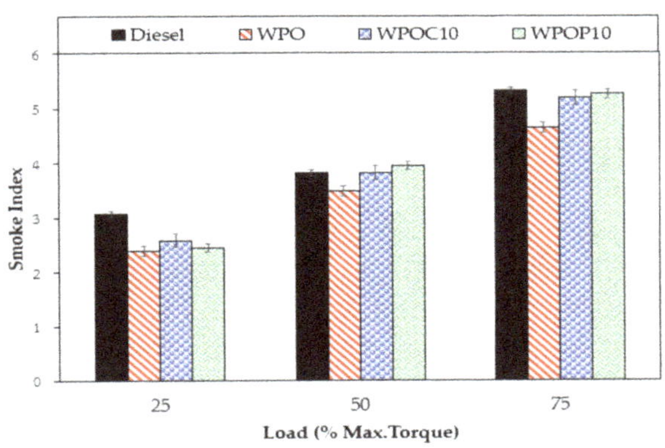

Figure 11. Smoke emissions.

4. Conclusions

The effects of biodiesel addition on fuel properties, combustion characteristics, engine performance, and exhaust emissions of crude waste plastic oil were studied. Two types of biodiesel, palm oil biodiesel and castor oil biodiesel, were selected as components for blending with waste plastic oil. The findings can be summarized as follows:

- Considering fuel lubricity and viscosity, the presence of 10% (v/v) biodiesel was the optimal ratio for improving the waste plastic oil.

- The brake thermal efficiency of the engine was slightly improved with the addition of biodiesel to waste plastic oil.
- The addition of castor oil as compared to palm oil biodiesel caused a delay in the start of the combustion, resulting in a higher peak of heat release rate.
- The reduction in the level of hydrocarbon- and oxide-containing nitrogen emissions was found with the addition of biodiesel, while carbon monoxide and the smoke emissions were increased.
- Comparing the two tested biodiesels, the presence of castor oil in waste plastic oil showed lower carbon monoxide and smoke emissions without any penalty regarding the levels of hydrocarbon- and oxide-containing nitrogen emissions when the engine was operated at a high load.
- Future studies on engine modification, such as to the piston, injection timing, and injection pressure, can be considered for further improvements through the reduction of exhaust emissions by the use of biodiesel as a blend component in waste plastic oil.

Author Contributions: Conceptualization, E.S.; investigation, C.K., S.M., and N.K.; writing—Original draft preparation, C.K.; writing—Review and editing, C.K. and E.S.; supervision, J.S. and W.A.; resources, K.W. and P.L. All authors have read and agreed to the published version of the manuscript.

Nomenclature

BSFC	brake-specific fuel consumption
BTE	brake thermal efficiency
CO	carbon monoxide
COME	castor oil methyl ester
HC	hydrocarbon
HDPE	high-density polyethylene
HFRR	high-frequency reciprocating rig
KOH	potassium hydroxide
LDPE	low-density polyethylene
MBT	mechanical biological treatment
NO_X	nitrogen oxides
PE	polyethylene
POME	palm oil methyl ester
PP	polypropylene
PS	polystyrene
WPO	waste plastic oil
WPOC10	fuel blend of 10% castor oil biodiesel and 90% waste plastic oil
WPOP10	fuel blend of 10% palm oil biodiesel and 90% waste plastic oil
GC-MS	gas chromatography–mass spectrometry

References

1. Energy Policy and Planning Office, Ministry of Energy, Thailand. Energy Statistics of Thailand. Available online: http://www.eppo.go.th/index.php/th/informationservices/ct-menu-item-56?orders[publishUp]=publishUp&issearch=1 (accessed on 28 July 2019).
2. Gabbar, H.A.; Aboughaly, M.; Stoute, C.A. DC thermal plasma design and utilization for the low density polyethylene to diesel oil pyrolysis reaction. *Energies* **2017**, *10*, 784. [CrossRef]
3. Damodharan, D.; Sathiyagnanam, A.P.; Rana, D.; Kumar, B.R.; Saravanan, S. Extraction and characterization of waste plastic oil (WPO) with the effect of n-butanol addition on the performance and emissions of a DI diesel engine fueled with WPO/diesel blends. *Energy Convers. Manag.* **2016**, *131*, 117–126. [CrossRef]
4. Kaimal, V.K.; Vijayabalan, P. A detailed study of combustion characteristics of a DI diesel engine using waste plastic oil and its blends. *Energy Convers. Manag.* **2015**, *105*, 951–956. [CrossRef]

5. Areeprasert, C.; Asingsamanunt, J.; Srisawat, S.; Kaharn, J.; Inseemeesak, B.; Phasee, P.; Khaobang, C.; Siwakosit, W.; Chiemchaisri, C. Municipal plastic waste composition study at transfer station of Bangkok and possibility of its energy recovery by pyrolysis. *Energy Procedia* **2017**, *107*, 222–226. [CrossRef]
6. Baskaran, R.; Sathish Kumar, P. Evaluation on performance of CI engine with waste plastic oil-diesel blends as alternative fuel. *Int. J. Res. Appl. Sci. Eng. Technol.* **2015**, *3*, 642–646.
7. Syamsiro, M.; Saptoadi, H.; Kismurtono, M.; Mufrodi, Z.; Yoshikawa, K. Utilization of waste polyethylene pyrolysis oil as partial substitute for diesel fuel in a DI diesel engine. *Int. J. Smart Grid Clean Energy* **2018**, *8*, 38–47. [CrossRef]
8. Sachuthananthan, B.; Reddy, D.R.; Mahesh, C. Production of diesel like fuel from municipal solid waste plastics for using in CI Engine to study the combustion, performance and emission characteristics. *Int. J. Pure Appl. Math.* **2018**, *119*, 85–98.
9. Kalargaris, I.; Tian, G.; Gu, S. Combustion, performance and emission analysis of a DI diesel engine using plastic pyrolysis oil. *Fuel Process. Technol.* **2017**, *157*, 108–115. [CrossRef]
10. Mani, M.; Nagarajan, G.; Sampath, S. Characterisation and effect of using waste plastic oil and diesel fuel blends in compression ignition engine. *Energy* **2011**, *36*, 212–219. [CrossRef]
11. Ramesha, D.K.; Kumara, G.P.; Lalsaheb; Mohammed, A.V.T.; Mohammad, H.A.; Kasma, M.A. An experimental study on usage of plastic oil and B20 algae biodiesel blend as substitute fuel to diesel engine. *Environ. Sci. Pollut. Res.* **2015**, *23*, 9432–9439. [CrossRef]
12. Senthilkumar, P.; Sankaranarayanan, G. Effect of Jatropha methyl ester on waste plastic oil fueled DI diesel engine. *J. Energy Inst.* **2016**, *89*, 504–512. [CrossRef]
13. Keera, S.T.; El Sabagh, S.M.; Taman, A.R. Castor oil biodiesel production and optimization. *Egypt. J. Pet.* **2018**, *27*, 979–984. [CrossRef]
14. ISO12156-1: Diesel Fuel—Assessment of Lubricity Using the High Frequency Reciprocating Rig. 2016. Available online: https://www.iso.org/standard/65227.html (accessed on 1 June 2020).
15. Wathakit, K.; Sukjit, E.; Maithomklang, S.; Srisertpol, J. Fuel properties, performance and emission of alternative fuel from pyrolysis of waste plastics. In Proceedings of the 2019 the 6th International Conference on Mechatronics and Mechanical Engineering, Wuhan, China, 9–11 November 2019; IOP Publishing Ltd.: Bristol, UK, 2020.
16. Shirneshan, A.; Almassi, M.; Ghobadian, B.; Borghei, A.M.; Najafi, G. Brake specific fuel consumption of diesel engine by using biodiesel from waste cooking oil. *World Sci. J.* **2013**, *1*, 45–52.
17. Kumar, S.; Prakash, R.; Murugan, S.; Singh, R.K. Performance and emission analysis of blends of waste plastic oil obtained by catalytic pyrolysis of waste HDPE with diesel in a CI engine. *Energy Convers. Manag.* **2013**, *74*, 323–331. [CrossRef]
18. Tarabet, L.; Loubar, K.; Lounici, M.S.; Hanchi, S.; Tazerout, M. Experimental evaluation of performance and emissions of DI diesel engine fueled with eucalyptus biodiesel. In Proceedings of the Internal Combustion Engines: Performance, Fuel Economy and Emissions, London, UK, 29–30 November 2011; pp. 167–176.
19. Adaileh, W.M.; Alqdah, K.S. Performance of diesel engine fuelled by a biodiesel extracted from a waste cocking oil. *Energy Procedia* **2012**, *18*, 1317–1334. [CrossRef]
20. Yusop, A.F.; Mamat, R.; Yusaf, T.; Najafi, G.; Yasin, M.H.M.; Khathri, A. Analysis of particulate matter (PM) emissions in diesel engines using palm oil biodiesel blended with diesel fuel. *Energies* **2018**, *11*, 1039. [CrossRef]
21. Arato, K.; Takashima, T. A study on reduction of heat loss by optimizing combustion chamber shape. *SAE Int. J. Engines* **2015**, *8*, 596–608. [CrossRef]
22. Pumpuang, A.; Maithomklang, S.; Sukjit, E.; Dejvajara, D.; Samaiklang, P.; Sanluecha, S. *Utilization of Castor Oil-Based Ethyl Ester Biodiesel in a Diesel Engine*; No. 2019-32-0606; SAE Technical Paper: 2019; SAE International: Pennsylvania, PA, USA, 2019.
23. Kaimal, V.K.; Vijayabalan, P. A study on synthesis of energy fuel from waste plastic and assessment of its potential as an alternative fuel for diesel engines. *Waste Manag.* **2016**, *51*, 91–96. [CrossRef]
24. Mrityunjaya, S.K.M.; Ramesha, D.K.; Premkumara, G.; Ranaprapareddy, N. Performance and emission characteristics of a compression ignition engine using blends of bio-oil as a fuel. *J. Middle Eur. Constr. Des. Cars* **2011**, *9*, 40–44. [CrossRef]
25. Das, M.; Sarkar, M.; Datta, A.; Santra, A.K. An experimental study on the combustion, performance and emission characteristics of a diesel engine fuelled with diesel-castor oil biodiesel blends. *Renew. Energy* **2018**, *119*, 174–184. [CrossRef]

26. Ramadhas, A.S.; Muraleedharan, C.; Jayaraj, S. Performance and emission evaluation of a diesel engine fueled with methyl esters of rubber seed oil. *Renew. Energy* **2005**, *30*, 1789–1800. [CrossRef]
27. Sukjit, E.; Herreros, J.M.; Dearn, K.; Tsolakis, A. *Improving Ethanol-Diesel Blend through the Use of Hydroxylated Biodiesel*; No. 2014-01-2776; SAE Technical Paper: 2014; SAE International: Pennsylvania, PA, USA, 2014.
28. Panwar, N.L.; Shrirame, H.Y.; Rathore, N.S.; Jindal, S.; Kurchania, A.K. Performance evaluation of a diesel engine fueled with methyl ester of castor seed oil. *Appl. Therm. Eng.* **2010**, *30*, 245–249. [CrossRef]
29. Sukjit, E.; Liplap, P.; Maithomklang, S.; Arjharn, W. *Experimental Investigation on a DI Diesel Engine Using Waste Plastic Oil Blended with Oxygenated Fuels*; No. 2017-24-0116; SAE Technical Paper: 2017; SAE International: Pennsylvania, PA, USA, 2017.
30. Azad, A.K.; Rasul, M.G.; Khan, M.M.K.; Sharma, S.C.; Bhuiya, M.M.K. Recent development of biodiesel combustion strategies and modelling for compression ignition engines. *Renew. Sustain. Energy Rev.* **2016**, *56*, 1068–1086. [CrossRef]
31. Niculescu, R.; Clenci, A.; Iorga-Siman, V. Review on the use of diesel-biodiesel-alcohol blends in compression ignition engines. *Energies* **2019**, *12*, 1194. [CrossRef]
32. Sukjit, E.; Maithomklang, S.; Trirak, N.; Klinkaew, N. Emission benefits from the use of castor oil in a compression ignition engine fuelled with diesel-ethanol blends. *Int. J. Electr. Energy* **2018**, *6*, 57–63. [CrossRef]
33. Sukjit, E.; Herreros, J.M.; Piaszyk, J.; Dearn, K.D.; Tsolakis, A. Finding synergies in fuels properties for the design of renewable fuels-hydroxylated biodiesel effects on butanol-diesel blends. *Environ. Sci. Technol.* **2013**, *47*, 3535–3542. [CrossRef]

6

Energy, Economic and Environmental Evaluation of a Proposed Solar-Wind Power On-grid System using HOMER Pro®

Farid Antonio Barrozo Budes [1], Guillermo Valencia Ochoa [1], Luis Guillermo Obregon [2], Adriana Arango-Manrique [3] and José Ricardo Núñez Álvarez [4,*]

1 Mechanical Engineering Department, Universidad del Atlántico, Carrera 30 Número 8-49, Puerto Colombia, Barranquilla 080007, Colombia; fbarrozo@mail.uniatlantico.edu.co (F.A.B.B.); guillermoevalencia@mail.uniatlantico.edu.co (G.V.O.)

2 Research Group on Sustainable Chemical and Biochemical Processes, Chemical Engineering Department, Universidad del Atlántico, Carrera 30 Número 8-49, Puerto Colombia, Barranquilla 080007, Colombia; luisobregon@mail.uniatlantico.edu.co

3 Electrical and Electronic Engineering Department, Universidad del Norte, Km 5 Vía Puerto Colombia, Barranquilla 080007, Colombia; adrianaarango@uninorte.edu.co

4 Energy Department, Universidad de la Costa, Calle 58 Número 55- 66, Barranquilla 080002, Colombia

* Correspondence: jnunez22@cuc.edu.co.

Abstract: The electrical sector in the Caribbean region of Colombia is currently facing problems that affect its reliability. Many thermo-electric plants are required to fill the gap and ensure energy supply. This paper thus proposes a hybrid renewable energy generation plant that could supply a percentage of the total energy demand and reduce the environmental impact of conventional energy generation. The hybrid plant works with a photovoltaic (PV) system and wind turbine systems, connected in parallel with the grid to supply a renewable fraction of the total energy demand. The investigation was conducted in three steps: the first stage determined locations where the energy system was able to take advantage of renewable sources, the second identified a location that could work more efficiently from an economic perspective, and finally, the third step estimated the number of PV solar panels and wind turbines required to guarantee optimal functioning for this location using, as a main method of calculation, the software HOMER pro® for hybrid optimization with multiple energy resources. The proposed system is expected to not only limit environmental impacts but also decrease total costs of electric grid consumption from thermoelectric plants. The simulations helped identify Puerto Bolivar, Colombia, as the location where the hybrid plant made the best use of non-conventional resources of energy. However, Rancho Grande was found to offer the system more efficiency, while generating a considerable amount of energy at the lowest possible cost. An optimal combination was also obtained—441 PV arrays and 3 wind turbines, resulting in a net present cost (NPC) of $11.8 million and low CO_2 production of 244.1 tons per year.

Keywords: solar energy; wind energy; energy efficiency; environmental impact; economic evaluation; on-grid system; HOMER Pro software

1. Introduction

The continuous increase in greenhouse effects in the energy field [1,2], the potential danger that represents the future of this trend, and the continuous rise in this kind of energy production boost the development of new trends of energy generation (NTEG), which lead to energy transition [3–5]. NTEG will be independent of hydrocarbons and fossil fuels because research on clean energy acquisition methods is on the rise, offering reliable solutions to the current problem [6,7]. Solar and wind energy systems are the most selected methods for clean energy production because of their viability and easy acquisition [8]. In 2006, the World Energy Outlook estimated that energy production could be duplicated in 25 years. Furthermore, the publication also expects a growth of 57% in renewable energy production [9]. The boost from new trends in renewable energy generation takes into consideration the current development of technologies that can be used to obtain energy from the movement of the sea waves, as well as ocean currents. There are other ways to obtain renewable energy, one of which is the photovoltaic (PV) system. Asia Pacific is estimated to remain the global PV leader in 2025 with the largest installed solar PV facility in the world [10]. The PV market has grown in both developed and developing countries, implying that renewable energy is a viable global resource. The world's largest PV energy production from an installed plant is located in Pavagada solar park in India. The park, which can generate up to 2 GW, was fully functional in 2019 [11]. A report in the Journal of Geophysical Research estimated that the highest reachable capacity of wind energy around the world is approximately 72 million GW, which corresponds to 500% more than the energy consumption of every kind of power [12].

Hybrid energy, which is the use of different kinds of energy, is more efficient than conventional energy generation. The availability of wind energy in Colombia, combined with biomass energy, has had a significant influence on the Caribbean region [13]. The exploitation of this source of energy can be an excellent solution to the energy problems prevalent in the region [14]. This solution lies in the design of a hybrid renewable energy plant that has the capacity to use all the renewable energy resources existing in this region [15].

However, fossil fuels are still considered the main energy source in the region, although they cause considerable damage to the environment by the high generation of greenhouse gases [6,16]. The use of fossil fuels by "fracking" increases greenhouse gases and other gases like arsenic and mercury [10]. Greenhouse gas emissions increased by at least 70% in the period between the 1970s to the beginning of the 21st century, with the energy sector being the main responsible factor [17]. An analysis by the Oak Ridge National Laboratory (Tennessee, United States) on carbon dioxide emissions states that greenhouse gasses in Colombia increased from 16 megatons to 84 megatons (80%) in the 1960s to 2014; it was compared to Portugal, Finland, Chile, Austria, Sweden, Ireland, and Hungary [18].

At present, energy potential (wind and solar energy) in the Caribbean Colombian region is going to waste, more specifically in La Guajira department; these forms of renewable energy can help mitigate greenhouse emissions and increase electricity generation. With hybrid systems, operating costs are reduced because they do not require as much maintenance as conventional energy generation methods, and owing to a learning curve that helps people understand how this technology operates [19].

The Hybrid Optimization for Multiple Energy Resources (HOMER) software provides the necessary tools to establish different simulations with multiple energy resources [20] and study behavior over time. This enables the simulation of a hybrid power plant with twice or more renewable resources. This academic tool can be used to determinate viability from economic and environmental perspectives and/or energy generation systems. Therefore, HOMER is ideal for this research work.

The main contribution of this research is to highlight the renewable energy potential in the Caribbean region of Colombia (more specifically in La Guajira), showcase the possibilities to meet the growing energy demand, and offer renewable energy as a great solution to the region's problems. The following sections of the paper are organized as follows. Section 2 provides the legal and regulatory framework for renewable energy projects, a description of the available sources in different locations of La Guajira, a scheme of the proposed system, and the planning undertaken for this research. Section 3 provides the theory and equations required to implement the study. Section 4 presents the results of the simulations with a complete analysis.

2. Contextualization and Required Information

This section highlights the Colombian policy associated with renewable integration in the electrical grid. It also gives geographical details of the different weather stations located in La Guajira, Colombia. The tables and graphs presented in this section contain relevant data obtained by the meteorological stations over 20 years. In addition, this research work's planning process is described.

2.1. Political Context

Colombia has a legal and policy framework that helps justify the development of this research work; it is shown in Figure 1 [21]. It is important to note that the legal framework is responsible for establishing limits and penalties for all future projects. An essential point of the policy is to provide economic benefits for projects that meet their guidelines and contribute to the development of new trends—in this case, the energy sector. The benefits include reduction in taxes, such as the cost of importing energy generating devices from renewable sources, and return on investment in net income coming directly from the state. In some countries, the production of clean energy in homes is promoted through incentives for those who provide this type of resource as a surplus in the community power grid.

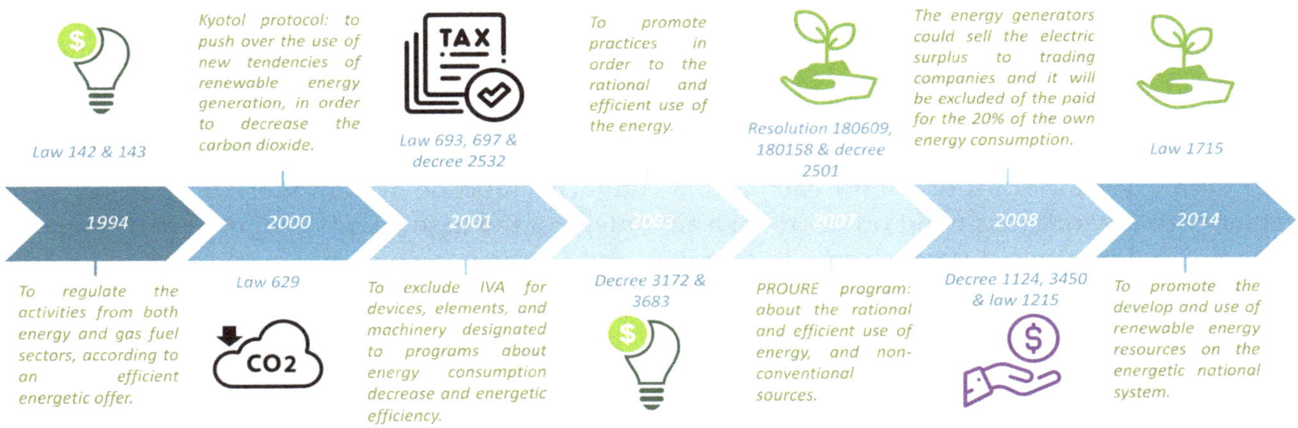

Figure 1. Timeline of the legal and regulatory framework for energy management in Colombia [20,21].

2.2. Energy System Scheme

The data were obtained from different meteorological stations located in La Guajira department, Colombia, as shown in Figure 2. There are nine stations dedicated to collecting temperature, wind speed, and solar radiation data. The data of pressure, relative humidity, and temperature in Figure 2 correspond to average yearly values.

In order to provide accurate results and establish the ideal location to use solar and wind energy, it was necessary to obtain data from each of the meteorological stations and pinpoint their specific locations [22].

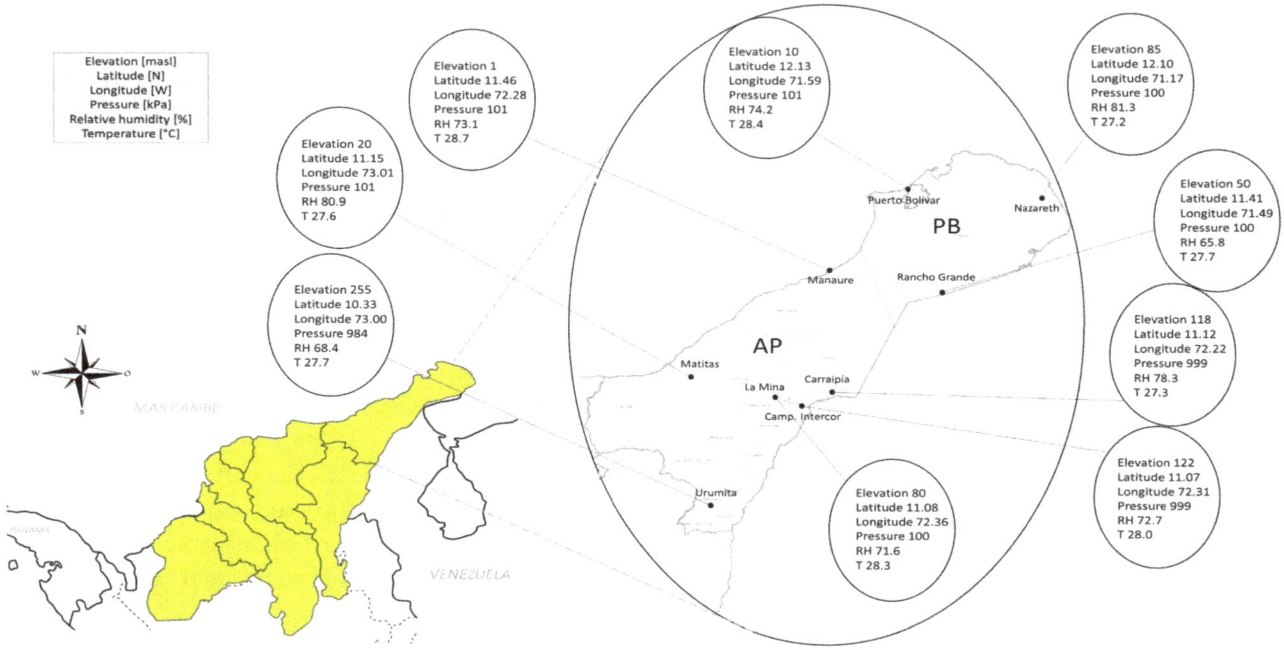

Figure 2. Geographic locations of the different meteorological stations in La Guajira, Colombia.

After studying the problem mentioned in this study, the creation of a hybrid wind and solar power generation plant was proposed to tap into the energy potential (Figure 3). This plant is proposed to have wind turbines 80 m in rotor height and with output power of 1.5 MW each, and an inverter module system like the Goldwind PMDD 1.5 MW Wind turbine [23]. A set of PV arrays (1kW per array) with 4 PV modules of 250 W each and a converter module system were proposed to work together with the turbine's wind power to take advantage of the high energy potential in the area. The map is divided into two different zones to establish areas influenced by wind speed and zones where potential renewable energy is higher and meteorological stations are located.

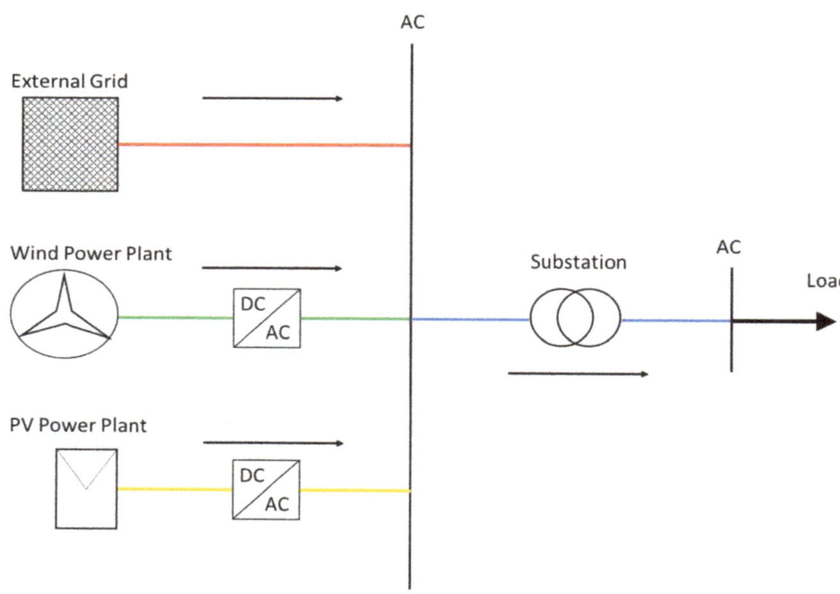

Figure 3. Proposed energy system scheme.

2.3. *Planning*

The investigation was carried out in three stages. The first stage involved the analysis of data collected from the weather stations over 10 years. The data concerning energy production and renewable fraction were studied without considering the costs on the system in an effort to estimate locations where renewable energy sources could be used in higher proportions.

The second stage sought to determine the most efficient location for a hybrid energy system that uses both wind and PV systems, working at the same time. This stage focuses on an economic perspective to study the results of total energy production, fraction of renewable energy, and aspects such as total net present cost (NPC) and CO_2 production to determine a location with optimal behavior.

The third stage determined the most efficient arrangement of wind and PV technologies working together, that is, the ideal number of wind turbines and PV panels depending on the energy demand and characteristics of the selected location. Finally, the grid power that the plant could develop, and its optimal composition was determined.

The first and second stages were calculated using simulations made in HOMER Pro software. The third stage used an optimization process through the MATLAB optimization function called Optimtool and the TOPSIS method for Pareto optimization. The data was used in the MATLAB curve fitting complement to determine the corresponding function for two main optimization objectives (energy cost and CO_2 emissions).

2.4. *Renewable Energy Resource Data*

Figure 4, and Tables 1 and 2 show the curves corresponding to wind speed in different locations at an altitude of 80 m with the corresponding temperature and solar radiation matrices for nine measurement points during a year, where AP means "Almirante Padilla" influence zone, and PB means "Puerto Bolivar" influence zone.

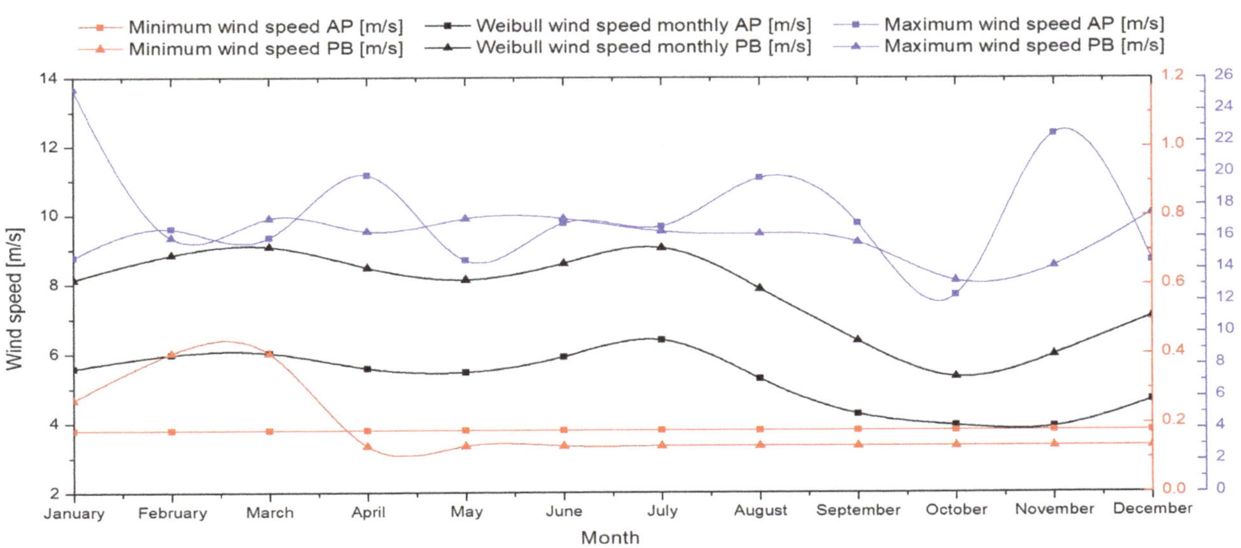

Figure 4. Wind speed graph for different measurement points in La Guajira [13].

Table 1. Temperature [°C] matrix for the different measurement points in La Guajira [22].

Locations	Jan.	Feb.	Mar.	Apr.	May	Jun.	Jul.	Aug.	Sep.	Oct.	Nov.	Dec.
Camp. Intercor	26.9	27.3	28.2	28.5	28.6	28.7	29.2	29.0	28.2	27.6	27.2	27.0
Carraipía	25.9	26.3	27.0	27.7	27.9	28.2	28.4	28.6	27.9	27.2	26.7	26.1
La Mina	27.0	27.6	28.3	28.7	28.9	29.3	29.5	29.4	28.4	27.7	27.3	26.9
Manaure	27.8	27.8	27.9	28.4	29.1	30.0	29.9	29.5	29.1	28.6	28.6	28.1
Matitas	26.7	26.8	27.2	27.7	27.9	28.8	28.9	28.4	27.7	27.1	27.1	26.6
Nazareth	25.7	25.8	26.3	27.0	27.6	28.0	27.9	28.5	28.5	27.8	27.1	26.2
Puerto Bolivar	27.0	26.9	27.4	28.2	29.0	29.5	29.4	29.5	29.3	28.8	28.3	27.5
Rancho Grande	25.9	26.4	27.3	28.1	28.5	28.7	28.3	28.3	28.3	27.7	27.5	27.2
Urumita	27.3	28.0	28.4	28.5	27.9	28.0	28.5	28.3	27.4	27.0	26.8	26.7

Table 2. Solar radiation [Wh/m^{2}day] matrix for the different measurement points in La Guajira [17].

Locations	Jan.	Feb.	Mar.	Apr.	May	Jun.	Jul.	Aug.	Sep.	Oct.	Nov.	Dec.
Camp. Intercor	4184	4368	4570	3622	3709	3970	4730	4769	4471	3469	3814	3495
Carraipía	3600	3648	3635	2986	2818	3563	4164	4152	3572	3226	3053	3037
La Mina	4258	4380	4350	4053	3637	4308	4646	4550	3904	3587	3382	3761
Manaure	3716	3917	4245	3827	3931	4516	4524	4866	4327	3525	3346	3146
Matitas	3751	3815	3526	3045	3208	4022	4451	4365	367	3337	3301	3355
Nazareth	3452	3788	3978	3443	3417	4146	4690	4819	4137	3302	3093	2851
Puerto Bolivar	4381	4973	5254	5050	4784	4692	5287	5536	4747	4004	3783	3899
Rancho Grande	4441	4772	5072	4465	4327	4630	5097	5261	4601	3943	3683	3802
Urumita	4400	4642	4509	3664	3391	3463	3994	3903	3377	3290	3369	3625

It is important to highlight the need to consider temperature in this case study. La Guajira, being a coast, is one of the warmest places in the Colombian territory. In addition, PV panels have a deficit when the temperature over them is too high. Furthermore, the physical properties of wind see a negative change with temperature increase. This reduces the amount of energy generated from PV arrays and wind turbines. Therefore, the temperature factor in the simulations was not considered, as it could represent possible incorrect results in this investigation.

2.5. Forecasting of Energy Demand

Thermoelectric plants are essential in the Colombian energy dispatch. However, large quantities of fossil fuels are required in thermoelectric plants, which means high and continuous operating costs, in addition to the high production of polluting gases and the legal consequences [24].

For this reason, it is necessary to implement a hybrid renewable energy generation plant that can replace a high percentage of the energy produced in thermoelectric plants. It would help in reducing the use of thermoelectric plants and facilitate their operation. If optimal operation of the hybrid plant is found, it could guarantee supply security and even enable selling of the remaining energy to the electric grid.

It was necessary to determine energy demands as a function of time, as shown in Figure 5. The energy that can be produced in thermoelectric plants is then compared with that produced by wind and solar systems. The effect of temperature is taken into account. Table 3 shows the parameters used in this research.

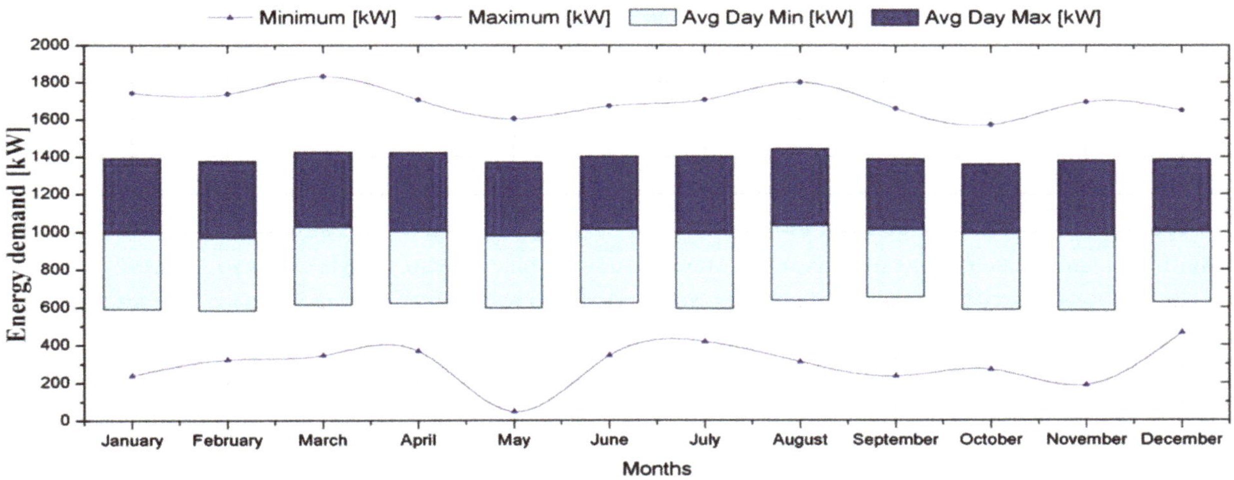

Figure 5. Seasonal profile of energy demand used in this study.

Table 3. Different parameters requested by HOMER Pro® for the specific economic and power calculations, where O&M means Operation & Maintenance and * indicates an estimated value.

Parameter	Description	Value	Unit
Grid power price *	Price that the electric utility charges for energy purchased from the grid	0.2	USD/kWh
Simulation period	Time considered for the study	10	Years
Annual scaled average	Parameter used to scale the whole array of hourly data up or down	24,000	kWh/day
Pike	Energy top value	1833.2	kW
Discount rate *	The rate considered to borrow money	8	%
Rate of inflation *	The percentage at which money is devalued along time	2	%
Wind turbine hub height	Height of the rotor measured from the ground	80	Meters
Grid CO_2 *	CO_2 emission factor of the energy generated by the grid	0.1	kg/kWh
PV O&M cost per unit	Operation and maintenance cost per PV unit	10	USD/year
Wind O&M cost per unit	Operation and maintenance cost per wind turbine unit	30,000	USD/year

3. Methodology

3.1. Wind Speed Estimation

The Weibull probability distribution was implemented to estimate the most probable wind speed in different locations using maximum and minimum values. A random variable x has a Weibull distribution if its probability density function is given as shown in Equation (1) [25].

$$f(x;\alpha,\theta) = \begin{cases} \frac{\alpha}{\theta^{\alpha}} \cdot x^{\alpha-1} \cdot exp\left[-(x/\theta)^{\alpha}\right] & x \geq 0 \\ 0 & x < 0 \end{cases} \quad (1)$$

The parameters α and θ [26] are estimated with experimental data. Depending on their values, Equation (1) can obtain the form of Equation (2), called the function of the probability density of the Rayleigh distribution:

$$f\left(x;\sigma^2\right) = \frac{x}{\sigma^2} \cdot \exp\left(-x^2/2\sigma^2\right) \quad x > 0 \quad (2)$$

In this case study, the factors of the different types of distribution expressed monthly for the locations of Puerto Bolívar and Almirante Padilla are presented in Table 4. On the other hand, Figure 6 shows the different velocity distributions of wind speed.

Table 4. Multi-annual adjusted distributions of wind speed for Puerto Bolivar (PB) and Almirante Padilla (AP), from 2003 to 2013 [22].

Distribution			**Weibull PB**						**Weibull AP**			
Month	**Jan.**	**Feb.**	**Mar.**	**Apr.**	**May**	**Jun.**	**Jan.**	**Feb.**	**Mar.**	**Apr.**	**May**	**Jun.**
Form	3.62	3.94	4.03	3.84	3.06	3.63	1.84	2.05	2.06	1.87	1.89	20.88
Scale	7.12	7.66	7.70	7.56	7.14	7.89	3.51	3.76	3.80	3.52	3.45	3.73
Month	**Jul.**	**Aug.**	**Sep.**	**Oct.**	**Nov.**	**Dec.**	**Jul.**	**Aug.**	**Sep.**	**Oct.**	**Nov.**	**Dec.**
Form	3.88	3.06	2.31	2.21	2.61	3.11	2.14	1.75	1.69	1.77	-	1.75
Scale	8.11	7.13	5.73	5.11	5.44	6.42	4.02	3.31	2.68	2.48	-	2.94
Distribution			**Rayleigh PB**						**Rayleigh AP**			
Month	**Jan.**	**Feb.**	**Mar.**	**Apr.**	**May**	**Jun.**	**Jan.**	**Feb.**	**Mar.**	**Apr.**	**May**	**Jun.**
Scale	6.64	6.94	7.15	7.04	6.71	7.37	3.66	3.74	3.76	3.64	3.55	3.67
Lower threshold	0.01	0.30	0.01	0.10	0.10	0.10	−0.09	0.01	0.02	−0.09	−0.07	0.03
Month	**Jul.**	**Aug.**	**Sep.**	**Oct.**	**Nov.**	**Dec.**	**Jul.**	**Aug.**	**Sep.**	**Oct.**	**Nov.**	**Dec.**
Scale	7.57	6.69	5.55	4.97	5.15	6.01	3.95	3.57	2.93	2.59	-	3.13
Lower threshold	0.09	0.10	0.05	0.05	0.10	0.10	0.03	−0.17	−0.15	−0.04	-	−0.10

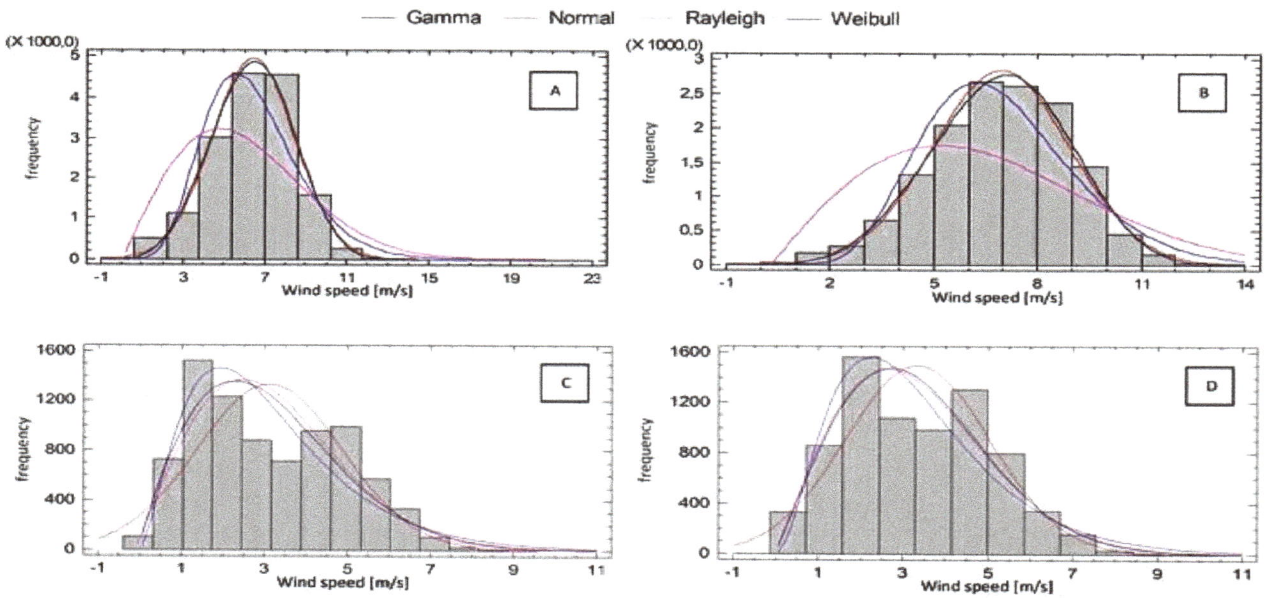

Figure 6. Adjusted frequency distributions of wind speed for Puerto Bolivar, (**A**) January, (**B**) February; and Almirante Padilla, (**C**) January, and (**D**) February.

The data for wind velocity were taken at a height of 10 m. Hellman's exponential law was used to determine the average wind velocity at any altitude:

$$V_h = V_{10}\cdot(h/10)^{\mu} \tag{3}$$

where, V_h is wind velocity at the required altitude h, V_{10} is the wind velocity at altitude of 10 m, and μ is the Hellman exponent, which varies with the roughness of the terrain [27]. These values were found to be 0.28 and 0.14 for the locations of Almirante Padilla and Puerto Bolívar airport, respectively [13].

3.2. HOMER Economic Analysis

In the principal cost analysis, total net present cost *(NPC)* and cost of energy *(COE)* are determined using Equation (4).

$$\boldsymbol{NPC}(\$) = \boldsymbol{TAC/CRF} \tag{4}$$

where, *TAC* is the total annualized cost and *CFR* is the capital return factor, calculated using Equation (5).

$$\boldsymbol{CRF}(\$) = \boldsymbol{i}\cdot(\mathbf{1}+\boldsymbol{i})^{N}/\left[(\mathbf{1}+\boldsymbol{i})^{N}-\mathbf{1}\right] \tag{5}$$

where, N is the number of years, and i is the annual range of real interest [%]. The cost of energy COE is the average unit cost of energy produced [$/kWh], and is determined using Equation (6).

$$\boldsymbol{COE}(\$/\boldsymbol{kWh}) = \boldsymbol{C_{tot.ann}/E} \tag{6}$$

where, $C_{tot.ann}$ is the total annual cost, and E is the total energy consumption per year.

3.3. HOMER Estimation of the Output Power of PV Panels

Equation (7) is used to determine the output power of PV panels:

$$\boldsymbol{P_{PV}} = \boldsymbol{Y_{PV}}\cdot\boldsymbol{f_{PV}}\cdot\left(\frac{\boldsymbol{G_T}}{\boldsymbol{G_{T,STC}}}\right)\cdot\left[\mathbf{1}+\boldsymbol{\alpha_P}\cdot\left(\boldsymbol{T_c}-\boldsymbol{T_{c,STC}}\right)\right] \tag{7}$$

where, Y_{PV} is the nominal capacity of the panel matrix, f_{PV} is the reduction factor of the panels, G_T is the incident solar radiation in the PV matrix in the current time step, $G_{T,STC}$ is the radiation incident in standard conditions, α_P is the power temperature coefficient, T_c is the temperature of the PV cell in the current time step, and $T_{c,STC}$ is the temperature of the PV cell under conditions of a standard test.

3.4. HOMER Estimation of the Output Power of Wind Turbines

Equation (8) is used to determine the output power of wind turbines.

$$\boldsymbol{P_{WTG}} = \left(\frac{\rho}{\rho_0}\right)\cdot\boldsymbol{P_{WTG,STP}} \tag{8}$$

where P_{WTG} is the output power of the wind turbine, $P_{WTG,STP}$ is the output power of the wind turbine at standard conditions, ρ is the actual density of air, and ρ_0 is the density of air at standard conditions.

3.5. Curve Fitting and Multi-Objective Optimization of the Forecast

With tabulated data, curve fitting using regressions is necessary; thus, it is important to have the best mathematical function to make forecasts. Optimtool was thus implemented to create an optimization process using the previously found function and determine Pareto's efficiency as a way of plotting the results. Pareto's efficiency shows a set of solutions delimited by the values closest to the origin coordinate if it is a minimum optimization; or, on the contrary, if it is a maximum optimization, the solution is delimited with the farthest values. The study then proceeded to use the multiple criteria method called Technique for Order of Preference by Similarity to Ideal Solution (TOPSIS) [28]. This method selects one of the values obtained in the Pareto efficiency, using the closest distance of any data from the lower vertex delimited by both ends of the graph. This technique uses the following equation:

$$\begin{aligned}\boldsymbol{d_{ix}} &= \sqrt{\sum_{j=1}^{n}\left(\boldsymbol{t_{ij}}-\boldsymbol{t_{xj}}\right)^2}, \quad \boldsymbol{i} = \mathbf{1},\mathbf{2},\ldots,\boldsymbol{m}\\ \boldsymbol{d_{iy}} &= \sqrt{\sum_{j=1}^{n}\left(\boldsymbol{t_{ij}}-\boldsymbol{t_{yj}}\right)^2}, \quad \boldsymbol{i} = \mathbf{1},\mathbf{2},\ldots,\boldsymbol{m}\end{aligned} \tag{9}$$

where d_{ix} and d_{iy} are the distances from the selected point t_{ij} to the ideal positive point t_{xj}, and the ideal negative point t_{yj}, respectively.

The relative proximity to the ideal solution (Siy) is determined by Equation (10):

$$S_{iy} = \frac{d_{iy}}{\left(d_{iy} + d_{ix}\right)} \qquad 0 \leq S_{iy} \leq 1 \tag{10}$$

This method allows to determine the best operating conditions for a system based on mathematical tools. However, it is also possible to achieve optimal conditions of power generation systems through advanced exergetic analysis, which are formulations based on the physical phenomena involved in each piece of equipment in the system [29,30].

4. Results and Discussions

This section presents the energy results, economic perspective, and multi-objective optimization.

4.1. Energy Results

Figure 7 shows the profile of the energy production obtained by PV systems, their nominal output power, and equivalent PV hours per year in multiple locations in La Guajira, Colombia. In this step, device replacement and capital costs are not considered, because the objective of this section is to evaluate the energy generation potential of the hybrid system.

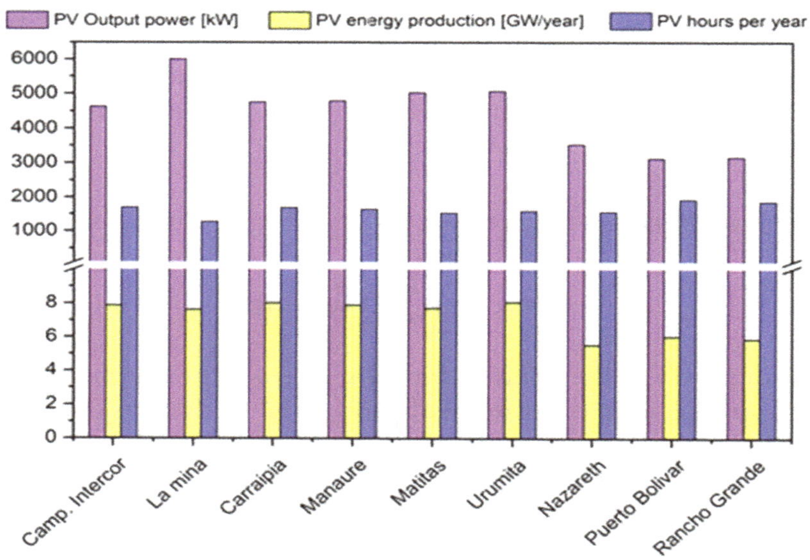

Figure 7. Comparison between the rated power and energy production for the photovoltaic (PV) system.

Figure 7 shows that the most significant annual energy production is in the town of Urumita. However, this does not mean that Urumita is the best location. It is necessary to quantify the efficiency of the devices to determine an optimal location for the PV system. The fraction between the nominal PV power and the energy generated per year gives a specific number of hours for each location in one year. Therefore, it estimates how PV arrays take advantage of solar light during operation, which means that locations where PV arrays have the longer number of hours per year will take more advantage of solar radiation.

On the other hand, it is necessary to highlight that PV output power indicates the nominal power for all the set of PV arrays connected between them, which are made of 4 panels with 250 W each. On this basis, Puerto Bolívar was identified as the location where the system works with the highest number of hours, despite having less energy generation compared to the other locations. This means that Puerto Bolívar is the best place to use our PV system. In percentage terms, the efficiency of this system is the fraction between the worked hours per year and the hours of one year, which gives an efficiency of 21.9%.

Subsequently, the optimal location for the wind system was determined. Considering that the use of wind turbines with hub height of 80 m and nominal power of 1.5 MW is fixed, a comparison was made between the number of wind turbines (given by the software for each location) and the energy production generated per year, which points towards the ideal location (Figure 8).

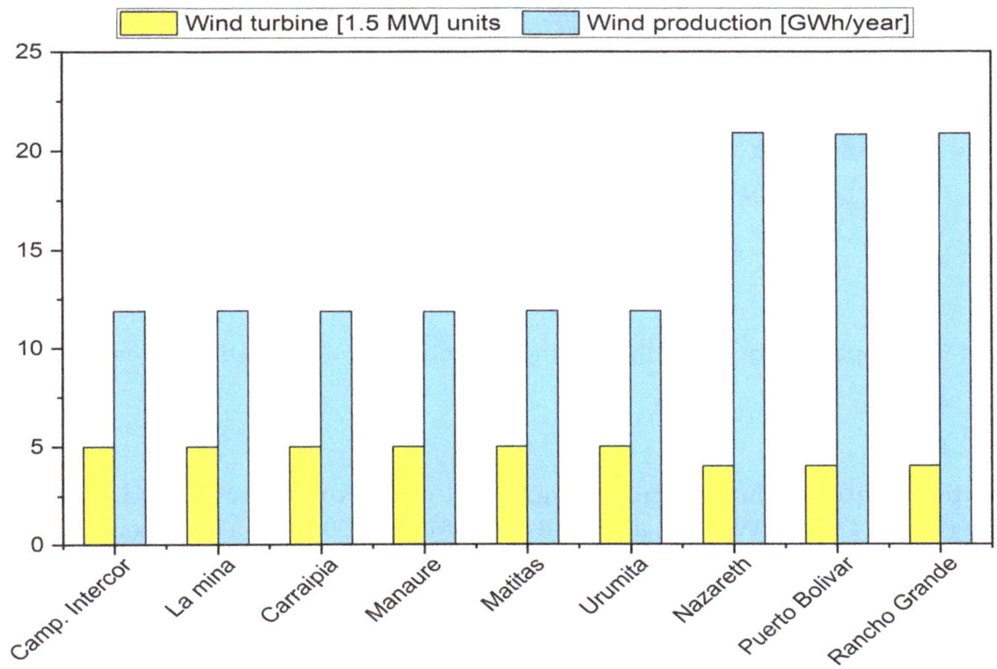

Figure 8. Comparison between the number of wind turbines and the annual energy production for each location.

Figure 8 shows six locations where the system considered five 1.5 MW turbines to reach the energy production plotted in the previous graphic for each location, while the other three only needed four turbines to reach a higher amount of energy. Thus, these three locations were analysed to find the area with the highest generation of wind power: Nazareth was identified as the location with 0.4% and 0.18% more wind power, compared to Puerto Bolívar and Rancho Grande, respectively.

After the best location for installation of the wind generation system was identified, the optimal general location was determined since the optimal sites for both types of devices are different. Therefore, it was necessary to analyse the percentage of renewable fraction present in each of the locations to identify which of these two locations generates the highest amount of renewable energy. The renewable fraction in a system is the ratio of the amount of clean energy produced and the total energy demand of our system.

Once the optimal location was determined, the parameters of the hybrid renewable energy generation plant were modified to increase power generation, resulting in a decrease in costs.

Figure 9 shows that Nazareth, Puerto Bolivar, and Rancho Grande are the locations with the highest RF percentage to produce energy by renewable resources. However, these values have a minimal difference, which means that it is necessary to verify the second step of the simulation; this refers to an economic perspective to determine the optimal location.

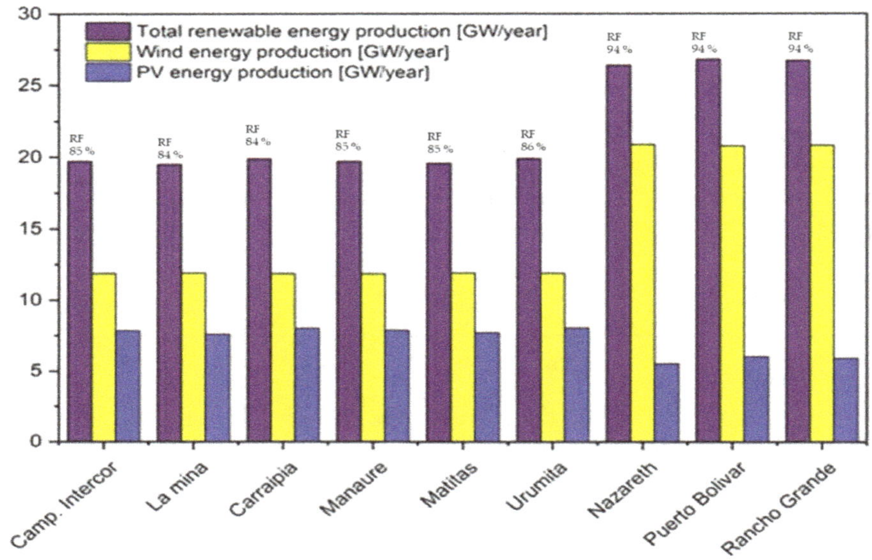

Figure 9. Comparison between the production and renewable fraction for each location.

4.2. Economic Perspective

The simulation requires capital and replacement costs for PV systems and wind turbines. A generic PV array (1 kW; 4 × 250 W) with a capital cost of USD 3000 and a replacement cost of USD 3000 was used. Generic wind turbines of 1.5 MW with a capital cost of USD 3,000,000 and a replacement cost of USD 3,000,000 were used. The effects of temperature were considered in the simulation.

Using generic energy demand based on the requirements of La Guajira department, simulation at this stage resulted in a single generic wind turbine for each location. Therefore, Figures 10 and 11 show the relationship of the number of PV panels with net present cost (NPC), total energy production, a fraction of renewable energy, and CO_2 emissions for the following locations: Nazareth, Port Bolívar, and Rancho Grande.

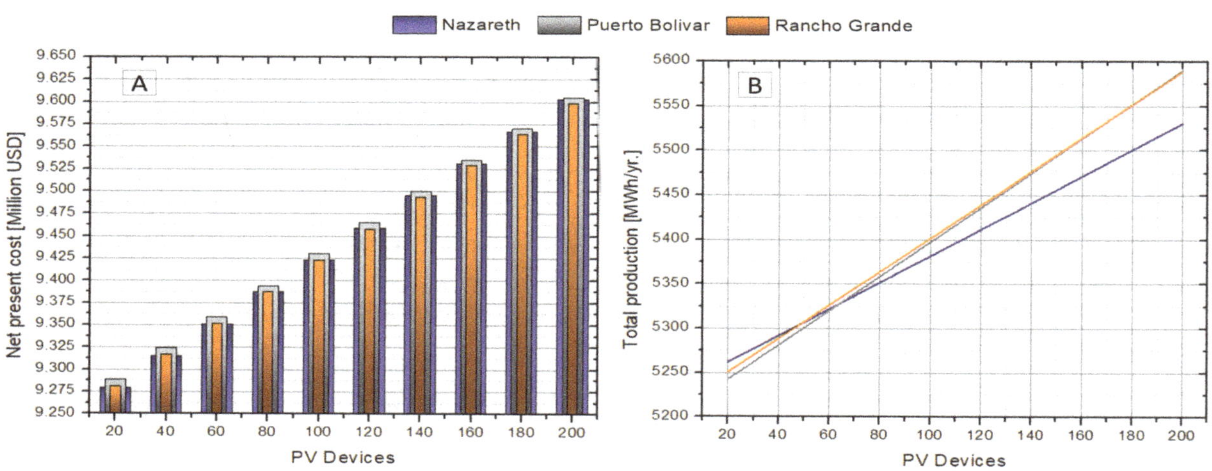

Figure 10. Comparative results for Nazareth, Puerto Bolivar, and Rancho Grande: (**A**) PV units and net present cost, (**B**) comparison between PV units and total production.

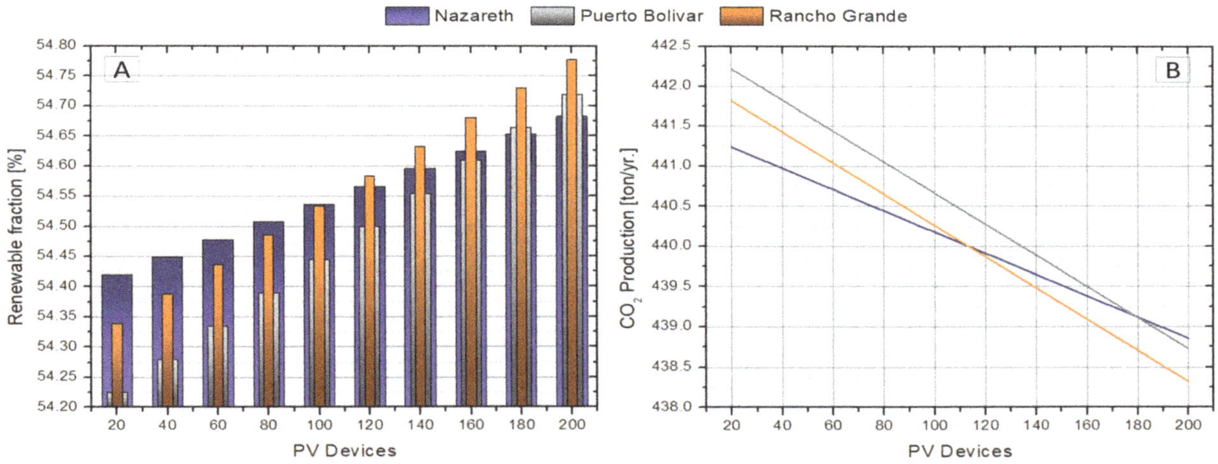

Figure 11. (**A**) Comparison between PV units and renewable fraction for Nazareth, Puerto Bolivar, and Rancho Grande, (**B**) comparison between PV units and CO_2 production for Nazareth and Rancho Grande.

Figure 10A shows that Nazareth and Rancho Grande are the locations where the NPC of the entire project, based on the number of PV panels implemented, is the lowest. Practically, the NPC for Nazareth and Rancho Grande are the same, which means that it is necessary to consider the following criteria to make the right decision about which place is optimal. In the town of Nazareth, the reason for total energy production and the number of PV panels is the lowest, compared to the localities of Rancho Grande and Puerto Bolívar. In addition, Figure 10B shows that the total energy production in Rancho Grande remains the best for at least 200 PV arrays. Another essential criterion considered was the renewable fraction. It is critical to highlight that a 1 PV device or 1 PV unit is equal to a 1 PV array for this simulation analysis.

Figure 11A shows that Nazareth and Rancho Grande have almost the same fraction values to produce energy by renewable energy, which are higher than those of Puerto Bolívar. This means that Puerto Bolívar is not the optimal place to implement a hybrid PV and wind power plant. Even if the location has good production values, its renewable fraction is the lowest, resulting in a high NPC.

The last criterion to study is the CO_2 emission produced by thermoelectric power plants and the acquisition of other non-renewable energy sources. Figure 11B shows that CO_2 emissions in Nazareth are lower than those of Rancho Grande when there are 110 PV arrays. For large numbers of PV units, lower CO_2 production is located in Rancho Grande, making it the best possible location.

4.3. Multi-Objective Optimization of Rancho Grande for Location of a Hybrid System

Considering that Rancho Grande was identified as the optimal location to implement a hybrid wind and PV system, the next stage sought to determine, through the optimization of multiple objectives, the most efficient combination of the number of PV devices and wind turbines.

Wind turbines in the range of 1 to 10 and the number of PV arrays in the range of 50 to 500 were implemented. A matrix was then developed for current NPC and CO_2 production based on PV devices and wind turbines.

A mathematical function for each criterion, NPC (Equation (11)) and CO_2 (Equation (12)), was determined using the MATLAB curve fitting tool. Figures 12 and 13 show the corresponding polynomial regressions with their respective functions.

$$f_1(x,y) = \begin{array}{l} \mathbf{9.523 + 8.646{\cdot}10^{-6}{\cdot}x - 0.8988{\cdot}y + 0.0006295{\cdot}x{\cdot}y + 0.6652{\cdot}y^2} \\ \mathbf{-8.478{\cdot}10^{-5}{\cdot}x{\cdot}y^2 - 0.06331{\cdot}y^3 + 3.936{\cdot}10^{-6}{\cdot}x{\cdot}y^3 + 0.002267{\cdot}y^4} \end{array} \quad (11)$$

$$f_2(x,y) = \begin{array}{l} 602.8 - 0.1141 \cdot x - 198 \cdot y + 0.0199 \cdot x \cdot y + 39.17 \cdot y^2 \\ -0.0009659 \cdot x \cdot y^2 - 3.691 \cdot y^3 - 1.175 \cdot 10^{-5} \cdot x \cdot y^3 + 0.1323 \cdot y^4 \end{array} \quad (12)$$

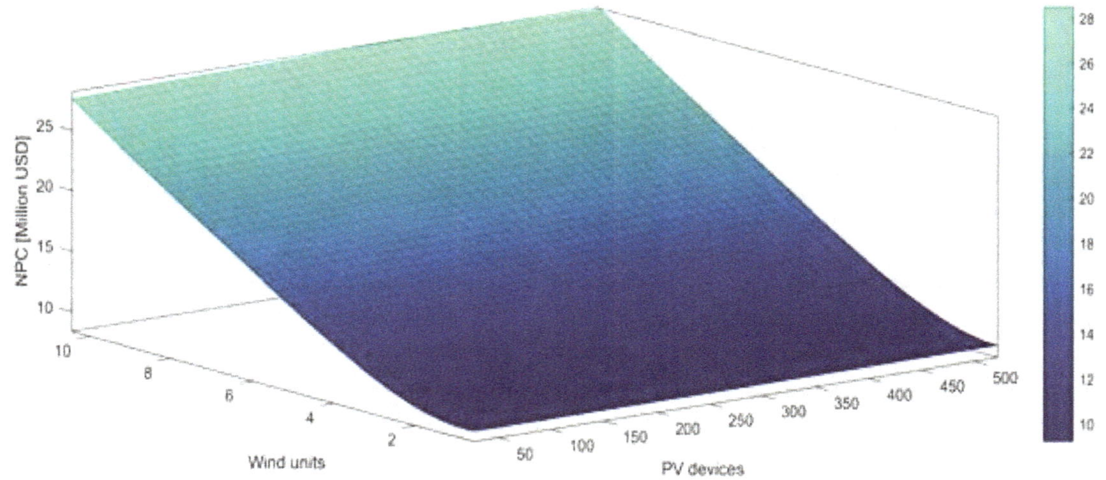

Figure 12. Polynomial regression for NPC (Rancho Grande).

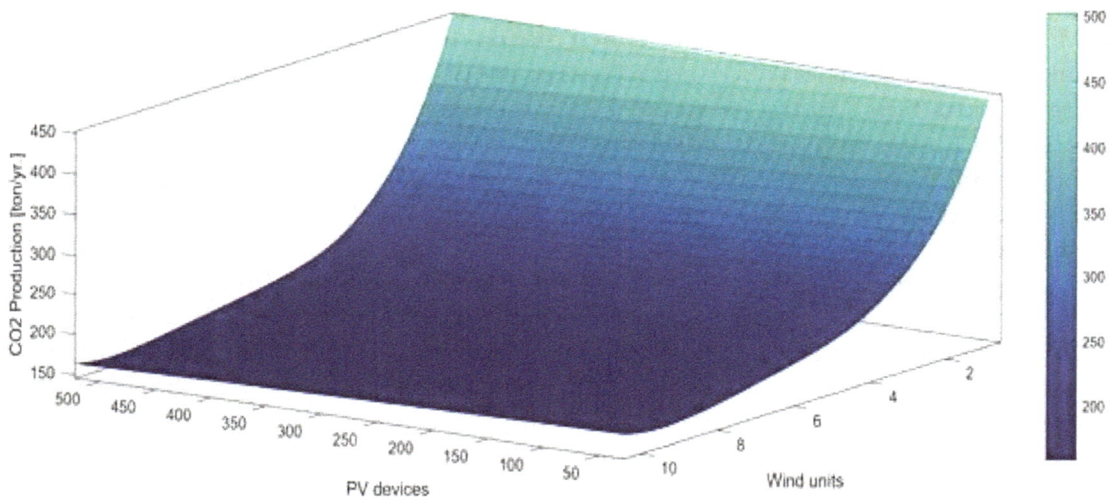

Figure 13. Polynomial regression for CO_2 emission (Rancho Grande).

These functions were useful in developing a MATLAB code for a multi-objective optimization process. Figure 12 shows the behavior of the NPC as a function of the PV devices and the wind units. It was necessary to use fourth-degree polynomial regression to obtain the minimum percentage of error, 0.0001%. Equation (11) is the mathematical function of the net present cost (NPC), where "x" is the number of PV units and "y" is the number of wind turbines.

Figure 13 shows the behaviour of CO_2 production as a function of the number of wind units and PV devices. Equation (12) is the corresponding mathematical function for CO_2 emissions, where "x" is the number of PV units, and "y" is the number of wind turbines. As can be seen, the effect of PV devices on CO_2 production is negligible, compared to the wind unit.

The optimization was conducted considering Objective 1 (the current NPC), and Objective 2 (the production of CO_2) with the two functions mentioned above. The purpose was to minimize both criteria using Pareto's efficiency. For each data of PV arrays and wind turbines, there are values of current NPC and CO_2 Production, as given in Table 5 (only a range of values was placed, since it can be more extensive).

Table 5. Data collected from previous multi-objective optimizations.

Index	Population $f1$	Population $f2$	$f_1(x,y)$	$f_2(x,y)$
1	**162.35**	**1.00**	**9.3208**	**424.2034**
2	455.86	2.84	11.5792	250.6008
3	374.16	1.58	9.8313	342.0218
4	458.33	7.68	21.8166	169.9665
5	450.50	7.50	21.4097	171.6429
6	190.93	1.09	9.3685	411.2848
7	463.61	8.10	22.7594	166.4565
8	464.54	7.08	20.4920	174.8915
9	182.34	1.16	9.3980	402.9515
10	466.82	9.27	25.4634	159.7775
11	**459.10**	**9.01**	**24.8412**	**160.7772**
12	252.28	1.31	9.5272	380.0974
13	170.97	1.47	9.5804	370.2378
14	460.99	5.58	17.1489	188.2983
15	204.76	1.51	9.6331	363.7958
16	**162.35**	**1.00**	**9.3208**	**424.2034**
17	467.55	8.58	23.8424	162.9501
18	289.97	1.20	9.4879	388.5064
19	456.46	6.39	18.9383	181.0377
20	452.18	6.71	19.6354	178.4340
21	460.93	4.41	14.6000	203.6394
22	**435.81**	**7.99**	**22.4599**	**167.9836**
23	443.14	4.65	15.0778	200.5169
24	447.40	3.76	13.2597	218.3149
25	436.18	3.23	12.2364	235.6264
26	294.45	1.59	9.7729	348.1610
27	453.42	5.85	17.7357	185.9505
28	456.58	4.38	14.5426	204.2703
29	292.95	2.85	11.4068	260.6003
30	451.30	6.14	18.3635	183.4294
31	390.02	2.24	10.5965	288.6604
32	460.17	3.87	13.5000	214.7941
33	**440.81**	**3.00**	**11.8483**	**244.0983**
34	212.64	1.61	9.7231	353.2899
35	407.42	7.34	20.9747	174.1048
36	466.60	8.46	23.5835	163.7199
37	434.90	2.31	10.7404	280.7996
38	458.70	9.03	24.8764	160.7231
39	**208.45**	**2.99**	**11.5348**	**259.4823**
40	295.98	2.07	10.2843	307.6137
41	208.34	1.17	9.4170	399.9407
42	430.48	2.02	10.3587	300.6018
43	**391.20**	**4.99**	**15.7236**	**197.8553**
44	461.77	5.12	16.1006	193.4751
45	445.79	5.75	17.4817	187.2504
46	731.67	2.49	10.9957	270.3299
47	443.03	3.41	12.5793	228.9088
48	404.32	5.30	16.4306	193.4263
49	468.14	9.52	26.0469	159.4742
50	162.38	1.02	9.3261	422.2238

$f_1(x,y)$ and $f_2(x,y)$ from Table 5, which are the values for NPC and CO_2 production, respectively, were then plotted (Figure 14).

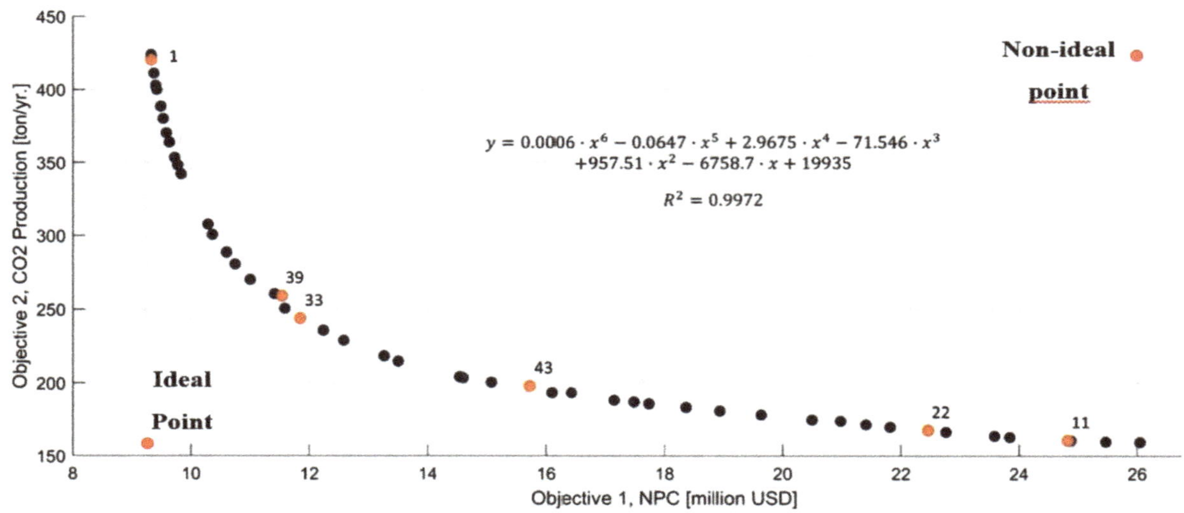

Figure 14. Pareto front for multi-objective optimization of NPC and CO_2 emissions.

It is important to note that Table 5 shows values that are not feasible due to decimal values; for example, index 17 gives 8.58 wind turbines. Decimal values are out of range because it is impossible to install 8.58 wind turbines of 1.5 MW. When choosing the nearest integer values, a considerable error was generated. On the other hand, PV arrays do not have this problem because their units are high, making it possible to approximate decimals to the nearest integer.

For this reason, it was necessary to determine the index that has the nearest integer for wind turbine devices, with at least 0.01 difference of any integer, as shown in Table 6.

Table 6. Feasible values to consider as an option, obtained from Table 4.

Index	PV Devices	Wind Units	NPC [million USD]	CO_2 Production [ton/yr]
1	162	1	9.3208	424.2034
11	459	9	24.8412	160.7772
22	436	8	22.4599	167.9836
33	**441**	**3**	**11.8483**	**244.0983**
39	**208**	**3**	**11.5348**	**259.4823**
43	391	5	15.7236	197.8553

While it is necessary to choose one of these values, it is important to note that all the previous data were optimal combinations, that is, data that is the best possible option for a specific number of selected devices. Figure 14 shows the locations of these indexes (in orange dots).

Figure 14 is a Pareto front composed of values of $f_1(x,y)$ and $f_2(x,y)$ shown in Table 5, which represents the net present cost (NPC) and CO_2 emissions, respectively. In addition, the characteristic equation is presented, where "x" is the number of PV arrays, and "y" is the number of wind turbines. Orange dots are the possible values given in Table 6, which represent specific configurations for the purposed hybrid plant. The ideal point is described as the interception of the vertical axis in the lowest value of CO_2 production with the horizontal axis in the lowest value of the Net Present Cost. The non-ideal point is described as the interception of the horizontal axis generated from the higher amount of CO_2 production with the vertical axis generated from the higher value of the Net Present Cost. Both ideal points are represented with red dots.

The next step is to implement the Technique for Order of Preference by Similarity to Ideal Solution (TOPSIS) method, which states that the selected point (taken in Figure 14) should have the shortest geometric length from the positive ideal solution and the most extended geometric length from the perfect negative solution. This optimization method has been widely used to improve the thermal and economic performance of energy systems [31–33]. Indexes 39 and 33 are the closest to the desired condition, as shown in Table 7.

Table 7. Results from TOPSIS analysis for index 33 and 39.

Index	Dx	Dy	Sy
33	**89.1292**	**181.4544**	**0.6706**
39	104.5021	166.1486	0.6139

Equation (10) is used to choose between the two values. The index with the value of Sy closest to 1 is the right point. The index 33 is closer to 1, making it the best possible option; therefore, a total of 441 PV arrays (1764 modules of 250 W) and 3 wind turbines of 1.5 MW should be used, resulting in an estimated NPC cost of 11.8 million dollars, and estimated CO_2 emission of 244.1 tons per year.

In this way, the best parameters that offer minimum energy costs to the residents, minimum emissions of pollutant gases, and maintain a reasonable energy production rate are obtained.

5. Conclusions

In this paper, an optimization process was developed to install a hybrid PV and wind power plant in La Guajira, Colombia. The study was done in three stages, with specific characteristics and region conditions. The first stage focused on an energy perspective, where it was found that three of the nine measurement locations—Nazareth, Puerto Bolivar, and Rancho Grande—were the best locations to take advantage of the available energy with 95% percentage each to produce energy using renewable energy. The renewable fraction for the other locations was close to 87%.

The second stage focused on an economic perspective, considering prices and taxes. This stage was developed only for the three best locations, finally identifying Rancho Grande as the optimal place to set up a hybrid energy plant. This location had an advantage over other areas in terms of renewable fraction, total production, and estimated CO_2 reduction.

The third and final stage focused only on the town of Rancho Grande, intending to determine the optimal combination of PV panels and wind turbines. This stage identified, for the plant, an optimal combination of 441 PV units and 3 wind turbines, giving an estimated minimum NPC of $11.8 million, and low CO_2 production of 244.1 tons per year.

Author Contributions: Conceptualization, G.V.O., F.A.B.B., and A.A.-M.; Methodology, G.V.O. and J.R.N.Á.; Software, G.V.O., L.G.O., and A.A.-M.; Validation, G.V.O., L.G.O., and J.R.N.Á.; Formal Analysis, G.V.O., J.R.N.Á., and F.A.B.B.; Investigation, G.V.O., L.G.O., and F.A.B.B.; Resources, J.R.N.Á. and A.A.-M.; Writing—Original Draft Preparation, F.A.B.B.; Writing—Review & Editing, L.G.O., A.A.-M., and G.V.O.; Funding Acquisition, G.V.O., and J.R.N.Á. All authors have read and agreed to the published version of the manuscript.

Acknowledgments: This research was supported by the Mechanical Engineering Program of Universidad del Atlántico. The Kai Research Group supports G. Valencia and F. Barrozo.

References

1. McCormick, R.L.; Tennant, C.J.; Hayes, R.R.; Black, S.; Ireland, J.; McDaniel, T.; Williams, A.; Frailey, M.; Sharp, C.A. Regulated emissions from biodiesel tested in heavy duty engines meeting 2004 emission standards. In Proceedings of the 2005 SAE Brasil Fuels & Lubricants Meeting, Rio De Janeiro, Brazil, 11–13 May 2005.
2. Valencia Ochoa, G.; Cárdenas Gutierrez, J.; Duarte Forero, J. Exergy, Economic, and Life-Cycle Assessment of ORC System for Waste Heat Recovery in a Natural Gas Internal Combustion Engine. *Resources* **2020**, *9*, 2. [CrossRef]

3. Valencia Ochoa, G.; Piero Rojas, J.; Duarte Forero, J. Advance Exergo-Economic Analysis of a Waste Heat Recovery System Using ORC for a Bottoming Natural Gas Engine. *Energies* **2020**, *13*, 267. [CrossRef]
4. Boca, G.D.; Saraçlı, S. Environmental Education and Student's Perception, for Sustainability. *Sustainability* **2019**, *11*, 1553. [CrossRef]
5. Noh, C.-H.; Kim, I.; Jang, W.-H.; Kim, C.-H. Recent Trends in Renewable Energy Resources for Power Generation in the Republic of Korea. *Resources* **2015**, *4*, 751–764. [CrossRef]
6. Valencia, G.; Duarte, J.; Isaza-Roldan, C. Thermoeconomic Analysis of Different Exhaust Waste-Heat Recovery Systems for Natural Gas Engine Based on ORC. *Appl. Sci.* **2019**, *9*, 4017. [CrossRef]
7. Núñez, A.J.; Benítez, P.I.; Proenza, Y.R.; Vázquez, S.L.; Díaz, M.D. Metodología de diagnóstico de fallos para sistemas fotovoltaicos de conexión a red. *Rev. Iberoam. Automática Inf. Ind.* **2020**, *17*, 94–105. [CrossRef]
8. Ueckerdt, F.; Brecha, R.; Luderer, G. Analyzing major challenges of wind and solar variability in power systems. *Renew. Energy* **2015**, *81*, 1–10. [CrossRef]
9. International Energy Agency. *World Energy Outlook 2016*; IEA Publications: Paris, France, 2016.
10. Szymczak, P.D. Asia Pac Leads Global Solar Photovoltaic (PV) Market. 18 December 2018. Available online: https://oilandgaseurasia.com/2018/12/18/asia-pac-leads-global-solar-photovoltaic-pv-market/ (accessed on 30 December 2019).
11. Fairley Peter. The Pros and Cons of the World's Biggest Solar Park. 2020. Available online: https://spectrum.ieee.org/energy/renewables/the-pros-and-cons-of-the-worlds-biggest-solar-park (accessed on 21 December 2019).
12. Moemken, J.; Reyers, M.; Feldmann, H.; Pinto, J. Future changes of wind speed and wind energy potentials in EURO-CORDEX ensemble simulations. *J. Geophys. Res. Atmos.* **2018**, *123*, 6373–6389. [CrossRef]
13. Valencia Ochoa, G.; Vanegas Chamorro, M.; Polo Jiménez, J. *Análisis Estadístico de la Velocidad y Dirección Del Viento en la Región Caribe Colombiana con Enfasis en la Guajira*; Sello Editorial Universidad Del Atlántico: Barranquilla, Colombia, 2016; p. 51.
14. Valencia, G.; Nuñez, J.; Acevedo, C. Research Evolution on Renewable Energies Resources from 2007 to 2017: A Comparative Study on Solar, Geothermal, Wind and Biomass Energy. *Int. J. Energy Econ. Policy* **2019**, *9*, 242–253.
15. Ochoa, G.V.; Blanco, C.; Martinez, C.; Ramos, E. Fuzzy Adaptive Control Applied to a Hybrid Electric-Power Generation System (HEPGS). *Indian J. Sci. Technol.* **2017**, *10*, 1–9. [CrossRef]
16. Milanes Batista, C.M.; Planas, J.A.; Pelot, R.; Núñez, J.R. A new methodology incorporating public participation within Cuba's ICZM program. *Ocean Coast. Manag.* **2020**, *186*, 105101. [CrossRef]
17. Sinay, L.; Carter, R.W.B. Climate Change Adaptation Options for Coastal Communities and Local Governments. *Climate* **2020**, *8*, 7. [CrossRef]
18. Boden, T.A.; Marland, G.; Andres, R.J. *Global, Regional, and National Fossil-Fuel CO2 Emissions, Carbon Dioxide Information Analysis Center, Oak Ridge National Laboratory*; U.S. Department of Energy: Oak Ridge, TN, USA, 2017. Available online: https://cdiac.ess-dive.lbl.gov/trends/emis/meth_reg.html (accessed on 26 October 2019).
19. Valencia, G.; Benavides, A.; Cárdenas, Y. Economic and Environmental Multiobjective Optimization of a Wind–Solar–Fuel Cell Hybrid Energy System in the Colombian Caribbean Region. *Energies* **2019**, *12*, 2119. [CrossRef]
20. HOMER Pro. Available online: https://www.homerenergy.com/products/pro/index.html (accessed on 29 July 2017).
21. Valencia, G.; Vanegas, M.; Villicana, E. *Disponibilidad Geográfica y Temporal de la Energía Solar en la Costa Caribe Colombiana*; Sello editorial de la Universidad del Atlántico: Barranquilla, Colombia, 2016.
22. Valencia Ochoa, G.; Vanegas Chamorro, M.; Villicaña Ortiz, E. *Atlas Solar de la Costa Caribe Colombiana*; Sello Editorial Universidad Del Atlántico: Barranquilla, Colombia, 2016.
23. Goldwind. Goldwind PMDD 1.5 MW Wind Turbine Brochure. 2017. Available online: https://www.goldw indamericas.com/sites/default/files/Goldwind%20Americas_Goldwind%201.5MW%20Brochure%20%2820 17%29_0.pdf (accessed on 2 February 2020).
24. UPME. *Escenarios de Oferta y Demanda de Hidrocarburos en Colombia*; Ministerio de Minas y Energía: Bogota, Colombia, 2012. Available online: http://www.upme.gov.co/docs/publicaciones/2012/escenarios_oferta_demanda_hidrocarburos.pdf (accessed on 9 April 2018).
25. Papoulis, A. *Probability, Random Variables, and Stochastic Processes*, 3rd ed.; McGraw-Hill: New York, NY, USA, 1991.

26. Jafari, A.A.; Zakerzadeh, H. Inference on the parameters of the Weibull distribution using records. *SORT* **2015**, *39*, 3–18.
27. Jung, S.; Arda Vanli, O.; Kwon, S.-D. Wind energy potential assessment considering the uncertainties due to limited data. *Appl. Energy* **2013**, *102*, 1492–1503. [CrossRef]
28. Etghani, M.M.; Shojaeefard, M.H.; Khalkhali, A.; Akbari, M. A hybrid method of modified NSGA-II and TOPSIS to optimize performance and emissions of a diesel engine using biodiesel. *Appl. Therm. Eng.* **2013**, *59*, 309–315. [CrossRef]
29. Ochoa, G.V.; Isaza-Roldan, C.; Duarte Forero, J. Economic and Exergo-Advance Analysis of a Waste Heat Recovery System Based on Regenerative Organic Rankine Cycle under Organic Fluids with Low Global Warming Potential. *Energies* **2020**, *13*, 1317. [CrossRef]
30. Ochoa, G.V.; Peñaloza, C.A.; Rojas, J.P. Thermoeconomic Modelling and Parametric Study of a Simple ORC for the Recovery of Waste Heat in a 2 MW Gas Engine under Different Working Fluids. *Appl. Sci.* **2019**, *9*, 4526. [CrossRef]
31. Valencia, G.; Núñez, J.; Duarte, J. Multiobjective optimization of a plate heat exchanger in a waste heat recovery organic rankine cycle system for natural gas engines. *Entropy* **2019**, *21*, 655. [CrossRef]
32. Ochoa, G.V.; Isaza-Roldan, C.; Forero, J.D. A phenomenological base semi-physical thermodynamic model for the cylinder and exhaust manifold of a natural gas 2-megawatt four-stroke internal combustion engine. *Heliyon* **2019**, *5*, e02700. [CrossRef]
33. Valencia Ochoa, G.; Acevedo Peñaloza, C.; Duarte Forero, J. Thermoeconomic Optimization with PSO Algorithm of Waste Heat Recovery Systems Based on Organic Rankine Cycle System for a Natural Gas Engine. *Energies* **2019**, *12*, 4165. [CrossRef]

Experimental and Theoretical Study on the Critical Breaking Velocity of Marine Natural Gas Hydrate Sediments Breaking by Water Jet

Leizhen Wang and Guorong Wang *

School of Mechanical Engineering, Southwest Petroleum University, Chengdu 610500, Sichuan, China; 201511000057@stu.swpu.edu.cn

* Correspondence: 200331010023@swpu.edu.cn.

Abstract: Water jet technology is a key technology in the marine natural gas hydrate (NGH) solid fluidization mining method. As an important parameter in water jet breaking NGH sediments technology, the critical breaking velocity of NGH sediments is unknown. In the present research, an orthogonal design experiment is carried out to study the critical velocity of NGH breakage by water jet, using frozen soil and sand as experimental samples. First, the time it takes to reach maximum NGH breaking depth is determined. Then, ultimate breaking distance is studied with respect to the NGH saturation, jet pressure, and nozzle diameter. Following that, the variation of critical velocity with NGH saturation is analyzed. Eventually, a formula to calculate the critical velocity for marine NGH breakage by water jet process is established, and the undetermined coefficient (η) in the formula is calibrated with the experiment data. The results show that the ultimate breaking distance is mostly achieved within 63 s. The three experimental factors in order of the effect on the ultimate breaking depth (from high to low) are NGH saturation, jet pressure, and nozzle diameter. The critical velocities for marine NGH breakage corresponding to the NGH saturations of 20%, 40,%, 6%, and 80% are 5.71 m/s, 7.14 m/s, 9.60 m/s, and 10.85 m/s, respectively. The undetermined coefficient η in critical velocity formula is 1.44 m/s.

Keywords: natural gas hydrate; critical velocity; solid fluidization method; water jet; ultimate breaking distance

1. Introduction

Natural gas hydrate (NGH), also named as "Flammable Ice", is considered to be an alternative clean energy source in the 21st century because of its huge reserves [1]. The total amount of NGH resources in the world is about 3×10^{15} m^3, most of which is stored in the ocean [2–4]. According to the types of accumulation, marine NGH can be divided into four types: the sandstone NGH reservoir, non-sandstone NGH reservoir, seafloor mounds NGH reservoir, and disseminated NGH reservoir [5]. Sandstone NGH reservoirs, which have a relatively high porosity and are easy to exploit, are usually selected for production tests [6–8]. Disseminated NGH reservoirs, although accounting for more than 90% of the total NGH reserves, are difficult to exploit with the current technology because of low permeability [9]. The NGH resources in the South China Sea are about 85 trillion m^3 [10]. However, they belong to the disseminated NGH reservoir type, and its matrix is mainly composed of clay and silty sand. The characteristics of hydrate in the South China Sea can be summarized as low permeability, shallow burial, and poor cementation. Although so many methods, such as heat injection, depressurization, chemical injection, carbon dioxide replacement, and solid fluidization, have been proposed, and are expected to be used in hydrate production [11–14]. However, for marine disseminated NGH reservoirs, only the depressurization method and the solid fluidization method have been technically verified by the offshore production test in the South China Sea in 2017 [14,15]. In the process of the production test with depressurization method, hydraulic slotting technology is used to solve the problem of the low permeability of the disseminated NGH reservoir [16]. The solid fluidization method is a new NGH production method, and its process is shown in Figure 1 [17]. First, a small borehole is drilled with coiled tubing. Then the NGH sediment around the borehole is broken into fine particles by water jet technology. After mixing with water, particles of hydrate, clay, and sand form slurry. Under the action of in situ separator, clay, and sand particles are separated and backfilled. Hydrate particles are transported to the ocean surface through the riser for further treatment [18].

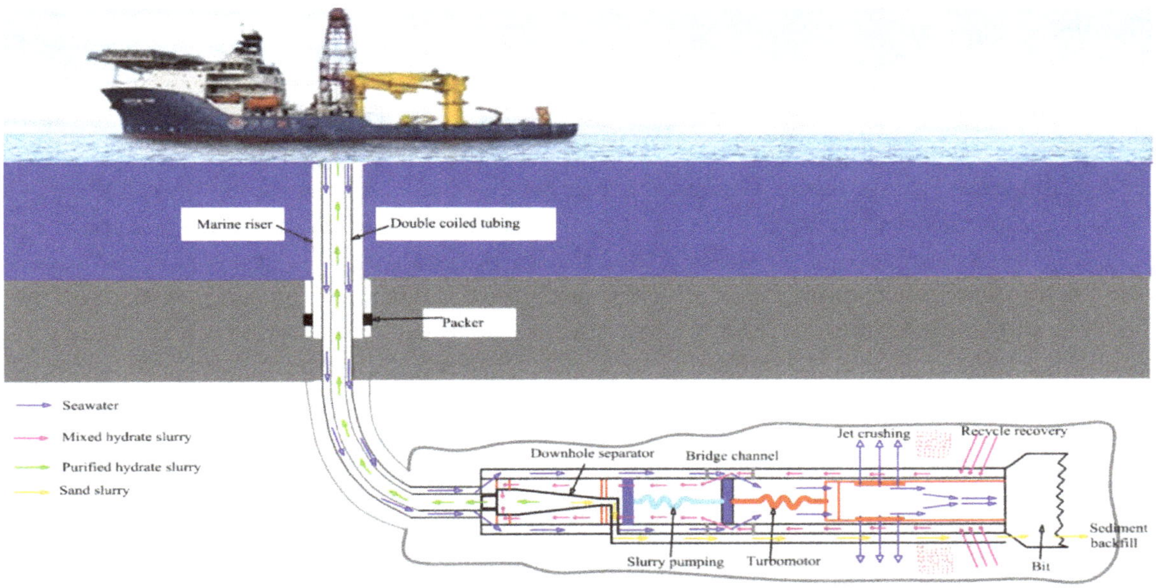

Figure 1. The exploitation process of natural gas hydrate (NGH) with the solid fluidization method. (Adapted from Qiu [17])

It can be seen that the technology of water jet breaking NGH is one of the core technologies of both the depressurization method and the solid fluidization method. Therefore, the technology of water jet breaking NGH has attracted the attention of some scholars. Pan [19] studied the effect of jet velocity on erosion depth and volume of NGH sediment by numerical simulation. Chen [20] studied the water jet erosion efficiency of hydrate-bearing sediment in the South China Sea under various work parameters, such as jet velocity, standoff distance, and nozzle diameter with the Arbitrary Lagrangian–Eulerian (ALE) method. Yang [21] carried out an experiment about the breaking of NGH with water jet. The influence of target distance, NGH saturation, and jet flow rate on the breaking depth is studied. Wang [22,23] carried out experiments of NGH breaking by water jet with frozen soil as experimental sample, developed a water jet mining tool, and optimized the water jetting parameters of the NGH production test in South China Sea. Wang [24] studied the effect of moving speed of mine head, jet flow rate, and jet action number on the jet breaking effect on four aspects: borehole shape, borehole diameter, breaking efficiency, and slurry concentration. The critical breaking velocity is an important parameter for evaluating the object's erodibility in water jet breaking technology. It means that only when the jet velocity on the contact surface of the jet and the object is higher than a certain value, will the object be damaged. In addition to evaluating the object's erodibility, the critical speed is also a key parameter to predict the breaking rate. However, the critical breaking velocity of NGH sediment under the action of the water jet is still unknown.

In this research, an orthogonal design experimental study of the critical velocity of NGH sediment under the action of water jet was carried out using frozen soil and sand as experimental samples. Firstly, the time it takes to reach maximum NGH breaking depth is determined. Then, the ultimate breaking distance is studied with respect to NGH saturation, jet pressure, and nozzle diameter. Following that, critical water jet breaking velocities corresponding to various sample saturations are calculated. Eventually, by combining the marine NGH strength expression with the critical breaking velocity semi-empirical formula, the formula to calculate the critical breaking velocity of marine NGH breaking by water jet is established. The characteristic velocity (η), an undetermined coefficient in the formula, is calibrated with the experiment data. This study will be helpful for the further research on the marine NGH breakage by water jet.

2. Experiment Equipment and Method

2.1. Experimemt Apparatus

As shown in Figure 2, the experiment apparatus mainly includes a water supply tank, pump, flowmeter, pressure gauge, experimental water tank, and nozzle. The volume of the water supply tank is 4.5 m^3. The pump is a plunger pump with the maximum working pressure and displacement of 30 MPa and 200 L/min, respectively. The measuring range of pressure gauge is 0–40 MPa at a 0.1 MPa precision, and the measuring range of flowmeter is 2–20 m^3/h at a 0.02 m^3/h precision. The function of the experimental water tank is to ensure that the jet is submerged. A movable baffle is installed in the water tank between the nozzle, and the experimental sample to avoid erosion of the sample while adjusting the test pressure. The type of nozzles used in the experiment is the cone straight type. The length of the cone angle section and the straight pipe section of a nozzle are the same, which are three times of nozzle diameter. The cone angle is 13°.

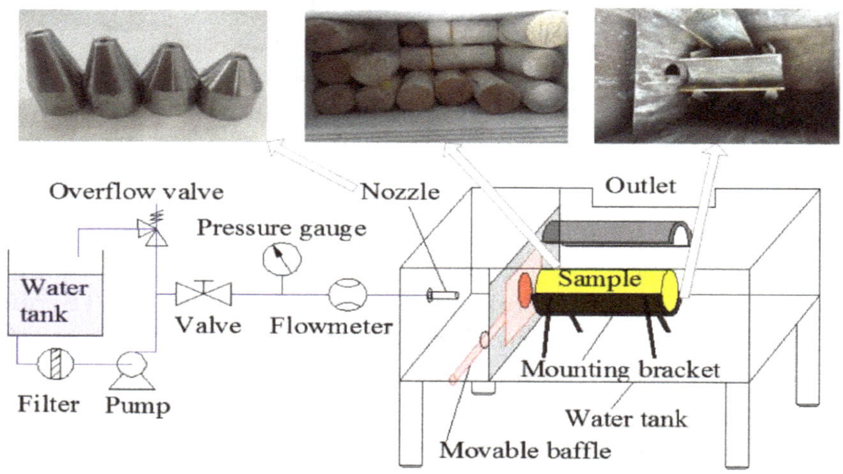

Figure 2. Schematic of the experimental system of the NGH breaking by water jet.

2.2. Sample Preparation

NGH is very easy to decompose at environmental temperature and pressure [25]. So, frozen sand and clay is used to instead of an NGH sediment as an experimental sample. In the early stage, it was generally believed that the mechanical properties of hydrate were similar to those of ice [26,27]. With the development of research, there are now different views on whether the mechanical properties of hydrate and ice, and the mechanical properties of hydrate sediment and frozen soil and sand, are the same. Stern [28] found that ice ordinarily exhibits a strength maximum before leveling off to steady flow stress, and methane hydrate exhibits monotonic work hardening (or strain hardening). Although the studies of Durham [29] have shown that methane hydrate is as much as 40 times stronger than ice, there is evidence that the tensile stress of methane hydrate was almost equal to that of ice, and the compressive stress was slightly higher than that of ice [30]. As for hydrate deposits and frozen soils, their differences mainly lie in the cohesion, while the internal friction angle is almost the same [31].

Winters [32] suggested that when porosity and saturation of the frozen soil (sand) and the NGH sediment are the same, their mechanical properties are consistent. Although the mechanical properties of frozen sand and clay are not exactly the same as that of NGH sediment, it can replace NGH sediment as an experimental sample to some extent. The experimental sample preparation is based on the matrix particle composition of NGH sediment samples in Shenhu area of South China Sea, which consists of clay (26.89%), silty sand (67.99%), and fine sand (5.12%). Porosity of the NGH core sample in the Shenhu area of South China Sea is 43.6%, and its matrix specific gravity is 2.7 [33]. Clay, silt, sand, and fine sand were blended uniformly in the same composition as that of the NGH sediments, to prepare the matrix sample. The mass of water can be calculated according to the following formula:

$$\frac{m_w}{m_s} = \frac{\xi}{1-\xi} \cdot \frac{\rho_w}{\rho_s} S_u \varepsilon \tag{1}$$

where: m_w is mass of water; m_s is mass of matrix; S_u is NGH saturation; ξ is porosity; ρ_w is density of water; ρ_s is density of matrix; ε is volume ratio of ice to water—because of the volume expansion of water during freezing, its value is 1.1.

For the experimental samples with saturation of 20%, 40%, 60%, and 80%, the mass ratio of water to matrix is 5.2%, 10.4%, 15.6%, and 20.8%, respectively. Evenly mixed water and matrix was poured into molds and frozen at −18 °C for seven days, after which the experiment samples were prepared.

2.3. Experimemt Procedure

A few hours before the start of the experiment, sufficient ice was added into the experimental water tank and water supply tank, to ensure the low temperature of experiment sample and jet water. Relative positions of the nozzle and the experimental sample were adjusted, so that they were on the same axis with a 20 mm distance. Prior to turning on the pump, the movable baffle was kept closed. The pump was turned on and the jet pressure was adjusted. When the jet pressure reached the experimental set point, the removable baffle was lifted to the open position, and the NGH water jet breaking experiment was timed as started. Every certain time duration, the removable baffle was closed to pause the jet breaking, and to record the jet breaking depth (the depth of the experimental sample that the water jet had broken into). When the jet breaking depth did not increase any more, an experiment of NGH jet water breaking was considered as completed. As shown in Table 1, a total of sixteen groups of orthogonal design experiments were designed and conducted with a variation of nozzle diameter, jet pressure, and NGH saturation.

Table 1. Orthogonal design experimental parameters of NGH breaking by water jet.

Experimental Serial Number	Saturation (%)	Nozzle Diameter (mm)	Jet Pressure (MPa)
1	20	1.5	2.5
2	20	2.0	5
3	20	2.5	7.5
4	20	3.0	10
5	40	1.5	5
6	40	2.0	2.5
7	40	2.5	10
8	40	3.0	7.5
9	60	1.5	7.5
10	60	2.0	10
11	60	2.5	2.5
12	60	3.0	5
13	80	1.5	10
14	80	2.0	7.5
15	80	2.5	5
16	80	3.0	2.5

3. Results and Discussion

3.1. Ultimate Breaking Time of the NGH Water Jet Breaking Process

For each jet breaking experiment, the sample breakage depth when the jet water could not break into the experiment sample any further is defined as maximum breaking depth. The time it takes to achieve the maximum breaking depth is the ultimate breaking time. Since the experimental sample is made of frozen soil, although the sample is submerged in the water tank to maintain low temperature, it should not be exposed to above zero degrees centigrade ambience for an excessive time. During the jet breaking process, the major breakage happens in the early stage for a short period of time, and, as the jet breaking proceeds, the breaking rate becomes low and breaking depth changes very little. Therefore, based on the correlation of breaking depth and time, the breaking depth that barely changes with time is approximately regarded as the maximum breaking depth, and the corresponding time is considered the ultimate breaking time. Figure 3 shows the variation of breaking depth with breaking time under various experimental conditions. As the jet breaking process proceeds, breaking depth increases with time, and breaking depth increase slows down with time. The first 0–15 s is the fast-breaking section, during which breaking depth increases rapidly. The following 16–63 s is the slow-breaking section during which breaking depth increases slowly. Comparing the breaking depth and time correlations under three experimental conditions in Figure 3, it was found that, given the same jet breaking time, the larger the breaking depth, the more significant breaking depth increases during the slow-breaking section, so a longer breaking time is needed to achieve the maximum breaking depth. Among all the NGH jets breaking experimental conditions, sample saturation of 20% (lowest), jet pressure of 10 MPa (highest), and nozzle diameter of 3.0 mm (largest) lead to the largest breaking depth, and thus results in the longest ultimate breaking time. Under this experimental condition (S_n = 20%, P = 10 MPa, D_n = 3.0 mm), average breaking rate (the rate of breaking depth increase is defined as breaking rate) during the fast-breaking section is 18.6 mm/s, and the average breaking rate of the slow-breaking section is as low as 1.4 mm/s. Breaking rate slows down further after the slow-breaking section, so the breaking depth increase after slow-breaking section is negligible. Therefore, based on the breaking depth vs. time curve of the experiment under sample saturation of 20%, jet pressure of 10 MPa, and nozzle diameter of 3.0 mm, 63 s is considered ultimate breaking time. The jet breaking process is conducted for 63 s for the following experiments.

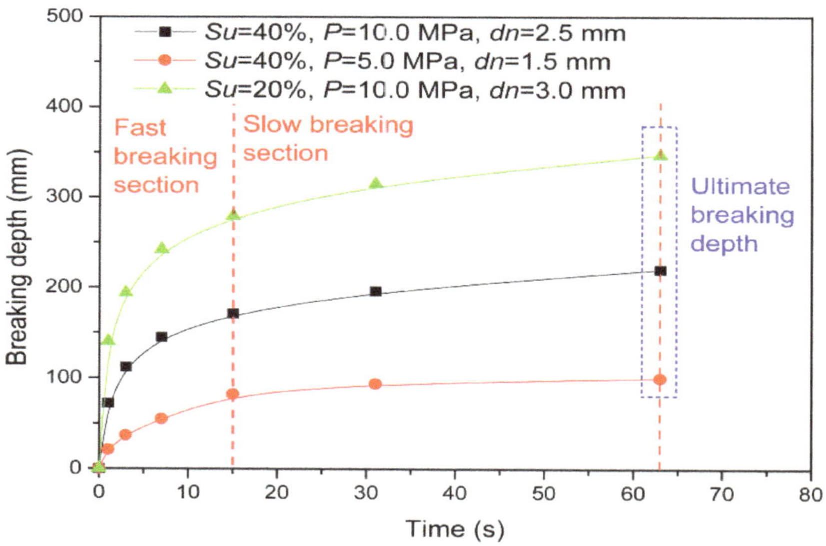

Figure 3. Variation of breaking depth with jetting time of water jet on NGH.

3.2. *Ultimate Breaking Distance of NGH Breaking by Water Jet*

Ultimate breaking distance, the summation of the target distance (the initial distance between the nozzle and the experimental sample prior to the NGH water jet breaking process, 20 mm) and the maximum breaking depth, is an important parameter to determine the critical water jet breaking velocity. Figure 4a,d shows the variation of the ultimate breaking distance with nozzle diameter and jet pressure when the sample saturation was 20%, 40%, 60%, and 80% respectively. The average ultimate breaking distances corresponding to 20%, 40%, 60%, and 80% sample saturations were 230.5 mm, 179 mm, 126.5 mm, and 109.25 mm, respectively, which suggests that the higher sample saturation, the shorter the ultimate breaking distance.

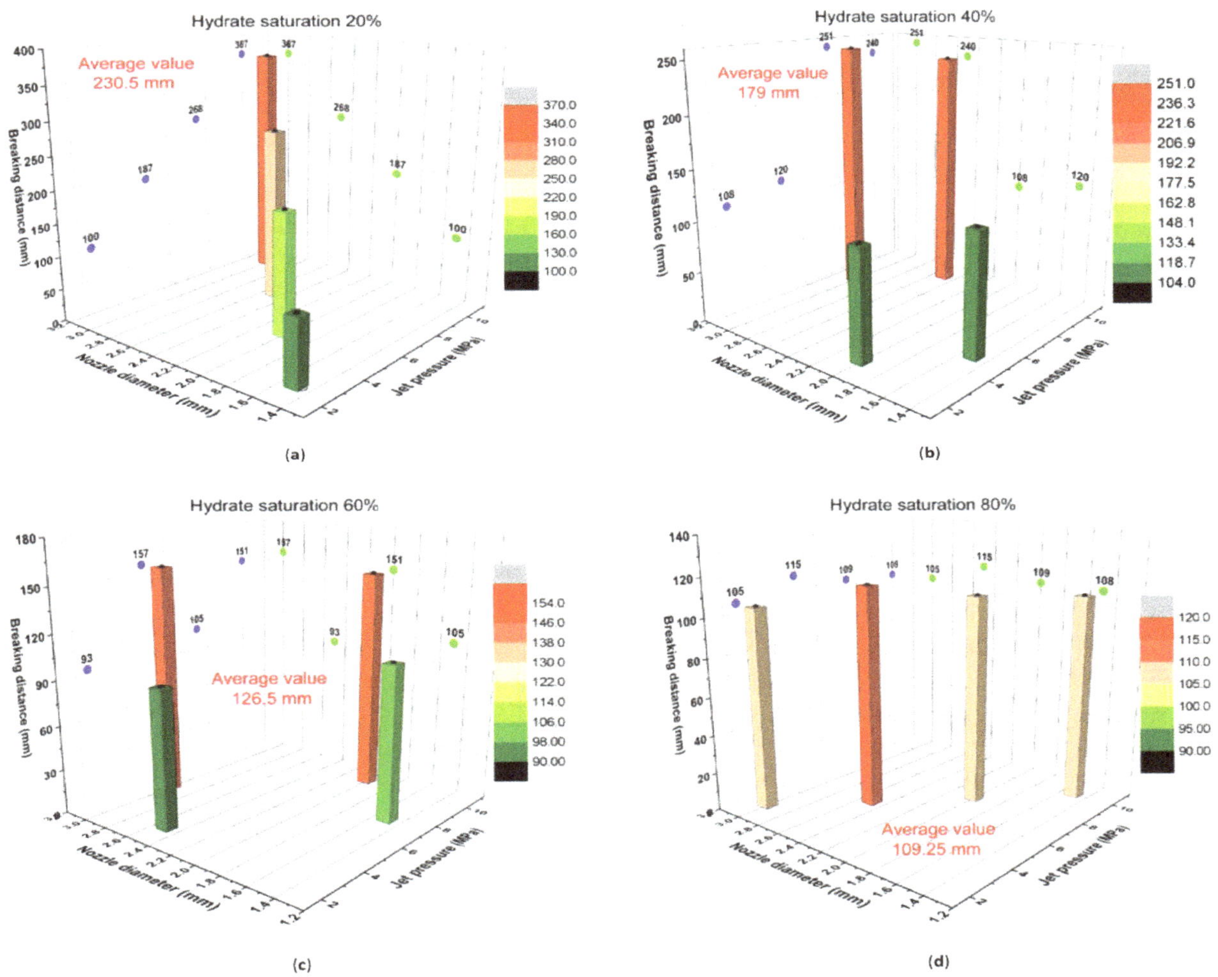

Figure 4. Variation of the breaking distances with nozzle diameter and jet pressure at various NGH saturations: (**a**) NGH saturation is 20%. (**b**) NGH saturation is 40%. (**c**) NGH saturation is 60%. (**d**) NGH saturation is 80%.

In orthogonal design test analysis, the range of a certain factor indicates the variation of the factor. A larger range indicates a higher influence that the specific factor has on the experimental index. As an example, the ultimate distance range of the sample saturation was calculated as the highest average ultimate distance (230.5 mm) among those corresponding to four various sample saturations, minus the lowest average ultimate distance (109.25 mm), equaling 121.25 mm. The ultimate breaking distance ranges of nozzle diameter and jet pressure were calculated using the same method, as shown in Table 2. In Table 2, *Ki* indicates the average ultimate breaking distance corresponding to Experiment Level *i*. Table 2 shows that the average ultimate breaking distance *Ki* increases with nozzle diameter and jet pressure increase. The average ultimate breaking distance ranges of the three experimental factors, sample saturation, jet pressure, and nozzle diameter, were 121.25 mm, 115 mm, and 112.5 mm, respectively. This shows that the sample saturation has the greatest effect on ultimate breaking distance. However, NGH saturation is a sediment property that could not be changed artificially. In summary, the three factors in order of the effect on the ultimate breaking depth (from high to low) are sample saturation, jet pressure, and nozzle diameter. There is a limitation in this orthogonal design experiment. Although the influence of a single parameter on the ultimate breaking depth is analyzed, the influence of interaction parameters on the ultimate breaking depth are not considered. The latter can more clearly reveal the relationship between the ultimate breaking distance and the experimental parameters. The main reason is that the main goal of this paper is to study the critical velocity of NGH sediment, which is a parameter to characterize the natural characteristics of NGH sediment. So, the critical velocity of NGH sediment is independent of jet pressure and nozzle diameter. Therefore, the interaction study is not considered in this orthogonal design experiment.

Table 2. Range analysis table of orthogonal design experiment.

Parameter	Factor A	Factor B	Factor C
	Saturation	Jet Pressure	Nozzle Diameter
K_1	230.5 (20%)	101.5 (2.5 MPa)	107.5 (1.5 mm)
K_2	179 (40%)	144 (5.0 MPa)	138.75 (2.0 mm)
K_3	126.5 (60%)	183.25 (7.5 MPa)	179 (2.5 mm)
K_4	109.25 (80%)	216.5 (10.0 MPa)	220 (3.0 mm)
R	121.25	115	112.5

3.3. *Critical Velocity of the NGH*

The critical velocity of the NGH breakage by the water jet is the axial water jet velocity at the sample breakage cross section when the ultimate breaking distance is achieved. The water jet flow velocity can be calculated based on the attenuation equation of the jet flow field. Although the water jet flow in this study is the non-free jet inside the enclosure, its axial jet velocity follows the same attenuation law as the free jet flow [34]. For the free jet flow, the jet flow field consists of the core area where axial velocity keeps the same, and the diffusion area where axial velocity declines with distance from the nozzle. Axial velocity can be expressed as Equations (2) and (3):

When $x < x_0$

$$v_x = v_0 \tag{2}$$

When $x > x_0$

$$v_x = \frac{\lambda d_n}{x} v_0 \tag{3}$$

where: x is the distance to nozzle, x_0 is the length of the potential core, $x_0 = \lambda d_\mathrm{n}$, v_0 is the initial exit velocity of the jet, v_x is the axial velocity of jet at distance x from nozzle, d_n is the nozzle diameter, λ is a dimensionless parameter to be measured experimentally, and Rajaratnam [35] recommends that the value of λ is 6.3.

Since the target distance between nozzle and experiment sample is larger than the length of the core area, the attenuation law of the axial velocity in the diffusion area, Equation (3), applies to velocity calculation for this study. As defined, when jet distance x increases to the ultimate breaking distance, the corresponding axial jet velocity is the critical water jet breaking velocity of the NGH breakage by the water jet process. Figure 5a–d shows the variation of the critical water jet breaking velocities with nozzle diameter and jet pressure when the sample saturations were 20%, 40%, 60%, and 80%, respectively. With the same sample saturation, experiments conducted under various jet pressures and nozzle diameters have almost the same critical water jet breaking velocity, which implies that critical velocity is a feature of the broken object, in this case NGH or the experimental sample, and is not affected by jet flow parameters. With the increase in the sample saturation, average critical velocity under various jet pressures and nozzle diameters increases. Average critical water jet velocities corresponding to the sample saturation of 20%, 40%, 60%, and 80% are 5.71 m/s, 7.14 m/s, 9.60 m/s, and 10.85 m/s, respectively.

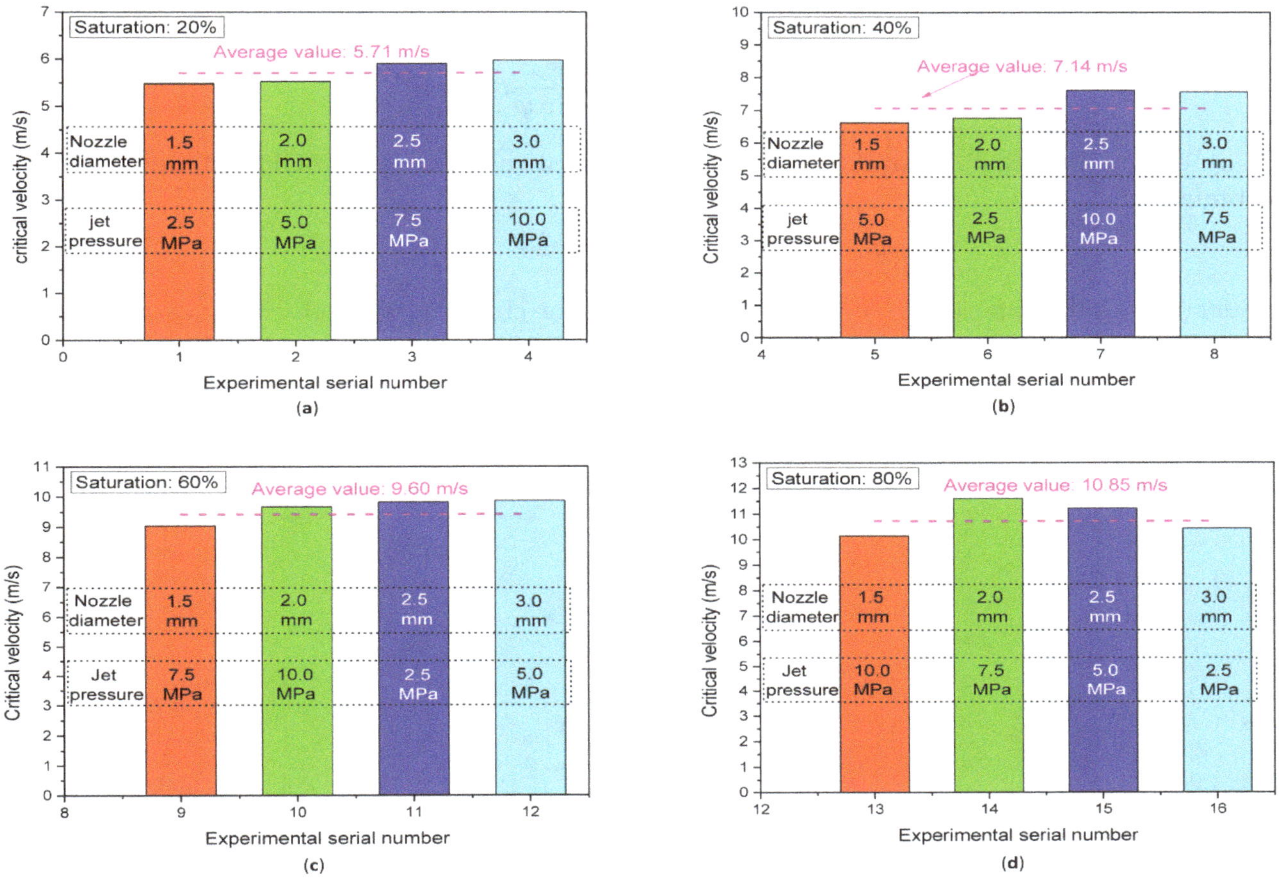

Figure 5. Variation of the critical breaking velocities with nozzle diameter and jet pressure under various NGH saturations: (**a**) NGH saturation is 20%. (**b**) NGH saturation is 40%. (**c**) NGH saturation is 60%. (**d**) NGH saturation is 80%. The different colors red, green, dark blue and light blue in the figure indicate that the nozzle diameters are 1.5 mm, 2.0 mm, 2.5 mm and 3.0 mm respectively.

To further theoretically analyze the correlation between critical velocity of NGH and NGH saturation, the critical breaking velocity formula established by Dabbagh [36] was introduced as Equations (4):

$$v_{cr} = \eta\left(\frac{q_u}{p_{atm}}\right)^k \tag{4}$$

where: η is a characteristic velocity, whose value equals to the critical velocity when the strength of NGH sediment is equal to the atmospheric pressure, η needs to be measured experimentally, k is a dimensionless exponent, its value is 0.5, P_{atm} is atmospheric pressure, v_{cr} is critical velocity, q_u is uniaxial compressive strength.

Because NGH sediment decomposes under normal pressure and temperature, its uniaxial compressive strength cannot be measured directly. Therefore, the expression of uniaxial compressive strength of NGH sediment is derived from maximum deviator stress, which is fitted with triaxial test data. Kuniyuki [37] recommends that maximum deviator stress of NGH sediment can be calculated by the following formula:

$$q_f = \frac{2\cdot\cos\varphi}{1-\sin\varphi}c_0 + \alpha S_h^{\beta} + \frac{2\cdot\sin\varphi}{1-\sin\varphi}\sigma_3 \tag{5}$$

where: q_f is maximum deviator stress of NGH sediment, in MPa; φ is the internal friction angle, 33.8°; c_0 is cohesion, 0.3 MPa; σ_3 is the effective confining pressures, in MPa; α and β are undetermined parameters, whose values are 4.64×10^{-3} and 1.58, respectively.

Regardless of the effective confining pressure, the uniaxial compressive strength of NGH sediment is equal to triaxial compressive strength, and its expression can be written as:

$$q_u = q_{f(\sigma_3 = 0)} = \frac{2 \cdot \cos\varphi}{1 - \sin\varphi} c_0 + \alpha S_h^{\beta} \tag{6}$$

Combining Equations (4) and (6), the formula for critical velocity of NGH sediment calculation was obtained:

$$v_{cr} = \eta \left[\frac{1}{p_{atm}} \cdot \left(\frac{2 \cdot \cos\varphi}{1 - \sin\varphi} c_0 + \alpha S_h^{\beta} \right) \right]^{0.5} \tag{7}$$

Substituting the experimental data to Equation (7), the characteristic velocity η was calculated, as shown in Figure 6. Corresponding to the NGH saturation of 20%, 40%, 60%, and 80%, η was calculated to be 1.41 m/s, 1.39 m/s, 1.52 m/s, and 1.44 m/s, respectively. Values of η calculated with various saturations are close, which proves that the critical water jet velocity formula is correct. Average critical water jet velocity under various saturations is 1.44 m/s.

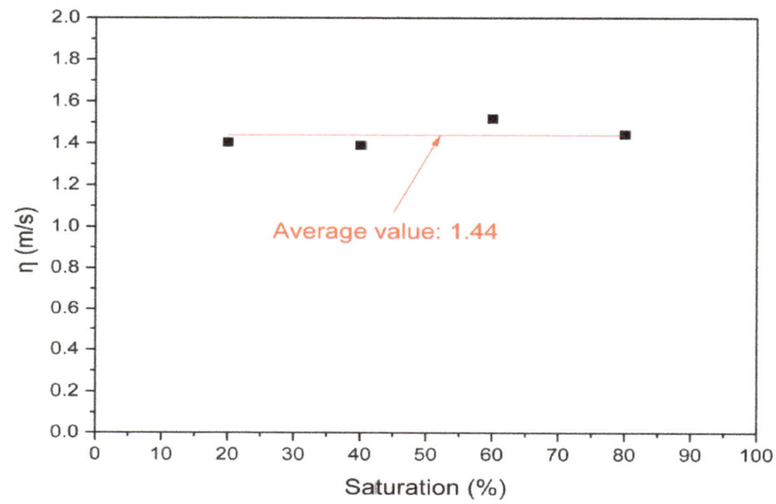

Figure 6. The relationship between the characteristic velocity η and the NGH saturation.

4. Conclusions

Based on the experimental and theoretical analysis, the critical water jetting breaking velocity for marine NGH breakage by water jet process was studied. Under various experimental conditions, the deeper the NGH breakage achieved through water jetting process, the greater the breaking ability of the water jet flow is within the same time, and therefore it takes longer to achieve the maximum breaking depth. Given the experimental conditions in this study, maximum breaking depth was achieved in 63 s of the NGH water jet breaking process.

Through orthogonal design experiment analysis, the ultimate breaking distance ranges of the three experimental factors, sample saturation, jet pressure, and jet nozzle diameter, were found to be 121.25 mm, 115 mm, and 112.5 mm respectively. Hence, the order (from high to low) of the effect of the three factors on the ultimate breaking depth is NGH saturation, jet pressure, and jet nozzle diameter.

Critical water jet breaking velocity of NGH breakage by the water jetting process was found to increase with NGH saturation increase. Average critical water jetting velocities corresponding with NGH saturations of 20%, 40% 60%, and 80% were 5.71 m/s, 7.14 m/s, 9.60 m/s, and 10.85 m/s, respectively. The formula was established to calculate the critical water jet breaking velocity of the marine NGH breakage by water jet process. Using the experiment data, the undetermined coefficient η in the critical water jet breaking velocity equation was calibrated to be 1.44 m/s. Considering that frozen clay and sand are used instead of NGH sediment as the experimental samples in the experiment, there will be some errors in this value η for real NGH sediment, which should be considered in practical engineering applications.

Author Contributions: Conceptualization, G.W.; Data curation, L.W.; Formal analysis, G.W.; Funding acquisition, G.W.; Investigation, G.W.; Methodology, L.W.; Project administration, G.W.; Resources, L.W.; Validation, L.W.; Writing—original draft, L.W.; Writing—review & editing, G.W. All authors have read and agreed to the published version of the manuscript.

References

1. Makogon, Y.F. Natural gas hydrates—A promising source of energy. *J. Nat. Gas. Sci. Eng.* **2010**, *2*, 49–59. [CrossRef]
2. Chong, Z.R.; Yang, S.H.B.; Babu, P.; Linga, P.; Li, X.-S. Review of natural gas hydrates as an energy resource: Prospects and challenges. *Appl. Energy* **2016**, *162*, 1633–1652. [CrossRef]
3. Boswell, R.; Collett, T.S. Current perspectives on gas hydrate resources. *Energy Env. Sci.* **2011**, *4*, 1206–1215. [CrossRef]
4. Pinero, E.; Marquardt, M.; Hensen, C.; Haeckel, M.; Wallmann, K. Estimation of the global inventory of methane hydrates in marine sediments using transfer functions. *Biogeosciences* **2013**, *10*, 959–975. [CrossRef]
5. Boswell, R.; Collett, T. The Gas Hydrates Resource Pyramid: Fire in the ice, methane hydrate newsletter. *US Dep. Energy* **2006**, *6*, 5–7.
6. Tamaki, M.; Fujii, T.; Suzuki, K. Characterization and Prediction of the Gas Hydrate Reservoir at the Second Offshore Gas Production Test Site in the Eastern Nankai Trough, Japan. *Energies* **2017**, *10*, 1678. [CrossRef]
7. Konno, Y.; Fujii, T.; Sato, A.; Akamine, K.; Naiki, M.; Masuda, Y.; Yamamoto, K.; Nagao, J. Key Findings of the World's First Offshore Methane Hydrate Production Test off the Coast of Japan: Toward Future Commercial Production. *Energy Fuels* **2017**, *31*, 2607–2616. [CrossRef]
8. Arata, N.; Sukizaki, S.; Denda, A.; Awashima, Y.; Okada, Y. Overview of the R&D for an Environmental Impact Assessment on Japan's Methane Hydrate R&D Program in Phase-1. *J. Jpn. Assoc. Pet. Technol.* **2009**, *74*, 350–359.
9. Wu, X.H.W.; Liu, W. Assessment of global gas hydrate resource potential and progress of production. *Mar. Geol. Front.* **2018**, *33*, 63–78.
10. Li, H.; Wei, N.; Jiang, L.; Zhao, J.; Cui, Z.; Sun, W.; Zhang, L.; Zhou, S.; Xu, H.; Zhang, X.; et al. Evaluation of Experimental Setup and Procedure for Rapid Preparation of Natural Gas Hydrate. *Energies* **2020**, *13*, 531. [CrossRef]
11. Wu, S.W.J. Reflections on the Successful Trial Production of Natural Gas Hydrate in Shenhu Sea Area of South China Sea. *Chin. Sci. Bull.* **2018**, *63*, 2–8. [CrossRef]
12. Li, G.; Moridis, G.J.; Zhang, K.; Li, X.-S. Evaluation of Gas Production Potential from Marine Gas Hydrate Deposits in Shenhu Area of South China Sea. *Energy Fuels* **2010**, *24*, 6018–6033. [CrossRef]
13. Moridis, G.; Reagan, M.T.; Boyle, K.L.; Zhang, K. Evaluation of the Gas Production Potential of Some Particularly Challenging Types of Oceanic Hydrate Deposits. *Transp. Porous Media* **2011**, *90*, 269–299. [CrossRef]
14. Zhou, S.C.W.; Li, Q. Research on the solid fluidization well testing and production for shallow non-diagenetic natural gas hydrate in deep water area. *China Offshore Oil Gas.* **2017**, *29*, 1–8.
15. Wu, X.; Liang, Q.; Ma, Y.; Shi, Y.; Xia, Z.; Liu, L.; Haeckel, M. Submarine Landslides and their Distribution in the Gas Hydrate Area on the North Slope of the South China Sea. *Energies* **2018**, *11*, 3481. [CrossRef]
16. First successful trial collection of natural gas hydrate in China. *Geol. China* **2017**, *44*, 620–621.
17. Qiu, S.; Wang, G.; Wang, L.; Fang, X. A Downhole Hydrocyclone for the Recovery of Natural Gas Hydrates and Desanding: The CFD Simulation of the Flow Field and Separation Performance. *Energies* **2019**, *12*, 3257. [CrossRef]
18. Zhou, S.; Chen, W.; Li, Q. Green production technology of deep water shallow gas hydrate with solid fluidization method. *China Offshore Oil Gas.* **2014**, *26*, 1–7.

19. Pan, D.C.C.; Yang, L.; Jin, C.; Li, X.; Liu, S.; Chen, Y.; Zhu, J. Physical simulation experiment system for jet erosion of natural gas hydrate. *Explor. Eng. Rock Soil Drill. Tunn.* **2018**, *45*, 35–39.
20. Chen, C.; Pan, D.; Yang, L.; Zhang, H.; Li, B.; Jin, C.; Li, X.; Cheng, Y.; Zhong, X. Investigation into the Water Jet Erosion Efficiency of Hydrate-Bearing Sediments Based on the Arbitrary Lagrangian-Eulerian Method. *Appl. Sci.* **2019**, *9*, 182. [CrossRef]
21. Yang, L. *Study on the Breaking Process of Marine Hydrate Subjected to High Pressure Water Jet and the Production Increase of Marine Hydrate Reservoirs Reconstruction*; Jilin University: Changchun, China, 2018.
22. Wang, G.; Zhong, L.; Zhou, S.; Liu, Q.; Li, Q.; Fu, Q.; Wang, L.; Huang, R.; Wang, G.; Li, X. Jet breaking tools for natural gas hydrate exploitation and their support technologies. *Nat. Gas. Ind. B* **2018**, *5*, 312–318. [CrossRef]
23. Wang, G.; Huang, R.; Zhong, L.; Wang, L.; Zhou, S.; Liu, Q. An optimal design of crushing parameters of marine gas hydrate reservoirs in solid fluidization exploitation. *Nat. Gas. Ind. B* **2019**, *6*, 257–261. [CrossRef]
24. Wang, L.; Wang, G.; Mao, L.; Zhou, S.; Zhong, L. Experimental research on the breaking effect of natural gas hydrate sediment for water jet and engineering applications. *J. Pet. Sci. Eng.* **2020**, *184*, 106553. [CrossRef]
25. Lv, X.; Shi, B.; Zhou, S.; Wang, S.; Huang, W.; Sun, X. Study on the Decomposition Mechanism of Natural Gas Hydrate Particles and Its Microscopic Agglomeration Characteristics. *Appl. Sci.* **2018**, *8*, 2464. [CrossRef]
26. Dvorkin, J.; Helgerud, M.B.; Waite, W.; Kirby, S.H.; Nur, A. Introduction to Physical Properties and Elasticity Models. In *Asia-Pacific Coasts and Their Management*; Springer Science and Business Media LLC: Berlin/Heidelberg, Germany, 2003; Volume 5, pp. 245–260.
27. Cox, J.L. *Natural Gas. Hydrates: Properties, Occurrence and Recovery*; US Department of Energy: Washington, DC, USA, 1983.
28. Stern, L.A.; Kirby, S.H.; Durham, W.B. Polycrystalline Methane Hydrate: Synthesis from Superheated Ice, and Low-Temperature Mechanical Properties. *Energy Fuels* **1998**, *12*, 201–211. [CrossRef]
29. Durham, W.B.; Stern, L.A.; Kirby, S.H. Ductile flow of methane hydrate. *Can. J. Phys.* **2003**, *81*, 373–380. [CrossRef]
30. Cao, P.; Wu, J.; Zhang, Z.; Fang, B.; Ning, F. Mechanical Properties of Methane Hydrate: Intrinsic Differences from Ice. *J. Phys. Chem. C* **2018**, *122*, 29081–29093. [CrossRef]
31. Zhang, X.; Wang, S.; Li, Q.; Zhao, J.; Wang, A. Experimental study of mechanical properties of gas hydrate deposits. *Rock Soil Mech.* **2010**, *31*, 3069–3074.
32. Winters, W.; Pecher, I.A.; Waite, W.; Mason, D.H. Physical properties and rock physics models of sediment containing natural and laboratory-formed methane gas hydrate. *Am. Min.* **2004**, *89*, 1221–1227. [CrossRef]
33. Kuang, Y.; Yang, L.; Li, Q.; Lv, X.; Li, Y.; Yu, B.; Leng, S.; Song, Y.; Zhao, J. Physical characteristic analysis of unconsolidated sediments containing gas hydrate recovered from the Shenhu Area of the South China sea. *J. Pet. Sci. Eng.* **2019**, *181*, 106173. [CrossRef]
34. Ho, C.E. *Turbulent Fluid Jet Excavation in Cohesive Soil: With Particular Application to Jet Grouting*; Massachusetts Institute of Technology: Cambridge, MA, USA, 2005.
35. Rajaratnam, N. Turbulent Jets. *Dev. Water Sci.* **1976**, *5*, 77–82.
36. Dabbagh, A.A.; Gonzalez, A.S.; Pena, A.S. Soil Erosion by a Continuous Water Jet. *Soils Found* **2002**, *42*, 1–13. [CrossRef]
37. Miyazaki, K.; Temma, N.; Aoki, K.; Yamaguchi, T. A Nonlinear Elastic Model for Triaxial Compressive Properties of Artificial Methane-Hydrate-Bearing Sediment Samples. *Energies* **2012**, *5*, 4057–4075. [CrossRef]

8

The Effect of CO_2 Partial Pressure on CH_4 Recovery in CH_4-CO_2 Swap with Simulated IGCC Syngas

Ya-Long Ding [1,2], Hua-Qin Wang [2], Chun-Gang Xu [1,3,4,5] and Xiao-Sen Li [1,3,4,5,*]

1 Guangzhou Institute of Energy Conversion, Chinese Academy of Sciences, Guangzhou 510641, China; dingyalong@huanghuai.edu.cn (Y.-L.D.); xucg@ms.giec.ac.cn (C.-G.X.)
2 College of Chemistry and Pharmaceutical Engineering, Huanghuai University, Zhumadian 463000, China; wanghuaqin880105@163.com
3 CAS Key Laboratory of Gas Hydrate, Guangzhou 510640, China
4 Guangdong Provincial Key Laboratory of New and Renewable Energy Research and Development, Guangzhou 510640, China
5 Guangzhou Center for Gas Hydrate Research, Chinese Academy of Sciences, Guangzhou 510640, China
* Correspondence: lixs@ms.giec.ac.cn.

Abstract: To investigate the influence of CO_2 partial pressure on efficiency of CH_4-CO_2 swap from natural gas hydrates (NGHs), the replacement of CH_4 from natural gas hydrate (NGH) is carried out with simulated Integrated Gasification Combined Cycle (IGCC) syngas under different pressures, and the gas chromatography (GC), in-situ Raman, and powder X-ray diffraction (PXRD) are employed to analyze the hydrate compositions and hydrate structures. The results show that with the P-T (pressure and temperature) condition shifting from that above the hydrate equilibrium curve of IGCC syngas to that below the hydrate equilibrium curve of IGCC syngas, the rate of CH_4 recovery drastically rises from 32% to 71%. The presence of water can be clearly observed when P-T condition is above the hydrate equilibrium curve of IGCC syngas; however the presence of water only occurs at the interface between gas phase and hydrate phase. No H_2 is found to present in the final hydrate phase at the end of process of CH_4-CO_2 swap with IGCC syngas.

Keywords: CH_4 hydrate; replacement; IGCC syngas; in-situ Raman

---

1. Introduction

Natural gas hydrates (NGHs) are ice-like compounds formed by water molecules and gas molecules under low temperature or/and high pressure. Water molecules form cavities with different morphology and dimension by hydrogen bonding and gas molecules fill the cavities by van de Walls force [1]. Gas hydrates in sea floor or permafrost usually exist in three structures such as structure I (sI), structure II (sII) and structure H (sH). According to literature reports, most NGH deposits are sI gas hydrates [2,3] which are composed of six $5^{12}6^2$ cages and two 5^{12} cages [1]. The NGHs, attracting much attention for their abundant reserves in the seabed and permafrost regions, have the potential to be an alternative energy resource in the future because it is estimated that the total amount of carbon resources contained in the NGHs is about twice as much as that in the proven fossil fuel reserves [4–9]. To date, many studies have been carried out to exploit CH_4 from NGHs deposits and several field trails have been conducted in last 40 years. The exploitation methods mainly include thermal stimulation, depressurization and chemical inhibitor injection, but the disadvantages such as enormous energy consumption, risks of the probably catastrophic landslide and serious environmental pollution are the main issues we are facing [10]. Notably, methane is an about 20 times more efficient greenhouse gas than CO_2 [11]. Therefore, effective and environmental friendly production technologies are expected, including CH_4-CO_2 swap, in-situ combustion and in-situ catalytic oxidation, etc.

The concept of CH_4-CO_2 swap in natural gas hydrates was firstly proposed by Ohgaki et al., [12–14] and this attracted worldwide attention for its ability to simultaneously produce CH_4 from NGHs and sequestrate CO_2 in sea sediments. Firstly, the researchers studied the thermodynamic feasibility of CH_4-CO_2 swap in gas hydrates, [15–18] and the results of Sivaraman's work showed that CO_2 hydrate could be formed under more moderate conditions of temperature and pressure than CH_4 hydrate and the heat of CO_2 hydrate formation (−57.98 kJ/mol) is larger than that of CH_4 hydrate dissociation (54.49 kJ/mol) [19]. Ors and Sinayuc [20] conducted an experimental study on the CH_4-CO_2 swap between gaseous CO_2 and CH_4 hydrate in porous media at 3.7 MPa and 277.15 K, and their results revealed that the CH_4-CO_2 swap process mostly took place at the gas-solid surface and the injection of gaseous CO_2 caused the dissociation of methane hydrate. Ota and Inomata et al. [21,22] studied the replacement of CH_4 in the hydrate by use of liquid CO_2 and the CH_4 recovery rate of 35% was obtained in 307 h, they also found that the CH_4 hydrate decomposed during the replacement process and the decomposition of the large cage (51262) in the CH_4 hydrate proceed faster than that of the small cage (512). They also suggested that CH_4 hydrate decomposition was probably dominated by rearrangement of water molecules in the hydrate whereas CO_2 hydrate formation seemed to be dominated by CO_2 diffusion in the hydrate phase. Lee and Ripmeester [23] used solid-state NMR methods to investigate the limiting equilibrium compositions and the distribution of guest molecules over different cages of the mixed hydrate formed from different CO_2 concentration gas mixture of CH_4 and CO_2, they suggested that the ratio of CH_4 in large cages to CH_4 in small cages of sI hydrate declined steadily to a value with increasing CO_2 concentration in the gas mixture. Schicks et al. [24] investigated the conversion of the primary CH_4 hydrate into a CO_2-rich hydrate using in situ microscopy, confocal Raman spectroscopy and powder XRD, and they suggested that the conversion process was induced by the gradient of the chemical potential between the hydrate phase and the environmental gas phase, and the conversion process could be described as a decomposition and reformation process. The conversion rate depended on the surface area of the hydrate phase and the concentration gradient of one component between the hydrate phase and the gas phase.

With further research, many researchers studied the feasibility of explore CH_4 from natural gas hydrates with gas mixture of CO_2 and N_2 which was the main component of flue gas. Kvamme [25] studied the feasibility of simultaneous CO_2 storage and CH_4 production from natural gas hydrate using mixtures of CO_2 and N_2, and suggested that the fast exchange between CH_4 and CO_2/N_2 mixtures was achieved through a new hydrate formation and the adding of N_2 into CO_2 is advantageous to gas permeability in the hydrates. Also, they suggested there were two primary mechanisms for the conversion of CH_4 hydrate into CO_2 hydrate, one was the direct solid state conversion [23] and the other was that new CO_2 hydrate formed from injected CO_2 and free water in the pores [26,27]. Lee et al. [28] explored the swap phenomenon occurred in sI and sII hydrates, and CH_4 recovery rates of 64% and 85% were obtained with CO_2 and CO_2/N_2, respectively, and the results showed that the sII hydrate was transformed into sI when sII hydrates were exposed to CO_2 and CO_2/N_2. Beyond that, they also investigated the recovery of CH_4 from gas hydrates intercalated within natural sediments using CO_2 and CO_2/N_2 gas mixtures, and they found that the recovery efficiency was nearly identical [2].

In addition to the above laboratory studies, there also were some field tests to produce CH_4 from CH_4 hydrate reservoirs in permafrost or subsea sediments [29–33]. In 2002 and 2008, the production tests were carried out in the permafrost reservoir of Mallik in northern Canada by injection of hot water and depressurization, and limited amounts of CH_4 was successfully exploited in a few days. After that, in 2012, the method of injection of 23% CO_2 and 77% N_2 was tested in the Alaska North Slope. The gas production were conducted above and near the P–T condition of CH_4 hydrate equilibrium respectively, and the total volume of produced gas mixture approached 30,000 m^3 over the whole test period though the mole fraction of CH_4 in the gas mixture were different at the different test period [31,34,35]. Then the depressurization technique was firstly used in offshore hydrate reservoirs at the eastern Nankai Trough in 2013, and the total gas production was around 120,000 m^3 in six days. Eventually the test was suspended due to the bad weather conditions and sand control problems.

Therefore, the commercial NGH exploitation is both highly complex and technologically challenging because the complex geologic structure, harsh engineering conditions and the risks of collapse of marine and potential serious damage to the marine ecosystem.

The changes of temperature and pressure have great influence on CH_4 recovery from NGHs via CH_4-CO_2 replacement. In order to clarify the influence of temperature and pressure and the possible influence of different added gas in CO_2, in this work, the changes of gas phase and hydrate phase during the replacement process under different pressure condition, using simulated IGCC syngas as displacement gas, are determined by in situ Raman spectroscopy, powder XRD and gas chromatography.

2. Experimental Section

2.1. Apparatus

As shown in Figure 1, the experimental apparatus consists of a gas supply system, a high-pressure vessel used as the hydrate formation or dissociation reactor, a cooling system and detection equipment. The stainless-steel reactor with an inner volume of 100 mL is embedded in the notch which is linked with the cooling system, and two quartz windows are mounted on front and back sides of the reactor for viewing the swap process and applying Raman measurement. In the cooling system, the ethylene glycol solution with the volume ratio of 1:3 are used as coolant, and the temperature can be controlled in the range of 253.15–303.15 K, and a Pt100 thermocouple (JM6081) supplied by Jiangsu Hongbo machinery manufacturing co. LTD (Nantong, Jiangsu, China) with uncertainties of ±0.1 K is employed to measure the temperature.

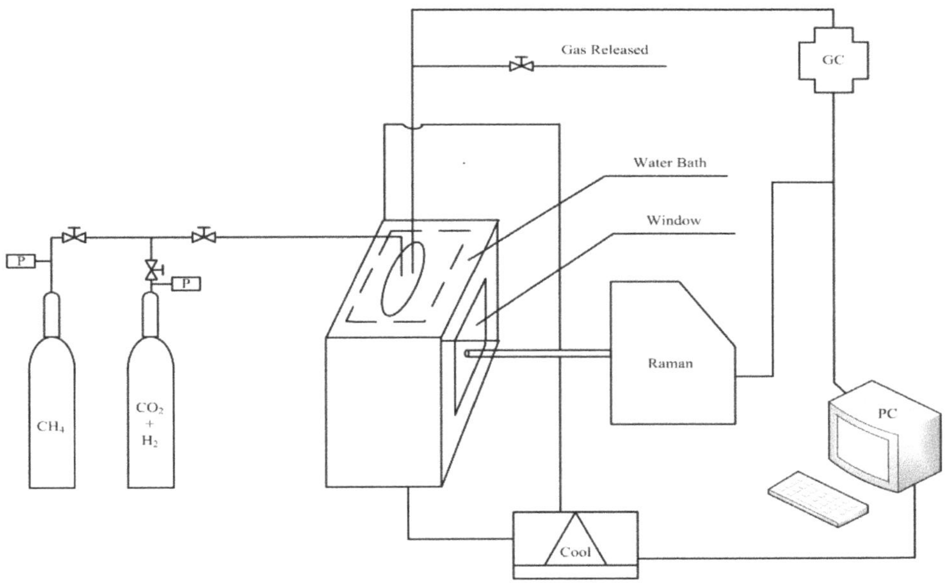

Figure 1. Schematic of experimental apparatus.

The gas component is determined by an Agilent 7890A gas chromatograph (GC, Agilent Technologies Inc., Palo Alto, CA, USA). The experimental CH_4 gas with a purity of 99.9% and 39.9% CO_2 +60.1% H_2 gas mixture were supplied by Foshan Huate Gas Co., Ltd. (Foshan, China). The deionized water with the resistivity of 18.25 mΩ cm^{-1} is produced with an ultra-pure water machine from Nanjing Ultrapure Water Technology Co., Ltd. (Nanjing, China).

The Raman spectra are obtained from on a LabRam Raman spectrometer (Jobin Yvon, Paris, France) with a 50 times tele lens and a single monochromator of 1800 grooves/mm grating and a multichannel air-cooled charged-coupled device (CCD) detector. In addition to this, the Raman spectrometer uses an Ar-ion laser source, which emits a 532 nm line with a power of 100 mW. The silicon (Si) crystal standard of 520.7 cm^{-1} is employed to calibrate the subtractive spectrograph.

The XRD patterns are recorded at 193 K on a D/MAX-2500 device (Rigaku, Tokyo, Japan) using graphite-monochromatized Cu Kα1 radiation (λ = 1.5406 Å) in the $\theta/2\theta$ scan mode. The XRD experiments are carried out in step mode with a fixed time of 3 s and a step size of 0.03° for 2θ = 10–60° for each hydrate sample.

The gas component in the collected gas samples and the final dissociated gas was analyzed on an Agilent 7890A GC (Agilent Technologies Inc., Palo Alto, CA, USA), equipped with a flame ionization detector (FID) and thermal conductivity detector (TCD). Besides that, the gas samples are detected with the method of uniform heating from 298.15–523.15 K, and H_2 (30 mL/min) is used as combustion gas, air (400 mL/min) is used as combustion-supporting gas and helium (25 mL/min) is used as make-up gas.

2.2. *Procedure*

Deionized water (60 mL) is injected into the reactor, followed by cooling down the system to 274.15 K and removing the air in the gas phase by injecting CH_4 gas continuously into the water from the bottom of reactor at a system pressure of 1.0 MPa, and after that the system is pressurized to 4.5 MPa. During the CH_4 hydrate sample preparation period, the intake of CH_4 gas and the agitation of the deionized water are maintained so that the injected CH_4 gas could dissolve well in the water and the experimental CH_4 hydrate sample forms evenly in the water phase. About five days later, once there is no water in the reactor as determined by Raman analysis, the CH_4 hydrate formation reaction is deemed finished. Hereafter, the simulated IGCC syngas of 39.9% CO_2 and 60.1% H_2 gas mixture is injected into the reactor from bottom of reactor, and the CH_4 gas above the hydrate is discharged at the same time under the stable pressure of 4.5 MPa. Once the CH_4 in gas phase is lower than 2%, the outlet valve is shut off and then the gas mixture of CO_2 and H_2 gradually displaces CH_4 from the hydrate. In the process of the replacement, the composition of the gas phase, gas-hydrate interface and the hydrate phase are determined through Raman spectroscopy every 24 h, and the component of the gas phase is also determined through GC analysis. At the end of replacement process, the hydrate is dealt with liquid nitrogen and the gas phase is removed, then the reactor is placed at room temperature for dissociation of the hydrate. Meanwhile, one same replacement reaction is conducted in another same reactor and at the end of the replacement process the hydrate is took out for powder XRD detection. These two experiments are labeled as experiment 1.

In addition to that, another two experiments, which are labeled as experiment 2, were conducted under the same conditions as above method, except for the fact the hydrate formation and the replacement processes are carried out at 6.0 MPa.

3. Results and Discussion

For monitoring the changes of the hydrate phase and gas phase during the replacement process, three fixed points are selected for Raman detection as point "A" in the hydrate phase, point "B" at the interface of gas phase and hydrate phase and point "C" in the gas phase.

After the formation of pure CH_4 hydrate, the Raman spectra and powder XRD pattern are obtained and shown in Figure 2. As shown in Figure 2a, all the hydrate samples have the representative Raman peaks of sI hydrate. The strong peak at 2905 cm^{-1} and relatively weak peak at 2915 cm^{-1} with the intensity ratio of 3:1 are attributed to the C-H symmetric stretching vibration of CH_4 molecules in large cages and small cages of sI hydrate, respectively. The peak at 2917 cm^{-1} which is detected in the gas phase is attributed to C-H symmetric stretching vibration of CH_4 gas. In addition, the powder XRD pattern of the hydrate samples are shown in Figure 2b, it can be seen that the hydrate samples show all peaks of the sI hydrate and little ice Ih. It can be concluded that all the formed hydrates are sI hydrates.

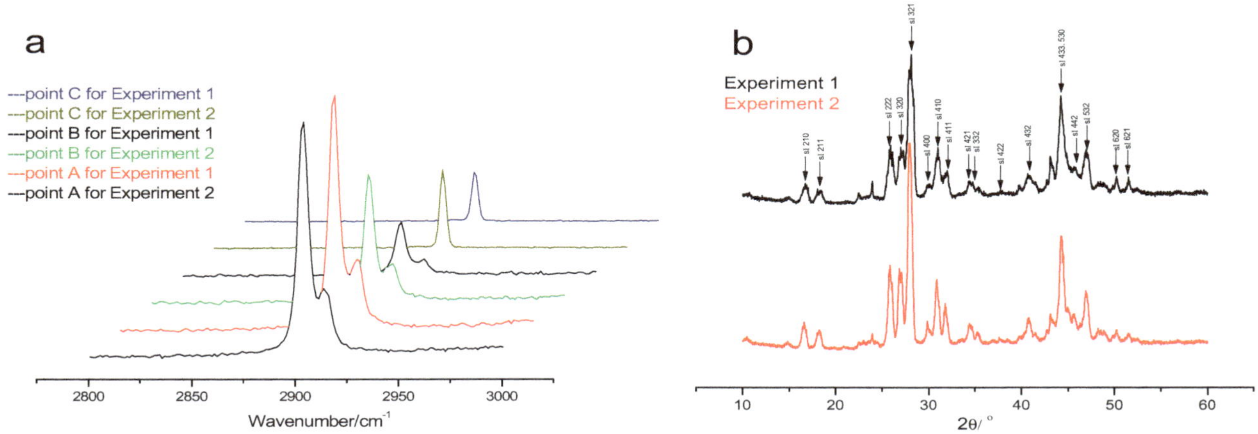

Figure 2. The Raman spectra for point "A", "B" and "C" (**a**) and powder XRD pattern for experiment 1 and 2 (**b**).

Once the water is fully converted into hydrate, the gas is quickly replaced by CO_2/H_2 under the same temperature and pressure conditions. During the process of replacement, the component changes in hydrate phase, gas-hydrate interface are determined by in situ Raman every 24 h until the intensities of Raman peaks obtained from point "A" show no changes.

Figure 3 shows the changes of Raman spectra of CO_2 hydrate at point "A" during the replacement process for both experiments, and the picture a and b represent the low pressure (4.5 MPa) and high pressure (6.0 MPa) experiments, respectively. The Raman peaks at 1277 and 1380 cm^{-1} correspond to the $C = O$ stretching vibration and bend vibration, respectively. Because of Fermi splitting, we cannot judge which cages are occupied by CO_2. It can be seen that the Raman peak intensity corresponding to CO_2 hydrate increases gradually for both low and high pressure experiments, but the growth rate for low pressure experiment is larger than that for high pressure experiment. The possible reason is that the partial pressure of CO_2 in the gas mixture of the low pressure experiment (1.8 MPa) is lower than that of high pressure experiment (2.4 MPa), bringing about the smaller driving force of CO_2 molecule in low pressure experiment, resulting in the retard of CO_2 hydrate formation. On the contrary, the driving force of the CO_2 molecules in the high pressure experiment is large enough to form CO_2 hydrate, and the formed CO_2 hydrate covers the CH_4 hydrate and impedes the gas mixture from contacting CH_4 hydrate.

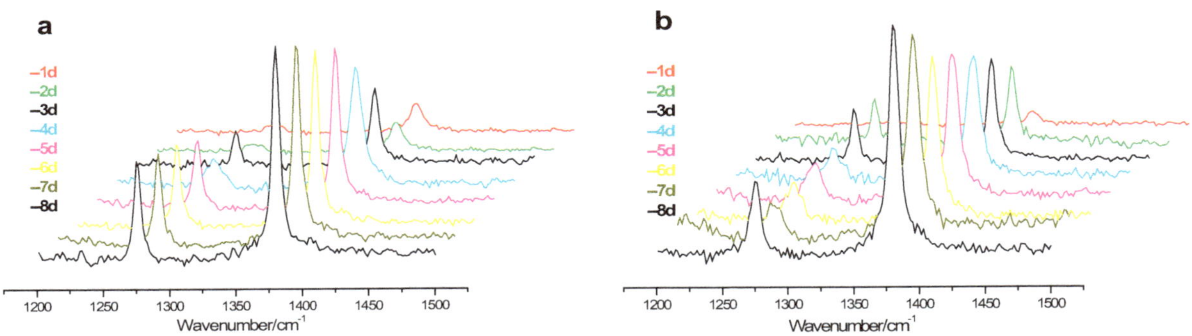

Figure 3. The Raman spectroscopy change of CO_2 hydrate at point "A" for experiment 1 (**a**) and experiment 2 (**b**), where the 1d, 2d, 8d represent the first, the second, the eighth day of the experiment process.

For the same smaller driving force reason more and more water results from the decomposition of the hydrate phase and is carried by the gas mixture up to the interface of the gas phase and hydrate phase in the low pressure experiment, and the more CO_2 hydrate is formed at the interface, as shown in Figure 4. The Raman peak of CO_2 hydrate for low pressure experiment increases faster than that for high pressure experiment, and finally, a stronger intensity of peak for low pressure experiment is obtained. Furthermore, it can be observed that there is significant water phase at the interface between gas phase and hydrate phase. However, a similar phenomenon has not been observed for the high pressure experiment in which the presence or not of the water from decomposition of hydrate can only be determined by in situ Raman, as shown in Figure 5, where the pictures a and b show the photo of vessel at the first and sixth day of replacement reaction of low pressure experiment, and the pictures c and d show the Raman peak change between 3100 and 3500 cm^{-1} which is attributed to O-H symmetric stretching vibration of H_2O molecules during the replacement reaction of high pressure experiment [36].

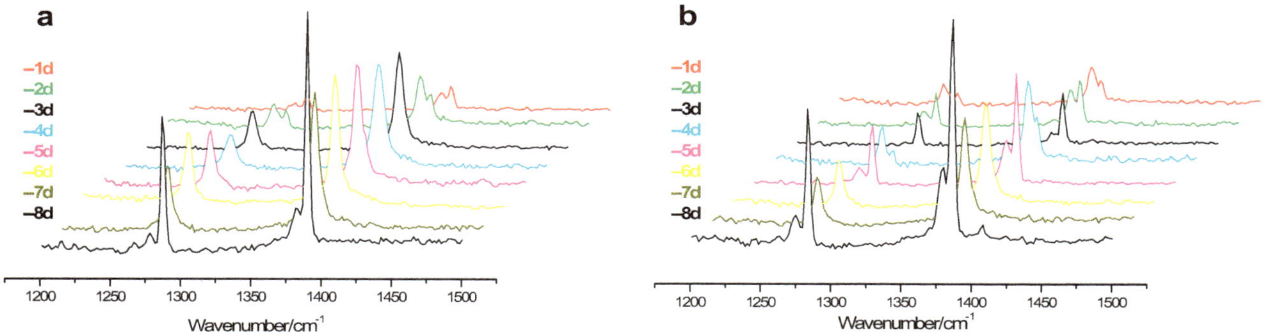

Figure 4. The Raman spectroscopy change of CO_2 hydrate at point "B" for experiment 1 (**a**) and experiment 2 (**b**), where the 1d, 2d, 8d represent the first, the second, the eighth day of the experiment process.

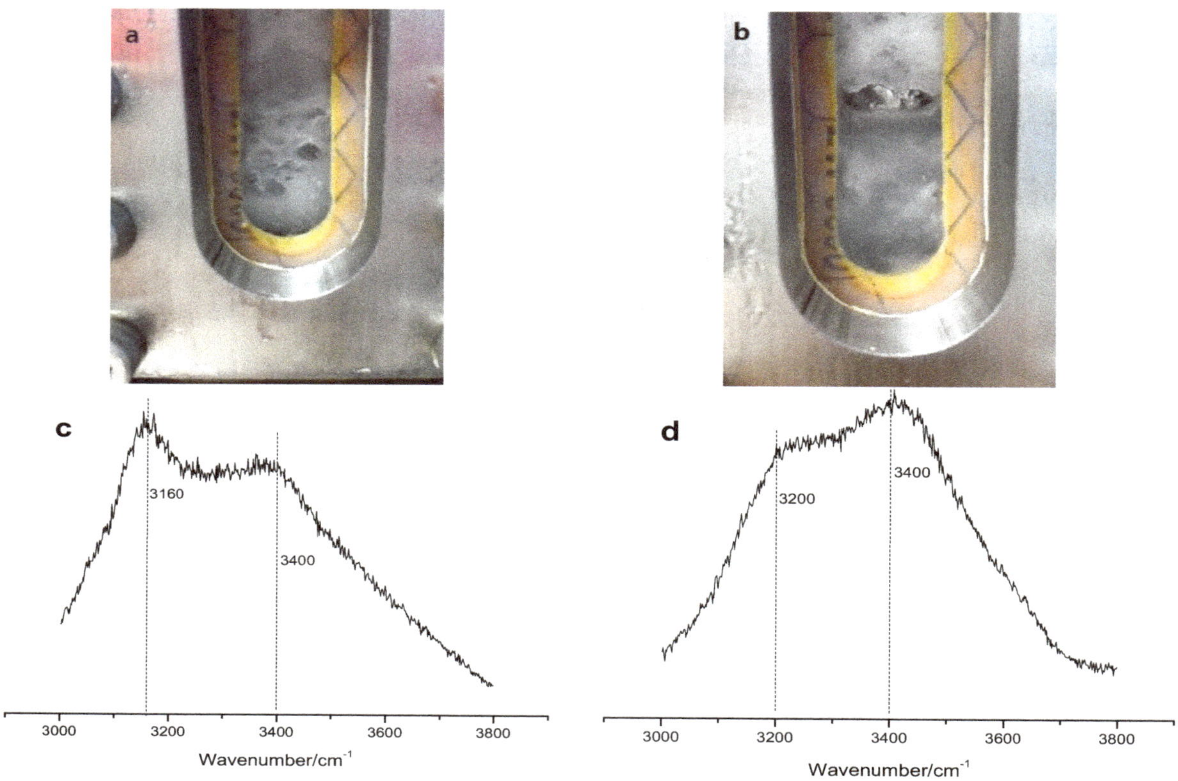

Figure 5. The picture of vessel at the first (**a**) and sixth (**b**) day of replacement reaction for experiment 1, and the Raman peak change between 3100 and 3500 cm^{-1} during experiment 2 (**c**,**d**).

It can be seen that the Raman peak shows sharp appearance at about 3160 cm^{-1} and short appearance at about 3400 cm^{-1} which are attributed to O-H symmetric stretching vibration of H_2O molecules in sI hydrate, then along with the replacement progress the peak appearance changes to low peak at 3200 cm^{-1} which is attributed to Fermi resonance between O-H stretch and bending mode of H_2O molecules in water phase, and high peak at 3400 cm^{-1} which is attributed to the O-H symmetric stretching vibration of H_2O molecules in water phase. This observation strongly verifies the decomposition of hydrate during the replacement process, though the degree of decomposition is different between low pressure experiment and high-pressure experiment. The presence of water from decomposition of hydrate phase may improve the permeability of hydrate phase and gives one reason for the higher CH_4 recovery yield in the low-pressure experiment.

During the replacement process, when the CH_4 hydrate is attacked by the CO_2/H_2 gas mixture, the hydrate lattice is decomposed and then the water from the decomposition of CH_4 hydrate forms CO_2 hydrate with the CO_2 gas molecules, so the amount of CH_4 hydrate at monitoring point "A" gradually decreases to a certain extent, as shown in Figure 6, which represents the changes of Raman spectra of CH_4 hydrate at point "A" during replacement process for both low and high pressure experiments. As can be seen that the Raman peaks of CH_4 hydrate gradually decrease during the replacement process, and the drop rate in the low pressure experiment is higher than that in the high pressure experiment, and the result is consistent with the rule of the increase of CO_2 hydrate in the Figure 3, where the increment of Raman peaks for CO_2 hydrate in low pressure experiment is higher than that in high pressure experiment. Apart from this, another result that can be seen from the picture *b* in Figure 6 is the changes of ratio of Raman peak intensity at 2905 cm^{-1} to that at 2915 cm^{-1} in the high pressure experiment. As known that the peak at 2905 cm^{-1} represents the CH_4 molecules in large cages of sI hydrate and the peak at 2915 cm^{-1} stands for the CH_4 molecules in small cages. During the high pressure replacement process, the intensity ratio of peak at 2905 cm^{-1} and peak at 2915 cm^{-1} changes gradually from premier 3:1 to final 2:1 and this decreasing ratio demonstrates that the replacement reaction only conducts in the large cage of the sI hydrate. The reason for this result may be the fast formation of CO_2 hydrate at the interface between gas phase and hydrate phase limits the injected gas mixture further contacting CH_4 hydrates.

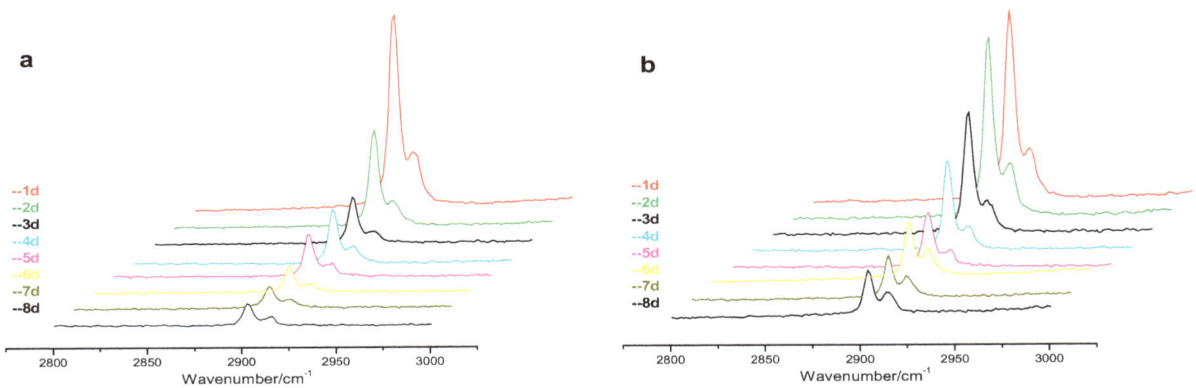

Figure 6. The Raman spectroscopy change of CH_4 hydrate at point "A" for experiment 1 (**a**) and experiment 2 (**b**), where the 1d, 2d, 8d represent the first, the second, the eighth day of the experiment process.

For preventing the released CH_4 from forming mixed hydrate with CO_2 gas, the gas injection and simultaneous discharge of gas mixture is conducted every 24 h, following the Raman detection, to decrease the CH_4 component concentration to less than 2% again, and the gas samples are collected at the beginning and end of the ventilation process and analyzed by gas chromatography. The results are shown in Table 1 and the data in Table 1 is plotted in Figure 7. It can be seen that in the low pressure experiment the CH_4 component concentration increases from less than 2% to 18.60, 17.18, and 7.48 mol% respectively in the first 3 days and tends to remain unchanged in the next 5 days till stabilizes at 1.0 mol% in the last 2 days. However, in the high-pressure experiment, the CH_4 component concentration only increases from less than 2% to 9.71 and 7.67 mol% in the first 2 days and the increment tends to remain unchanged in the following days till stabilizes at 0.5 mol% in the last 2 days. What should be noted is that the CO_2 percent decrement is a little larger than the of CH_4 percent increment in both experiments. This result could be attributed to the decomposition of CH_4 hydrate and the new formation of CO_2 hydrate, as well as the low-pressure experiment shows a greater extent of decomposition of CH_4 hydrate than the high pressure experiment.

Table 1. The increment of CH_4 fraction and the decrement of CO_2 fraction in the collected gas samples in experiment 1 and experiment 2.

Item / Time	1d	2d	3d	4d	5d	6d	7d	8d
CH_4 Increment in Experiment 1	18.60	17.78	7.48	3.10	2.80	1.53	1.03	1.04
CO_2 Decrement in Experiment 1	19.40	18.28	8.29	4.76	3.5	2.03	1.81	1.79
CH_4 Increment in Experiment 2	9.71	7.67	3.34	2.85	1.81	2.03	0.63	0.57
CO_2 Decrement in Experiment 2	10.14	8.24	3.45	2.94	2.05	2.11	0.62	0.61

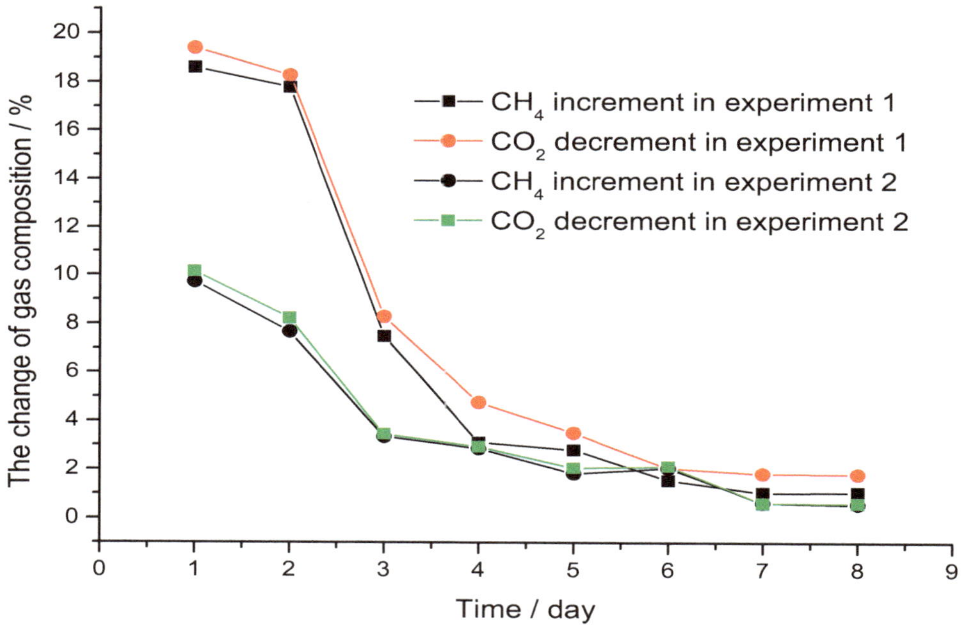

Figure 7. The increment of CH_4 fraction and the decrement of CO_2 fraction in the collected gas samples in experiment 1 and experiment 2.

When the Raman peak intensity of CH_4 hydrate and CO_2 hydrate at the point "A" and "B" show no changes and the CH_4 percent increment in the collected gas samples is less than 0.5, the replacement process is considered finished. Thereafter, the vessel is cooled with liquid nitrogen and the gas phase is withdrawn quickly by a vacuum pump, then the vessel is placed in the atmosphere for dissociation of the hydrate phase and the gas samples of decomposed hydrate phase are detected by gas chromatography. The gas chromatography results of the gas from decomposition of the final hydrate are shown in Table 2. As seen, the CH_4 residues in the hydrate phase after replacement process are 29.29% and 67.76%, indicating that the CH_4 recovery rates are 70.21% and 32.24% for low pressure experiment and high-pressure experiment, respectively. Also, the same low-pressure experiment and high pressure experiment are conducted, the final hydrate phases are removed from vessel and detected by powder XRD to determine the structure of the hydrates. The powder XRD patterns of the hydrates are shown in Figure 8, the peaks with blue arrows are attributed to the cubic crystal system which represents the sI hydrate. It can be concluded that the hydrate structure has no change during the replacement. In both low and high pressure experiments, there is no presence of H_2 in hydrate phase on the basis of Raman results, thus it can be concluded that the H_2 molecules does not occupy cages of the hydrate, they may only play a promotion role during the replacement process. Comparing with the results obtained previously, where the replacement process of CH_4 hydrate is conducted with gaseous CO_2, the obtained CH_4 recovery efficiency was 50%. Due to the adding of H_2, the CH_4 recovery efficiency was increased from 50% to 70.29%, it could be referred that the stability of the cages is broken and the CH_4 molecules in the cages could spilled over when the H_2 molecules contact with the hydrate lattices. However, the lattices could form again once CO_2 molecules step into the cages. The higher CH_4 yield for the low pressure experiment is due to the lower driving force of CO_2 in the gas mixture, and thus the disturbed hydrate lattices resulting from the connection of hydrate with H_2 molecules could not occupied by CO_2 molecules in good time, which causes the decomposition of hydrate. However, during the high-pressure replacement process, the CO_2 molecules can occupy the disturbed hydrate cages for the high CO_2 partial pressure, and the hydrate lattice stabilizes again.

Table 2. Gas compositions in the dissociated gases after the completion of replacements for experiment 1 and experiment 2.

Experimental Runs	Experiment 1		Experiment 2	
	CH_4	CO_2	CH_4	CO_2
Run 1	28.63	71.37	67.88	32.12
Run 2	29.94	70.06	67.64	32.36
Avg.	29.29	70.71	67.76	32.24

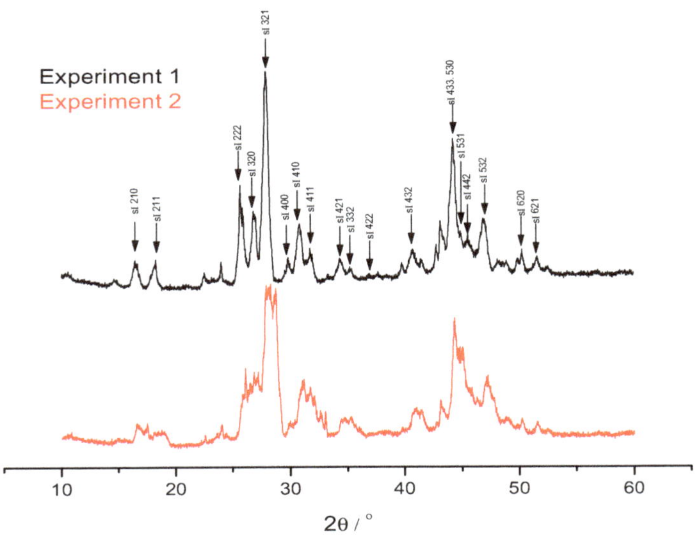

Figure 8. The PXRD pattern of gas hydrate phase after replacement process of experiment 1 and experiment 2.

4. Conclusions

The replacement of CH_4 from CH_4 hydrate with a gas mixture of CO_2/H_2 at the low pressure of 4.5 MPa and the high pressure of 6.0 MPa are investigated. The experimental results show that the CH_4 yield for the low-pressure experiment is 71% and the value for the high pressure experiment is 32% and H_2 molecules are not present in the final hydrate phase after the two replacement processes. Water appears obviously during the low-pressure experiment and the water from decomposed hydrate phase is not obvious during the high pressure experiment.

Author Contributions: Conceptualization, X.-S.L.; Investigation, Y.-L.D. and H.-Q.W.; Methodology, C.-G.X.; Supervision, X.-S.L.; Writing—original draft, Y.-L.D.; Writing—review & editing, C.-G.X. All authors have read and agreed to the published version of the manuscript.

References

1. Sloan, E.D.; Koh, C. *Clathrate Hydrates of Natural Gases*; CRC Press: Boca Raton, FL, USA, 2007.
2. Koh, D.Y.; Kang, H.; Kim, D.O.; Park, J.; Cha, M.; Lee, H. Recovery of methane from gas hydrates intercalated within natural sediments using CO_2 and a CO_2/N_2 gas mixture. *ChemSusChem* **2012**, *5*, 1443–1448. [CrossRef]
3. Ginsburg, G.; Soloviev, V. *Submarine Gas Hydrates*; VNIIOkeangeologia: St. Petersburg, Russia, 1998.
4. Baldwin, B.A.; Stevens, J.; Howard, J.J.; Graue, A.; Kvamme, B.; Aspenes, E.; Ersland, G.; Husebø, J.; Zornes, D.R. Using magnetic resonance imaging to monitor CH_4 hydrate formation and spontaneous conversion of CH_4 hydrate to CO_2 hydrate in porous media. *Magn. Reson. Imaging* **2009**, *27*, 720–726. [CrossRef]
5. Makogon, Y.F.; Holditch, S.A.; Makogon, T.Y. Natural gas-hydrates—A potential energy source for the 21st Century. *J. Pet. Sci. Eng.* **2007**, *56*, 14–31. [CrossRef]
6. Collett, T. *Natural Gas Hydrates: Vast Resources, Uncertain Future*; US Geological Survey: Reston, CA, USA, 2001.
7. Sloan, E.D. Fundamental principles and applications of natural gas hydrates. *Nature* **2003**, *426*, 353–363. [CrossRef]
8. Milich, L. The role of methane in global warming: Where might mitigation strategies be focused? *Glob. Environ. Chang.* **1999**, *9*, 79–201. [CrossRef]
9. Kvenvolden, K.A. Potential effects of gas hydrate on human welfare. *Proc. Natl. Acad. Sci. USA* **1999**, *96*, 3420–3426. [CrossRef]
10. Moridis, G.J. *Toward Production from Gas Hydrates: Current Status, Assessment of Resources, and Simulation-Based Evaluation of Technology and Potential*; Lawrence Berkeley National Laboratory: Berkeley, CA, USA, 2008.
11. Dlugokencky, E.; Houweling, S.; Bruhwiler, L.; Masarie, K.; Lang, P.; Miller, J.; Tans, P. Atmospheric methane levels off: Temporary pause or a new steady-state? *Geophys. Res. Lett.* **2003**, *30*. [CrossRef]
12. Hirohama, S.; Shimoyama, Y.; Wakabayashi, A.; Tatsuta, S.; Nishida, N. Conversion of CH_4-Hydrate to CO_2-Hydrate in Liquid CO_2. *J. Chem. Eng. Jpn.* **1996**, *29*, 1014–1020. [CrossRef]
13. Ohgaki, K.; Takano, K.; Sangawa, H.; Matsubara, T.; Nakano, S. Methane exploitation by carbon dioxide from gas hydrates. Phase equilibria for CO_2-CH_4 mixed hydrate system. *J. Chem. Eng. Jpn.* **1996**, *29*, 478–483. [CrossRef]
14. Ohgaki, K.; Takano, K.; Moritoki, M. Exploitation of CH_4 Hydrates under the Nankai Trough in Combination with CO_2 Storage. *Kagaku Kogaku Ronbunshu* **1994**, *20*, 121. [CrossRef]
15. Kvamme, B.; Tanaka, H. Thermodynamic stability of hydrates for ethane, ethylene, and carbon dioxide. *J. Phys. Chem.* **1995**, *99*, 7114–7119. [CrossRef]
16. Kang, S.-P.; Chun, M.-K.; Lee, H. Phase equilibria of methane and carbon dioxide hydrates in the aqueous $MgCl_2$ solutions. *Fluid Phase Equilibria* **1998**, *147*, 229–238. [CrossRef]
17. Anderson, R.; Llamedo, M.; Tohidi, B.; Burgass, R.W. Experimental measurement of methane and carbon dioxide clathrate hydrate equilibria in mesoporous silica. *J. Phys. Chem. B* **2003**, *107*, 3507–3514. [CrossRef]
18. Yezdimer, E.M.; Cummings, P.T.; Chialvo, A.A. Determination of the Gibbs free energy of gas replacement in SI clathrate hydrates by molecular simulation. *J. Phys. Chem. A* **2002**, *106*, 7982–7987. [CrossRef]
19. Sivaraman, R. The potential role of hydrate technology in sequestering carbon dioxide. *Gas Tips* **2003**, *9*, 4–7.
20. Ors, O.; Sinayuc, C. An experimental study on the CO_2–CH_4 swap process between gaseous CO_2 and CH_4 hydrate in porous media. *J. Pet. Sci. Eng.* **2014**, *119*, 156–162. [CrossRef]
21. Ota, M.; Morohashi, K.; Abe, Y.; Watanabe, M.; Smith, R.L., Jr.; Inomata, H. Replacement of CH_4 in the hydrate by use of liquid CO_2. *Energy Convers. Manag.* **2005**, *46*, 1680–1691. [CrossRef]
22. Ota, M.; Abe, Y.; Watanabe, M.; Smith, R.L.; Inomata, H. Methane recovery from methane hydrate using pressurized CO_2. *Fluid Phase Equilibria* **2005**, *228*, 553–559. [CrossRef]
23. Lee, H.; Seo, Y.; Seo, Y.T.; Moudrakovski, I.L.; Ripmeester, J.A. Recovering methane from solid methane hydrate with carbon dioxide. *Angew. Chem. Int. Ed.* **2003**, *42*, 5048–5051. [CrossRef]

24. Schicks, J.; Luzi, M.; Beeskow-Strauch, B. The conversion process of hydrocarbon hydrates into CO_2 hydrates and vice versa: Thermodynamic considerations. *J. Phys. Chem. A* **2011**, *115*, 13324–13331. [CrossRef]
25. Kvamme, B. Feasibility of simultaneous CO_2 storage and CH_4 production from natural gas hydrate using mixtures of CO_2 and N_2. *Can. J. Chem.* **2015**, *93*, 897–905. [CrossRef]
26. Baig, K.; Kvamme, B.; Kuznetsova, T.; Bauman, J. Impact of water film thickness on kinetic rate of mixed hydrate formation during injection of CO_2 into CH_4 hydrate. *AIChE J.* **2015**, *61*, 3944–3957. [CrossRef]
27. Kvamme, B.; Qasim, M.; Baig, K.; Kivelä, P.-H.; Bauman, J. Hydrate phase transition kinetics from Phase Field Theory with implicit hydrodynamics and heat transport. *Int. J. Greenh. Gas Control* **2014**, *29*, 263–278. [CrossRef]
28. Park, Y.; Kim, D.-Y.; Lee, J.-W.; Huh, D.-G.; Park, K.-P.; Lee, J.; Lee, H. Sequestering carbon dioxide into complex structures of naturally occurring gas hydrates. *Proc. Natl. Acad. Sci. USA* **2006**, *103*, 12690–12694. [CrossRef]
29. Dallimore, S.R.; Collett, T.S.; Taylor, A.; Uchida, T.; Weber, M.; Chandra, A.; Mroz, T.; Caddel, E.; Inoue, T. Scientific results from the Mallik 2002 gas hydrate production research well program, Mackenzie Delta, northwest territories, Canada: Preface. *Bull. Geol. Surv. Can.* **2005**, *585*, 1–16.
30. Yamamoto, K.; Dallimore, S. Aurora-JOGMEC-NRCan Mallik 2006–2008 gas hydrate research project progress. *Nat. Gas Oil* **2008**, *304*, 285–4541.
31. Schoderbek, D.; Boswell, R. Iġnik Sikumi# 1, Gas Hydrate Test Well, Successfully Installed on the Alaska North Slope. *Nat. Gas Oil* **2011**, *304*, 285–4541.
32. Nagakubo, S. Methane hydrate as a domestic energy resource: Japan's methane hydrate R&D program. *J. Geogr.* **2009**, *118*, 758–775.
33. Nagakubo, S.; Arata, N.; Yabe, I.; Kobayashi, R.; Yamamoto, K. Environmental impact assessment study on Japan's methane hydrate R&D program. *Fire Ice* **2011**, *11*, 4–11.
34. Koh, D.-Y.; Kang, H.; Lee, J.-W.; Park, Y.; Kim, S.-J.; Lee, J.; Lee, J.Y.; Lee, H. Energy-efficient natural gas hydrate production using gas exchange. *Appl. Energy* **2016**, *162*, 114–130. [CrossRef]
35. Schoderbek, D.; Farrell, H.; Hester, K.; Howard, J.; Raterman, K.; Silpngarmlert, S.; Martin, K.; Smith, B.; Klein, P. *Oil & Natural Gas Technology: ConocoPhillips Gas Hydrate Production Test: Final Technical Report*; DE-NT0006553-Final-Technical-Report; United States Department of Energy, National Energy Technology Laboratory: Albany, NY, USA, 2013.
36. Schicks, J.; Erzinger, J.; Ziemann, M. Raman spectra of gas hydrates—Differences and analogies to ice 1 h and (gas saturated) water. *Spectrochim. Acta Part A Mol. Biomol. Spectrosc.* **2005**, *61*, 2399–2403. [CrossRef]

9

Research on the Kinetics of Pyrolysis of Wood-Based Panels in Terms of Waste Management

Tomasz Jaworski and Małgorzata Kajda-Szcześniak *

Department of Technologies and Installations for Waste Management, Silesian University of Technology, 44-100 Gliwice, Poland; tomasz.jaworski@polsl.pl
* Correspondence: malgorzata.kajda-szczesniak@polsl.pl.

Abstract: Currently, there is a lot of interest in implementing the idea of a circular economy along with searching for optimal methods of waste management in terms of raw materials and energy. Waste wood-based floor panels are part of this discussion with regard to its management. The interest in this waste results from statistics and the prediction of its future quantities on the waste market. The separation and testing of individual layers of the waste floor panel was undertaken to answer the following question: Is it reasonable to mechanically separate the contaminated upper panel layer from the remaining part (which is suitable for material recycling) and subject it to thermal transformation methods? Thermogravimetric studies did not confirm the rationale of mechanical separation of layers for further management. Therefore, the use of pyrolysis was proposed as an alternative by showing the advantages of this process in the thermal transformation of the tested waste. The analyzed kinetics of this process included: mass loss, the influence of heating rate on the decomposition process, the impact of volatile parts in the substrate on the rate of mass loss, and the share of coke residue. Empirical formulas of the tested substrates in the molecular formula C–H–O–N (carbon-hydrogen-oxygen-nitrogen) were also proposed to assess its energy usefulness by entering the analyzed waste into a Van Krevelen diagram.

Keywords: kinetics; pyrolysis; wood-based panels; Van KREVELEN systematics; TG-MS/FTIR analysis

---

1. Introduction

It is estimated that more than 10 million m^3 of various types of wood materials are currently produced in Poland, over 3.5 million m^3 of which are MDF (medium-density fiberboard) and HDF (high-density fiberboard) [1].

At a time of great emphasis on the recovery of materials according to the idea of a circular economy (CE) and following the hierarchy of waste management (obligatory for business entities) with regard to the post-consumer wood waste, such as floor panels, two options for their management were considered as below [2–5]:

- mechanical and thermal method,
- thermal method.

The proposed methods fit into the idea of a circular economy, whose most important assumptions are presented below. The most popular definition of the circular economy was presented by the Ellen MacArthur Foundation: "A circular economy is one that is restorative and regenerative by design and aims to keep products, components, and materials at their highest utility and value at all times, distinguishing between technical and biological cycles" [6]. Another definition was provided in the publication [7], i.e., a circular economy is "a concept used to describe a zero-waste industrial economy that profits from two types of material inputs: (1) biological materials are those that can be reintroduced back into the biosphere in a restorative manner without harm or waste (i.e: they breakdown naturally); and, (2) technical materials, which can be continuously re-used without harm or waste" [8]. According to the authors of the publication [9], the circular economy shall be regarded as a "project and business modeling strategy that is, slowing, closing and narrowing the resource loop".

The circular economy in [10] is defined as "a regenerative system in which input resources and waste, emissions and energy leaks are minimized by slowing, closing and narrowing the material and energy loops. This can be achieved through long-term design, maintenance, repair, reuse, regeneration, renewal and recycling."

To a large extent, the circular economy in waste management is currently focused on resource recovery, the rational management of non-renewable resources, and environmental impact prevention. In the past, waste management was understood only in terms of storage and incineration [5,11].

The former of the aforesaid methods could include two processes. The first mechanical one involves the roughing (rupture, ...) of the upper layer, which is theoretically more contaminated with chemical impurities—mainly urea–formaldehyde resins—due to its exposure to direct use and requirements on abrasion, waterproofness, etc. This layer would require thermal transformation. The second process is related to the layers lying below the upper one. Since they seem to be less contaminated, they should be managed through material recycling. The latter method is to submit the entire waste to the process of thermal transformation. The choice of pyrolysis as a method of thermal transformation of waste is not accidental, since more and more attention is being paid to its nature and ability to gain products in the form of gas, combustible oils, and charcoal. Although the Polish legislation requires the combustion of pyrolysis products, there are other ways to use these raw materials, such as for example in the production of methanol: R3 Recycling or the recovery of organic substances that are not used as solvents (including composting and other biological transformation processes). This heading also includes gasification and pyrolysis using these components as chemical reagents [2]. This article does not consider the possibility (due to the lack of literature data) of the return of waste floor panels to the manufacturers of these products and their possible inclusion in the production cycle under the so-called internal recirculation.

A methodology for determining the kinetic parameters of waste pyrolysis for fuels with an unidentifiable chemical formula of the basic molecule (e.g. post-consumer floor panel) has been proposed. The methodology can be helpful for comparative studies between various solid fuels and also waste with the general formula CxHyOz. The location of the molecule, which was determined with this method, in the Van Krevelen diagram gives an excellent configuration of its energy status. This status can be changed subsequently through operations such as drying, mechanical drainage, heating, etc. in order to enhance energy properties (mainly calorific value by improving the ratio of fuel elements h/c (hydrogen/carbon) and o/c (oxygen/carbon)).

2. Thermodynamic and Kinetic Aspects of Pyrolysis

Pyrolysis is a process of thermochemical decomposition in an anaerobic atmosphere. It is a complex process of the decay of chemical compounds into smaller molecules under the influence of external thermal energy [12,13]. Most reactions that occur during pyrolysis are endothermic. In general, the pyrolysis process can be represented by reaction (1):

$$\text{CnHmOp} - \text{heat} - \text{CxHyOz} + \sum \text{CaHbOc} + H_2O + \text{C (charcoal) (liquid) (gas)} \quad (1)$$

As a result of the pyrolysis, three fractions are formed [12,13]:

- the charcoal, consisting largely of carbon, also contains the mineral parts included in the substrate pyrolyzed;
- liquid fraction, consisting of an organic part with a complex chemical composition and water part containing, among others, acetic acid or methanol;

- gas fraction, which is a mixture of CO_2, CO, H_2, and hydrocarbons, mainly methane.

The literature presents different conditions for the process and the impact of these parameters on the products obtained [14–16]. The basic division into types of pyrolysis is related to the rate of heating of the fuel molecule. Slow pyrolysis occurs when the time required to warm the molecule to the pyrolysis temperature is significantly longer than the characteristic pyrolysis reaction time (theating >> tr). When the situation is reversed, this process can be defined as fast pyrolysis. Other parameters that affect the quality and quantity of products are: a maximum temperature, the residence time of the primary decomposition products in the conversion zone, a degree of fuel fragmentation, the reactor heating method, and its construction. Table 1 presents types of pyrolysis and characteristic parameters influencing the decomposition, while Figure 1 shows the possibilities of the course of pyrolysis depending on the parameters in which the process is carried out.

Table 1. Types of pyrolysis and characteristic parameters influencing the decomposition [13].

Type of Pyrolysis	Residence time of Primary Decomposition Products in the Conversion Zone	Heating Rate [K/s]	Final Temperature [K]	Products
Carbonization	days	very slow	400	charcoal
Conventional	5–30 min	slow	600	charcoal, gas, liquid fraction
Fast	<2 s	very fast	500	liquid fraction
Very fast	<1 s	very fast	1000	reagents, gas
Underpressure	2–30 s	fast	400	liquid fraction

Pyrolysis as a method of obtaining fuels is a promising technology for the conversion of indirect wood-based waste such as floor panels. This type of waste can be found in the waste catalogue under the codes: 03 01 05, 17 09 04, and 20 01 38. The code allocation depends on the source of the waste floor panels [17].

Pyrolysis is a very complex process and, depending on the parameters of its operation, it is possible to obtain solid, liquid, and gas products in various weight ratios. The schematic diagram of the possibilities of the process, depending on the pressure, temperature, and speed of the process, was presented in work [13,18].

2.1. Heat Demand for the Pyrolysis Process

Although basically pyrolysis is an endothermic process, during its duration, there are effects of both heat extraction and its emission. At the time of pyrolysis, pyrogenetic water is formed, which is the result of the reaction of hydrogen with oxygen, as well as the result of the disintegration of hydroxyl group side chains. However, decomposition reactions are not the only chemical processes that occur during pyrolysis. The gases formed from the solid penetrate the remaining part of the solid phase (most often the char) and react with them. Therefore, it is difficult to determine clearly what is the thermal effect of the pyrolysis itself, and what are the effects of secondary reactions occurring between pyrolytic gases and the surface of the char [13]. The energy balance of the pyrolysis process is prepared in accordance with the first law of thermodynamics, from which it follows that if any transformation in the system changes its state and the system exchanges energy with the environment only through heat transport, then the relationship follows the equation below (2) [19]:

$$Q = \Delta h \tag{2}$$

where:

Δh – increase of enthalpy of the system
Q – heat exchanged with the environment

The above indicates that the change of enthalpy of the system is equal to the heat supplied, the quantity of which corresponds simultaneously to the heat demand for pyrolysis. In this case, possible heat losses to the environment are not taken into account. Thus, it is possible to use the calculation methodology proposed by [19], which is based on the balance of the enthalpy of substance formation, and takes into account the physical enthalpy of products in appropriate physical states.

Pyrolysis heat is determined on the basis of temperature measurements in DSC (Differential scanning calorimetry) scanning calorimeters. The results obtained in this type of device are a net measure of pyrolysis and secondary reactions. Pyrolysis heat is the difference in the energy of the formation of reagents and products at the reference temperature (298.15 K). The heat values of pyrolysis reported in the bibliography differ significantly from 750 kJ/kg (endotherm) to 130 kJ/kg (exothermic reaction). For example, pyrolysis heat in the range from 274 kJ/kg to 353 kJ/kg was determined using DSC [13]. Bibliography [13] presents thermogravimetric measurements for pine wood up to 700 °C, with a heating rate of 20 °C/min. After cooling, the samples of charcoal being still in the thermogravimeter were heated once again at the same rate. Shown the power consumed by pine wood and the charcoal during pyrolysis in TGA (Thermogravimetric analysis) as a function of temperature. After integration in time with the power difference consumed by the material, the thermal effect of the process is obtained. The difference among the thermal effects for wood and charcoal relates to pyrolysis heat, and it amounts to 1473 kJ/kg. This effect is clearly higher than the effect resulting from the above data.

2.2. Kinetics of the Pyrolysis Process

Fuels formed on the basis of waste floor panels are a complex solid substance, and there are thermal transformations in the hard-to-define structure of C–H–O–N molecules (which changes texture during the process) because it is difficult to implement a model description including reaction range, autocatalytic effects, diffusion effects, and structural changes. In the bibliography, one can find various approaches to pyrolysis kinetics; however, the basis for any analysis is the distribution reaction model, which occurs according to the following entry (3). This is as a single irreversible reaction of the thermal decomposition of carbon, but it is also helpful in case of a carbon-rich substance:

$$\text{Carbon} \xrightarrow{\quad} \text{k Xvolatile\_products} + \text{(1-X)charcoal} \tag{3}$$

This entry is often regarded in the literature of the subject as too simplified, because it omits important transitional stages of the substance distribution, including, first of all, subsequent and possibly parallel reactions of the distribution of primary products, such as coal tar (metaplast). For this reason, many forms of kinetic equations have been formulated, also by dividing pyrolysis into individual stages. However, a comparison of various schemes of the decomposition of organic carbon leads to the conclusion that regardless of the method of receiving final products, the formula expressed by the dependence (3) reflects the final result of pyrolysis. As a consequence, many authors approximate the process of decomposition of a substance rich in organic carbon with the first-order reaction occurring uniformly in the entire volume of the particle [20]. The speed of extracting volatile parts can be described in this case by the Formula (4):

$$dV/dt = k\,(V_0{}^{daf} - V) \tag{4}$$

The most common source of information on pyrolysis kinetics are data obtained from thermogravimetric and derivative thermogravimetric measurements (TG/DTG). The results of such measurements under dynamic conditions indicate the existence of several process stages characterized by different quantities of apparent activation energy. The stages of various kinetics are attributed to the pyrolysis of the individual components of which the floor panels are made, including: wood-based HDF (high-density fiberboard) boards, urea–formaldehyde resins, and laminates [21].

2.3. *Van KREVELEN Systematics and Wood-Based Waste*

The original intention and achievement of Van Krevelen [22] was the presentation of classical fuels in a plane coordinate system, where the atomic quotients of O/C were given on the abscissa x and on the ordinates y = H/C (excluding N, S, P, and micronutrients). Assuming that carbon is tetravalent, the maximum value of the ordinate is 4, and the abscissa x = O/C ≤2. The Van Krevelen diagram was later modified by Meunieur [23], who adopted the total chemical formula CHyOx (x, y coordinates from the diagram) for these fuels. Modification made by Meunieur showed the chemical structure more accurately than the percentage mass fraction of elements C, H, and O [24].

When analyzing wood-based waste (and knowing their chemical composition), it was possible to determine their picture in the Van Krevelen systematics. The aim was to define the place of this waste and its energy suitability in comparison to classical fuels. To determine the waste position in the Van Krevelen diagram, it is necessary to calculate the atomic quotients x and y for dry organic matter of waste without sulfur and phosphorus in accordance with the condition:

$$C + N + O + H = 100\% \quad (5)$$

where:

C, N, O, and H – content of elements in the sample in percentage by weight.

At the stage of laboratory tests, it was possible to determine the content of all four elements required by the procedure. To find an illustration/point in the diagram for waste, it is necessary to separate the content of these elements [24,25]. Determination of the empirical formula of wood-based waste ground on the percentage content of elements was based on the conversion of this content into molars of elemental atoms. The obtained values were divided by the lowest number in the ratio, which was in this case 1.071. Ultimately, the empirical formula is as follows:

$$C_4H_5NO_2 \quad (6)$$

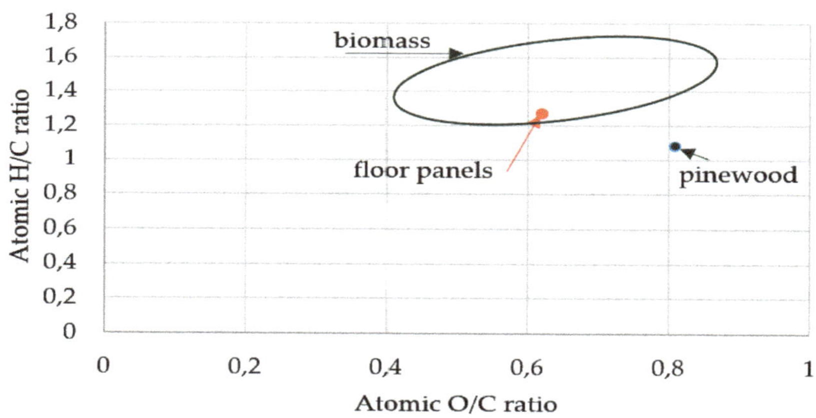

Figure 1. Illustration of the position of waste from floor panels in the Van Krevelen system (author's own study) [24,26].

Figure 1 presents the location of wood-based waste (floor panels) in the Van Krevelen diagram. The coordinates of the point characterizing wood-based waste were calculated according to the formulas [24,25]:

$$X = OM_c/CM_o = 0.62 \quad (7)$$

$$Y = HM_c/CMH = 1.27 \quad (8)$$

where:

MC, MH, and MO – molar masses of carbon, hydrogen, and oxygen;
C, H, O – content of elements in the dry organic mass of the residue in percentage by weight.

3. Materials and Methods

Tests TG-MS/FTIR were carried out on samples with the following assay [27]:

- ZMT/20/2019 (pinewood)
- ZMT/21/2019 (bottom layer of the floor panel)
- ZMT/22/2019 (middle layer made of HDF)
- ZMT/23/2019 (upper layer of the floor panel)
- ZMT/24/2019 (complete floor panel)
- ZMT/25/2019 (urea–formaldehyde resin)

These tests were carried out by means of the STA 409 PG Luxx thermogravimetric analyzer from Netzsch on a TG (Thermogravimetry) carrier coupled with a quadrupole mass spectrometer QMS Aeolos and a medium infrared spectrometer FTIR (Fourier Transform Infrared Spectroscopy) Tensor 27 from Bruker. The tests were carried out in an argon atmosphere (purity class 5.0, flow 25 mL/min), from 40 to 1000 °C, with a heating rate of 10 K/min. The sample weight was 10 ± 0.1 mg. The test was performed in crucibles with Al_2O_3 with a lid. In order to eliminate traces of oxygen in the thermal mixer, the OTS (Oxygen Trap System) system was used. The measurement was corrected with corrective measurements (measurement without a sample). The corrective measurements were carried out three times.

This paper presents an analysis of the physicochemical properties of selected post-consumer wood waste and urea–formaldehyde resin. The goal of the analysis was to determine specific fuel properties, such as moisture, ash content, content of volatile components, calorific value, and elemental composition (C, H, O, N, S, Cl).

The characteristics of the physicochemical properties of the tested samples are presented in Table 2.

Table 2. Physicochemical properties of samples.

Parameter	ZMT/20/2019	ZMT/21/2019	ZMT/22/2019	ZMT/23/2019	ZMT/24/2019
Total humidity, % mass	13.81	5.35	6.07	5.20	6.16
Combustible fraction, % mass	98.36	99.39	99.75	95.88	99.12
Ash, % mass	1.64	0.61	0.25	4.12	0.88
Volatile fraction, % mass	78.17	80.28	82.09	77.43	82.07
Heat of combustion, MJ/kg	20.23	19.63	18.70	18.88	19.89
Calorific value, MJ/kg	18.93	17.96	17.10	17.46	18.58
Carbon, % mass	44.94	53.23	51.72	46.51	48.34
Hydrogen, % mass	4.25	6.85	6.44	5.71	5.11
Oxygen, % mass	48.56	33.09	35.85	35.20	40.24
Nitrogen, % mass	0.61	5.99	5.46	8.08	5.19
Sulfur, % mass	0.003	0.23	0.28	0.38	0.24
Chloride, % mass	ppo	ppo	ppo	ppo	ppo

Ppo—below the limit of quantification.

The individual layers of the floor panel are characterized by similar humidity of 5–6%, while pine wood is characterized by humidity at the level of approximately 14%. All the samples had a low ash content of less than 4.12% and a high content of volatile matter in the range of 78–82%. The wastes are characterized by a calorific value at a similar level, i.e., above 17 MJ/kg. In all the layers of the panel, there was a high nitrogen content in the range from 5.19–8.08%, while a low nitrogen content was recorded in pine wood. The tested wastes had low sulfur and chlorine content.

4. Results and Discussion

Figure 2 shows the weight loss curves for individual samples. It reveals the unexpected similarity between the weight loss curves of the pure wood sample and the sample representing the upper layer of the panel, as well as three curves of samples: the bottom and the middle layers and the sample of the averaged complete panel composition. In the case of waste floor panel pyrolysis, usually three stages of thermal decomposition are identified: in the first stage (from environment temperature to about 140–180 °C), the samples lose moisture; in the second stage (from about 160 to 400 °C), the material pyrolysis takes place; the third stage, which is clearly slower than the preceding one, occurs at a temperature of about 400 to 700 °C. The divergence from the three-stage decomposition was observed at the sample of a urea–formaldehyde resin, which decomposed in five stages occurring in the following temperature ranges: 40–120 °C, 120–200 °C, 200–260 °C, 260–300 °C, and 300–450 °C.

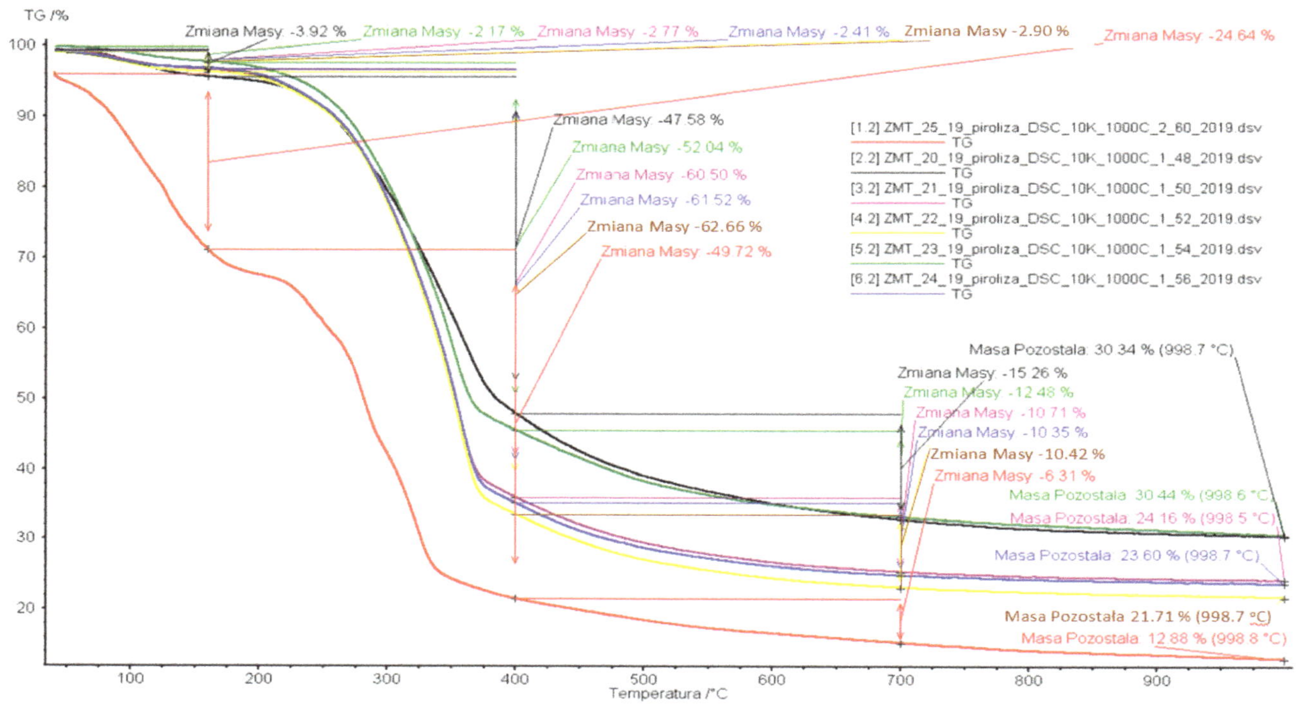

Figure 2. TG (thermogravimetry) curves of all the samples tested (own research).

Figure 3 shows the weight loss rates of individual samples. All the samples of different layers of the floor panel have a similar velocity of mass loss ranging from 6.145 to 7.792%/min, and the largest distribution occurs at the temperatures of 35.4 to 357.1 °C. This creates good conditions for simultaneous decomposition at almost the same temperature. A slightly different rate of decay characterizes clean wood and it amounts to 4.097%/min. The chemical additives in the panel increase the weight loss rate almost twice.

Figures 4–6 present representative results regarding TG, DTG curves, and ionic current (MS). Studies show that water (ion m/e = 18, blue line) is emitted from ZMT/20-24/2019 samples in two stages: when the samples lose moisture in the temperature range from the environment temperature to about 180 °C, and in the range from about 250 to 450 °C. The other main products of pyrolysis, such as: carbon monoxide (m/e = 28, purple line), carbon dioxide (m/e = 44, yellow line), methane (m/e = 16, red line), and nitric oxide (m/e = 30, green line) are emitted from samples in the range from about 200 to 500 °C. In the case of the ZMT/25/2019 sample (urea–formaldehyde resin), the temperature ranges in which individual products are released are slightly different, and also, ions were recorded: m/e = 17 (most likely ammonia – light green line) and m/e = 57 (light blue line).

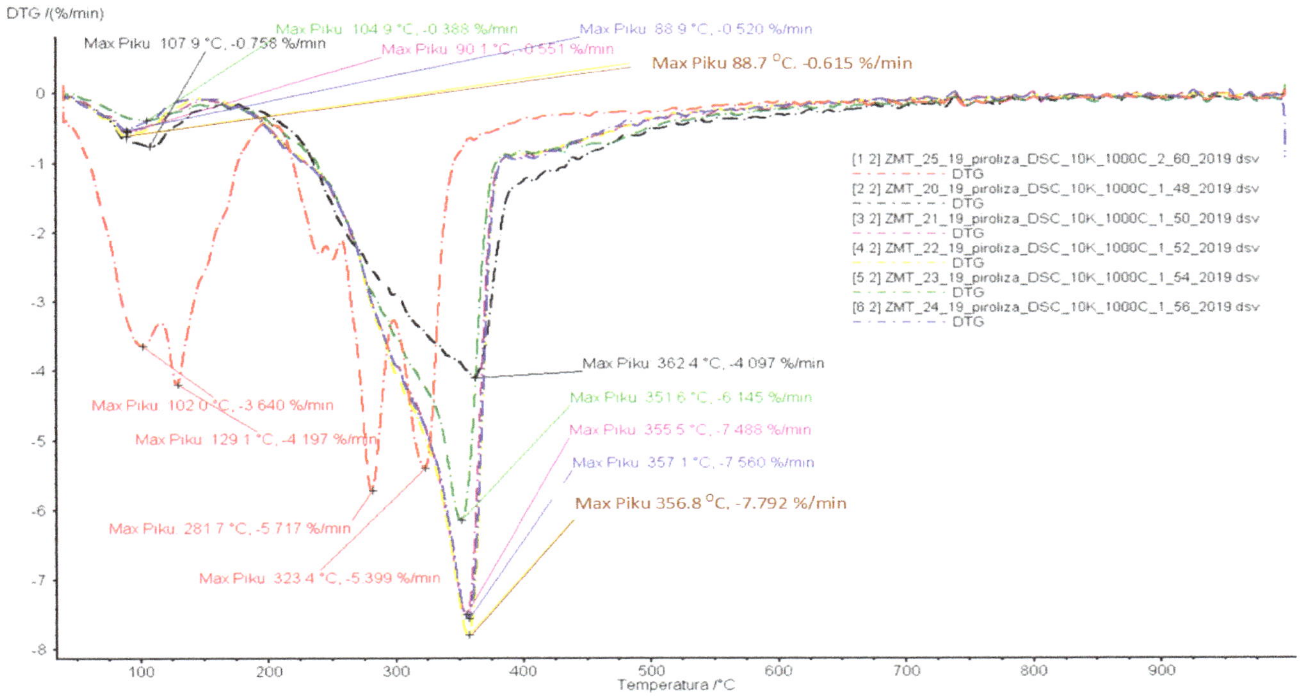

Figure 3. Derivative thermogravimetric measurements (DTG) curves of all samples tested (own research).

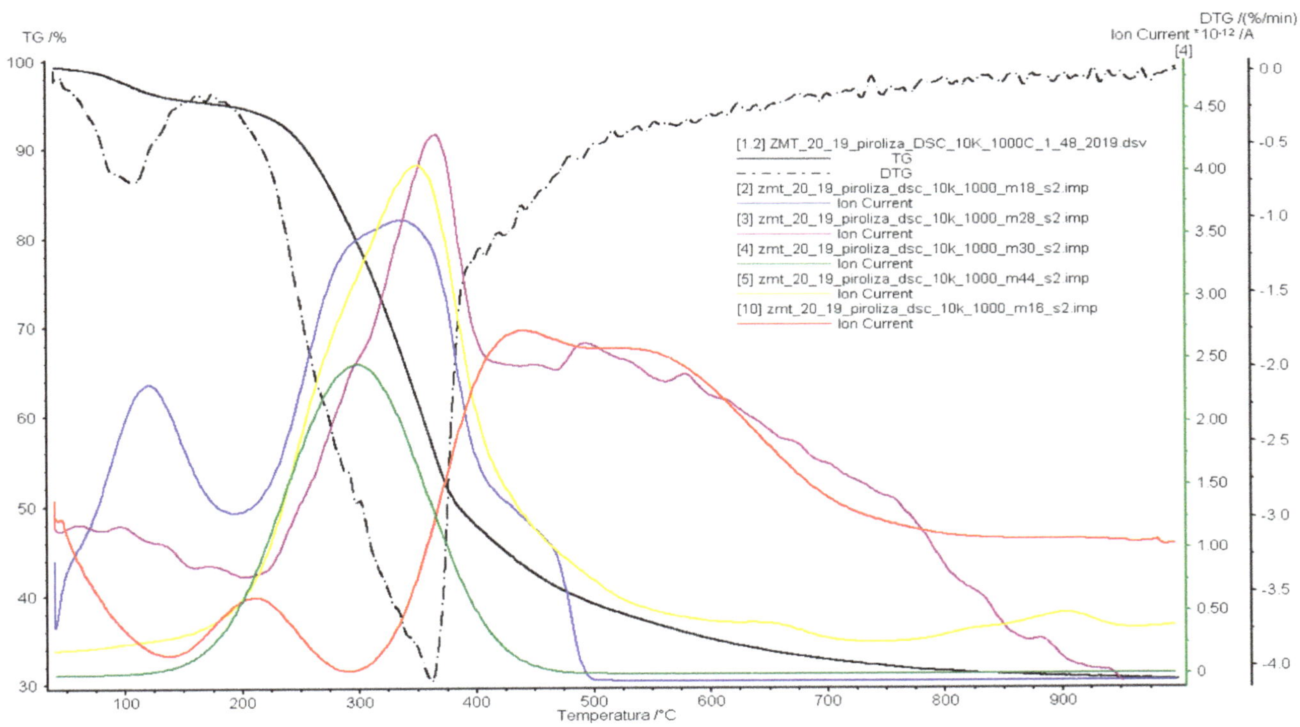

Figure 4. TG and DTG, curves and ion current intensity for the sample ZMT/20/2019 (own research).

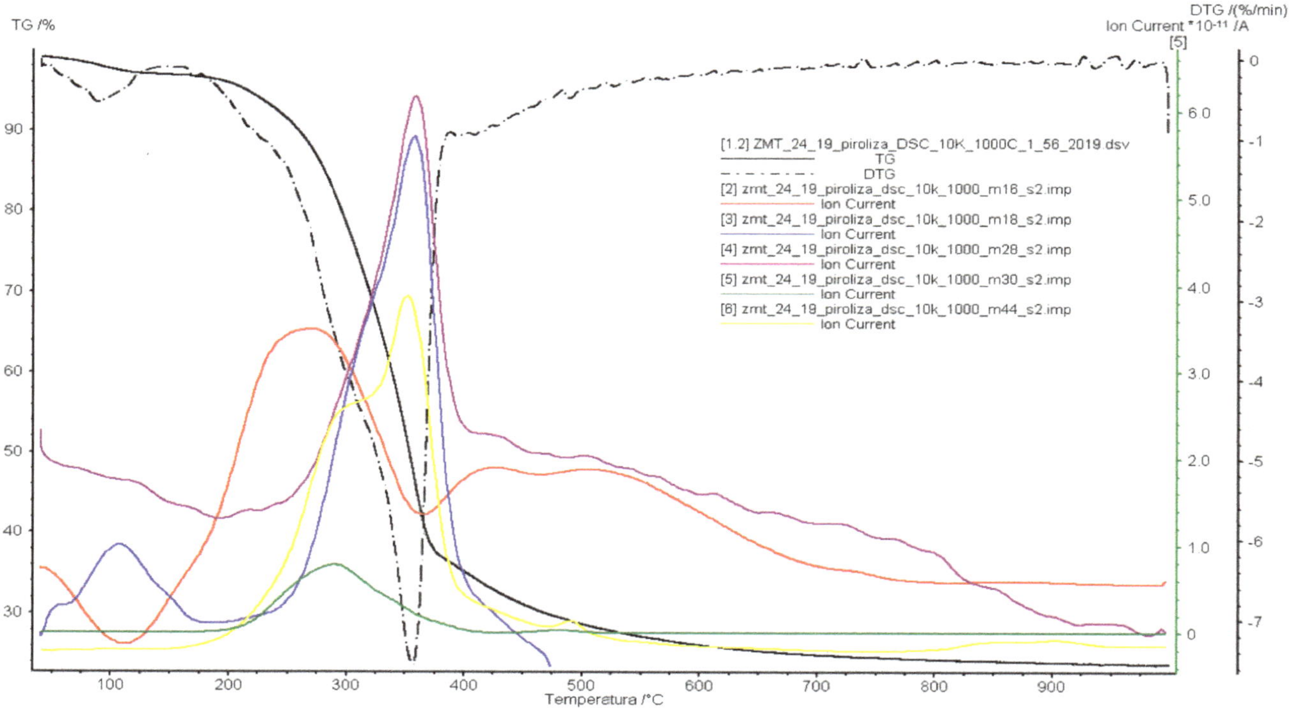

Figure 5. TG and DTG curves, and ion current intensity for the sample ZMT/24/2019 (own research).

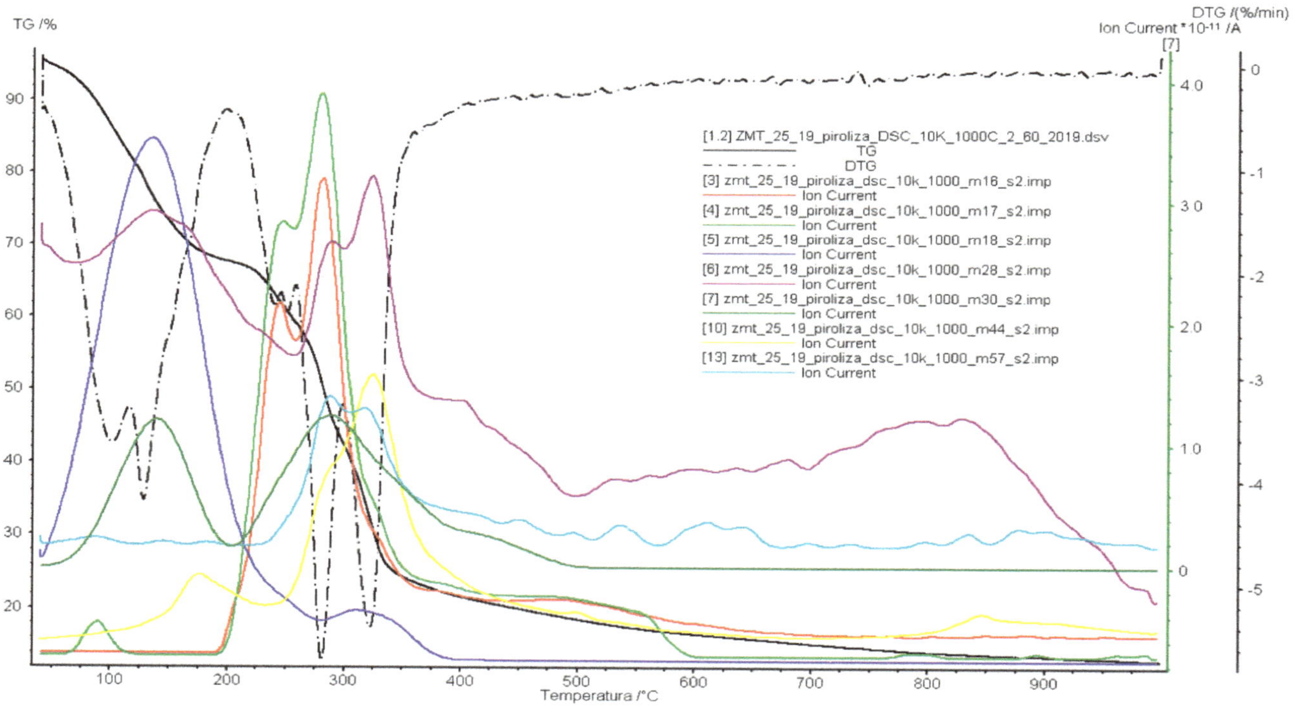

Figure 6. TG and DTG curves, and ion current intensity for the sample ZMT/25/2019 (own research).

The combination of thermoanalysis with the analysis of gaseous products has provided knowledge of the mechanisms that occur in the examined waste during thermal processes and will enable the qualitative assessment of gases emitted during heating with regard to the chemical composition.

Figure 7a,b show the course of the curves of the emission of selected gases as a function of temperature with the TG curves for the sample of the charcoal by using the FTIR spectrometer data during the pyrolysis process of the samples: ZMT/20/2019 and ZMT/24/2019. It is clearly evident that ammonia and nitrogen dioxide are emitted during the pyrolysis process of the waste floor panel. This phenomenon is not observed in the case of the pyrolysis of pure pine wood.

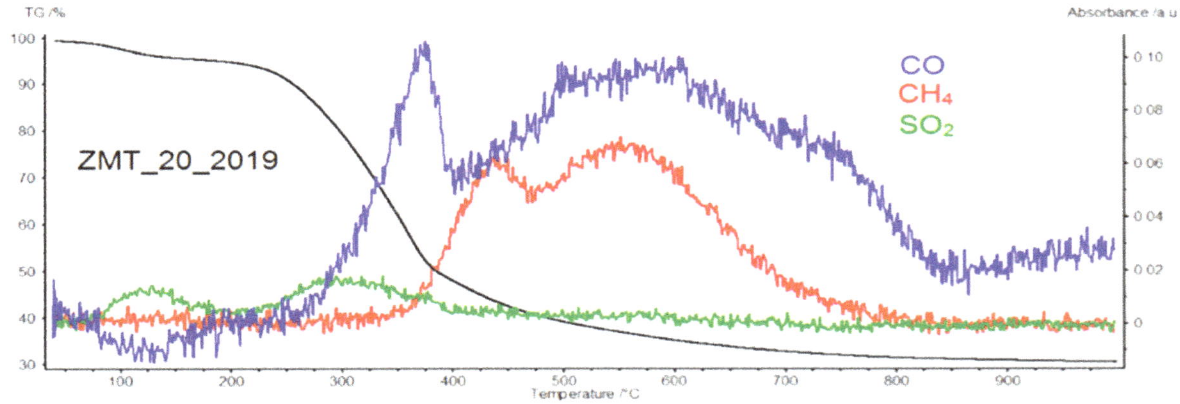

(**a**)

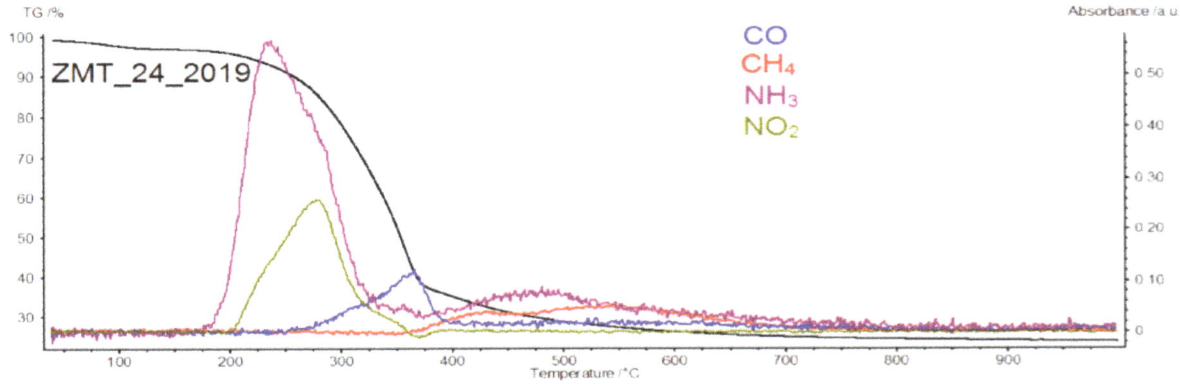

(**b**)

Figure 7. TG curve and curves of emissions of carbon monoxide, methane, nitrogen dioxide, and ammonia for the samples (**a**) ZMT/20/2019 and (**b**) ZMT/24/2019 (own research).

This phenomenon is caused by the addition of urea–formaldehyde resin to the floor panels, which contains about 37% nitrogen in the composition [28].

The products of pyrolysis depending on the conditions of the process (which include heating speed, temperature, residence time, granulation of the material, etc.) are as follows: gas (CO, CO_2, CH_4,...), pyrolysis oil, and carbonizate (carbonized solid). For example, increasing the heating rate directs pyrolysis products toward the production of bio-oil, and long-term heating is used to produce carbonized solids. All pyrolysis products have a positive chemical enthalpy (calorific value), so they can be used for energy, e.g., in the production of heat and electricity or cogeneration. However, pursuant to the Polish legislation, products of the waste pyrolysis process should ultimately be oxidized and incinerated. There is also a possible legal path for another use of pyrolysis gases for the production of chemicals, e.g., in methanol production.

5. Conclusions

Mechanical separation of the panel layers for the purpose of different management of the upper layer (thermal transformation) and lower layers (recycling) is not reasonable. This thesis is justified by the following:

(a) All samples, regardless of the location of the layer (upper/middle/lower), emit, according to the TG curve, high amounts of the NO_x and NH_3, which are derivatives of the decomposition of urea–formaldehyde resin, as shown in Figures 4–7;
(b) The sample of pure pine wood does not generate NO_x and NH_3 during thermal decomposition, as shown in Figures 4 and 7a;
(c) The addition of the hardeners (glues) in the form of the urea–formaldehyde resin efficiently contaminates the entire cross-section of the panel.

The thermal transformation (combustion/gasification/pyrolysis) shall be indicated as the only reasonable method of management of floor panels.. All these methods must deal with the issue of emissions of the contaminants as a result of decomposition or the secondary reaction mechanisms generated by the addition of urea–formaldehyde resin.

The problem of the combustion of floor panels in the layer on the grate of the waste thermal treatment installation (ITPOK) was the subject of description in the publication [21,29–31].

The proposed methods of thermal decomposition of floor panels using pyrolysis indicate the advantages of this process as evidenced by:

(a) Energy characteristics of the location of the formed fuel in the Van Krevelen diagram in the area of coal fuels and peat, as shown in Figure 1;
(b) The DTG distribution curves, and peaks of the maximum mass loss rate of the panel samples and the temperature at which they take place, are similar in the area from 6.145 to 7.792%/min., in temperatures: from 351.4 to 357.1 °C, as shown in Figure 3;
(c) The high content of volatile parts in Table 2 suggests a high tendency for thermal decomposition in accordance with Equation (4); a similar tendency is related to coals, which is documented in the publication [20]; and
(d) A factor promoting the rate of decomposition of matter during pyrolysis is also a lower oxygen content in the panel, which is 40.24% by weight, compared to its content in pure wood, which is 48.56% by weight, as shown in Table 2. Lower oxygen content improves the energy qualities of the fuel.

The research on pyrolysis should be further developed with solutions reducing the NO_x and NH_3 emissions from this process.

Post-consumer wood in the form of floor panels, MDF boards, etc., is a hazardous waste due to the content of chemical substances; however, these substances improve their functional qualities (hardness, water resistance, etc.). In this study, an attempt was made to check how contaminated the waste is and whether at least part of it is suitable for such management that is different from thermal management. Analyses of subsequent sections of the tested waste showed almost equal contamination with urea–formaldehyde resin in the panel cross-section. Therefore, it is recommended to direct this waste to organized thermal disposal. Pyrolysis seems to be a good solution here, especially since its products, depending on the mode of its conduction and the heating rate applied), can be also used as raw material for the production of chemicals.

Kinetic studies should be related to the thermodynamic analysis of pyrolysis. It is suggested to continue research on the basis of compounds: the enthalpy of creating substrates—which is an energy of process activation—oxygen content, and volatile parts in substrates, as well as the heating rate in the process (type of reactor for conducting pyrolysis).

Author Contributions: Conceptualization, T.J.; formal analysis, T.J.; writing, T.J.; funding acquisition for research, M.K.-S.; investigaton M.K.-S.; project administration M.K.-S.; resources M.K.-S.

References

1. Stowarzyszenie Producentów Płyt Drewnopochodnych w Polsce. Available online: http://sppd.pl/plyty-drewnopochodne-w-trendzie-ekoprodukcji.html (accessed on 8 July 2019).
2. Ustawa o odpadach (Dz. U. 2013 poz. 23 z póź. zm). Available online: http://prawo.sejm.gov.pl/isap.nsf/DocDetails.xsp?id=WDU20130000021 (accessed on 26 September 2019).
3. *Communication from the Commission to the European Parliament, the Council, the European Economic and Social Committee and the Committee of the Regions Closing the Loop—An EU Action Plan for the Circular Economy*; European Commission: Brussels, Belgium, 2015.
4. *Communication from the Commission to the European Parliament, the Council, the European Economic and Social Committee and the Committee of the Regions. The Role of Waste-to-Energy in the Circular Economy*; European Commission:: Brussels, Belgium, 2017.
5. Ghisellini, P.; Cialani, C.; Ulgiati, S. A review on circular economy: The expected transition to a balanced interplay of environmental and economic systems. *J. Clean. Prod.* **2016**, *114*, 11–32. [CrossRef]
6. Ellen MacArthur Foundation. *Towards a Circular Economy: Business Rationale for an Accelerated Transition*; The Ellen MacArthur Foundation, November 2015; Available online: https://www.ellenmacarthurfoundation.org/assets/downloads/publications/TCE_Ellen-MacArthur-Foundation_26-Nov-2015.pdf (accessed on 22 August 2019).
7. Lewandowski, M. Designing the Business Models for Circular Economy—Towards the Conceptual Framework. *Sustainability* **2016**, *8*, 43. [CrossRef]
8. Scott, J.T. *The Sustainable Business a Practitioner's Guide to Achieving Long-Term Profitability and Competitiveness*, 2nd ed.; Greenleaf Publishing: Sheffield, UK, 2015.
9. Bocken, N.M.P.; de Pauw, I.; Bakker, C.; van der Grinten, B. Product design and business model strategies for a circular economy. *J. Ind. Prod. Eng.* **2016**, *33*, 308–320. [CrossRef]
10. Geissdoerfer, M.; Savaget, P.; Bocken, N.M.P.; Jan Hultink, E. The Circular Economy—A new sustainability paradigm? *J. Clean. Prod.* **2016**, *143*, 757–768. [CrossRef]
11. Czop, M.; Poranek, N.; Czajkowski, A. Energetic usability and nuisance to the environment of selected fuels made of wastes. *Przem. Chem.* **2018**, *97*, 1460–1462. [CrossRef]
12. Basu, P. *Biomass Gasification and Pyrolysis Practical Design and Theory*, 2nd ed.; Elsevier: Oxford, UK, 2010.
13. Kardaś, D.; Kluska, J.; Klein, M.; Kazimierski, P.; Heda, Ł. *Modelowe Kompleksy Agroenergetyczne: Teoretyczne i Eksperymentalne Aspekty Pirolizy Drewna i Odpadów*; Monografia Publisher: Olsztyn, Poland, 2014.
14. Zolezzi, M.; Nicolella, C.; Ferrara, S.; Iacobucci, C.; Rovatti, M. Conventional and fast pyrolysis of automobile shredder residues (ASR). *Waste Manag.* **2004**, *24*, 691–699. [CrossRef] [PubMed]
15. Pütün, A.E.; Önal, E.; Uzun, B.B.; Özbay, N. Comparison between the "slow" and "fast" pyrolysis of tobacco residue. *Ind. Crops Prod.* **2007**, *26*, 307–314. [CrossRef]
16. Bridgwater, A.V.; Peacocke, G.V.C. Fast pyrolysis processes for biomass. *Renew. Sustain. Energy Rev.* **2000**, *24*, 1–73. [CrossRef]
17. Rozporządzenie Ministra Środowiska z dnia 9 grudnia 2014 r. w sprawie katalogu odpadów (Dz. U. 2014 poz. 1923). Available online: http://prawo.sejm.gov.pl/isap.nsf/DocDetails.xsp?id=WDU20140001923 (accessed on 26 September 2019).
18. Diebold, J.P.; Scahill, J. Ablative Fast Pyrolysis of Biomass in the Entrained-Flow Cyclonic reactor at SERI. In Proceedings of the 14th Biomass Termochemical Conversion Contractors' Review Meeting, Arlington, VA, USA, 22–25 June 1982; Solar Energy Research Institute: Golden, CO, USA, 1982.
19. Ściążko, M.; Więcek, P.; Sobolewski, A. Heat demand for the pyrolysis process and standard enthalpy of reaction. *Karbo* **2012**, *4*, 238–248.
20. Ściążko, M. Solid Fuels as CxHyOz Molecules—Thermodynamic and Kinetic Aspects. *Chemik* **2013**, *67*, 379–386.
21. Kajda-Szcześniak, M.; Jaworski, T. Analysis of the process of combustion of post-consumer wood using quantitative assessment indicators. *Wood* **2016**, *59*, 91–107. [CrossRef]
22. Van Krevelen, D.W. Graphical-statistical method for the study of structure and reaction processes of coal. *Fuel* **1950**, *29*, 269–284.
23. Meunieur, J. *Vergasung Fester Brennstoffe und die Oxidative Umwandlung von Kohlenwasserstoffen*; Verlag Chemie: Weinheim, Germany, 1962.

24. Ostojski, A. Sewage sludge classification based on elementary analysis. In *Nowe Metody Redukcji Emisji Zanieczyszczeń i Wykorzystnia Produktów Ubocznych Oczyszczalni Ścieków*; Obarska-Pempkowiak, H., Pawłowski, L., Eds.; Komitet Inzynierii Środowiska PAN: Lublin, Poland, 2009; Volume 4, pp. 41–51.
25. Kempa, E. Systematyka osadów ściekowych. In *Materiały Międzynarodowej Konferencji NaukowoTechnicznej "Osady Ściekowe: Odpad czy Surowiec?"*; Bień, J., Ed.; Wydawn. Politechniki Wrocławskiej: Częstochowa, Poland, 1997; pp. 7–35.
26. Heidenreich, S.; Müller, M.; Foscolo, P.U. *Advanced Biomass Gasification. New Concepts for Efficiency Increase and Product Flexibility*, 1st ed.; Chapter 3—Biomass Pretreatment; Academic Press: Cambridge, MA, USA, 2016; pp. 11–17.
27. *Research Report No: ZMT/2/2019*; ICHPW: Zabrze, Poland, 2019.
28. Wandrasz, J.W.; Wandrasz, A.J. *Formed Fuels. Bio-Fuels and Fuels from Waste in the Thermal Process*; Seidel-Przywecki Sp. z o.o.: Warszawa, Poland, 2006.
29. Kajda-Szcześniak, M.; Jaworski, T. Characteristics of the combustion process of woodwork waste in the installation of thermal treatment of municipal solid waste (tpok). *Wood Res.* **2018**, *63*, 15–23.
30. Pikoń, K.; Kajda-Szczesniak, M.; Bogacka, M. Environmental effect of co-combustion of coal with wood waste in low-power boilers. *Przem. Chem.* **2015**, *94*, 1548–1550.
31. Jaworski, T.; Pikoń, K.; Kajda-Szcześniak, M. Experimental and theoretical modeling of the waste combustion in the chamber with moving grate. *Chem. Process Eng.* **2018**, *39*, 3–14.

10

Evaluation of a Modified Equivalent Fuel-Consumption Minimization Strategy Considering Engine Start Frequency and Battery Parameters for a Plugin Hybrid Two-Wheeler

Naga Kavitha Kommuri *, Andrew McGordon, Antony Allen and Dinh Quang Truong

WMG, University of Warwick, Coventry CV4 7AL, UK; A.McGordon@warwick.ac.uk (A.M.); A.Allen@warwick.ac.uk (A.A.); T.Dinh@warwick.ac.uk (D.Q.T.)

* Correspondence: k.kommuri@warwick.ac.uk

Abstract: An appropriate energy management strategy is essential to enhance the performance of hybrid electric vehicles. A novel modified equivalent fuel-consumption minimization strategy (ECMS) is developed considering the engine operating point deviation from the optimum operating line. This paper focuses on an all-inclusive evaluation of this modified ECMS with other state-of-art energy management strategies concerning battery ageing, engine switching along with fuel economy and charge sustenance. The simulation-based results of a hybrid two-wheeler concept are analysed, which shows that the modified ECMS offers the highest benefit compared to rule-based controllers concerning fuel economy and reduction in engine switching events. However, the improvement in fuel economy using modified ECMS has significant negative potential effects on critical battery parameters influencing battery ageing. The results are analysed and found consistent for two different drive cycles and three different powertrain component configurations. The results show a significant reduction in fuel consumption of up to 21.18% and a reduction in engine switching events of up to 55% with modified ECMS when compared with rule-based strategies. However, there is a significant increase in battery temperature by 31% and battery throughput by 378%, which plays a major role in accelerating battery ageing. This paper emphasizes the need to consider battery-ageing parameters along with other control objectives for a robust assessment of energy management strategies. This study helps in laying down a foundation for future improvements in energy management development and it also aids in establishing a basis for comparing energy management controllers.

Keywords: parallel plugin hybrid; hybrid two-wheeler; load-levelling control (LLC); equivalent fuel-consumption minimization strategy (ECMS); battery ageing; fuel economy; engine switching and charge sustained

---

1. Introduction

Concerns over climate change, constraints on energy resources, stringent regulations on emissions and poor energy efficiency are all pushing the transportation industry to focus more on alternative technologies, such as Hybrid Electric Vehicles (HEVs) [1]. Plugin HEVs (PHEVs) have become the prime focus in recent times. However, battery cost is one of the major hurdles in a PHEV to limit its prevalence in the automotive industry [2]. The key point in using the full potential of hybridization is by designing a proper supervisory control strategy, which manages the split of demanded power between energy sources in the most efficient and optimum way. One of the important aspects of PHEVs is to optimize the performance of the hybrid energy system by developing an intelligent battery ageing-conscious energy management strategy.

Studies [3–5] suggested that hybrid electric controllers can be divided into rule-based controllers and optimization-based controllers. Rule-based controllers are based on predefined rules based on experience, calibration and tuning of controllers for the desired output without prior knowledge of the trip [6,7]. Optimization controllers overcome the inherent rigidity of rule-based controllers by using the optimal control strategy that minimizes a cost function [8,9]. The most powerful optimal control used for optimization controller is equivalent fuel-consumption minimization strategy (ECMS) and it is generally accepted to be a promising real-time controller because of its feasibility and optimality [5,10,11]. In [12], Paganelli et al. have developed optimal control for a parallel hybrid and compared the fuel consumption with a rule-based conventional strategy; the fuel consumption reduced by 17.5% for the test cycle considered. Similarly, in [10–15], the authors C. Sun et al., J. Han et al., G. Paganelli et al., C. Musardo et al., L. Serrao et al. and L. Namwook et al. have implemented ECMS and compared the results of fuel consumption and charge sustainability with rule-based strategies. In all these references considered, ECMS shows significant improvement in fuel economy along with charge sustenance when compared to rule-based strategies. However, the control objectives in these references remained limited to fuel economy, tailpipe emissions and charge sustainability. The fuel economy of any type of HEV is highly dependent on the energy capacity of the on-board energy storage system [2]. Battery replacement cost is one of the critical factors to be explored in depth in PHEVs concerning energy management strategies. Therefore, it is very important to estimate the effect of energy management on battery health alongside with fuel economy and other control objectives. The assessment of energy management strategies should always include an exclusive study of all the powertrain components and their critical parameters. However, in the considered literature, there is no all-inclusive study exploring the connection or real tradeoff between battery ageing, engine switching and fuel economy.

Batteries are the most dominant choice of an energy storage system in road transport. Nonetheless, batteries experience an irreversible degradation processes while in use, which affects the energy capacity and the internal resistance of the storage system [9]. Battery capacity loss is due to many factors, which include various environmental and operating conditions. High C rate operation, a low or high state of charge (SOC), high battery temperature and the excessive depth of discharge are a few critical factors contributing to battery health degradation [16–19]. Batteries represent a large part of vehicle cost in a PHEV [2]. Hence, studying the effect of energy management on battery ageing is an important criterion to consider. In [20], the authors proposed a multiobjective optimal control problem, which considered both fuel consumption and battery ageing by converting the battery pack replacement cost to an equivalent fuel cost. However, the limitation of [20] is that they have not exclusively studied the effect of energy management on battery parameters influencing battery ageing: battery temperature, battery throughput and charge/discharge C rate.

HEVs have multiple power sources, and their fuel economy and emissions can be optimized by energy management strategy. As a result, under the influence of the energy management strategy, HEVs frequently make transitions between different pre-defined modes to achieve an optimal instantaneous power distribution. However, HEVs tend to suffer from issues such as drivability, starting system reliability and excessive clutch wear during mode transition [21,22]. Unlike four-wheelers, two-wheelers are an inherently unstable vehicle. The vehicle drivability and safety are linked to each other in two-wheelers, especially during the vehicle cornering phases [23]. Thus, it becomes even more important to develop an energy management strategy considering engine switching events along with battery ageing for a hybrid two-wheeler.

None of the references considered have done a stipulated comprehensive assessment of energy management strategies considering the battery ageing effects, engine switching events along with fuel economy. All the studies in the considered references were limited to fuel economy and tailpipe emissions. The evaluation and comparison of energy management strategies are incomplete without exclusively studying the effect of energy management strategies on all the critical battery parameters influencing battery degradation. This paper emphasizes the assessment of rule-based and optimal control-based strategies by thoroughly studying their effect on battery ageing parameters, engine switching events along with fuel economy and charge sustenance.

The four controllers considered for this study are:

(1) The conventional Basic rule-based Energy Management Strategy (BEMS);
(2) The Modified rule-based Energy Management Strategy (MEMS);
(3) The conventional Equivalent fuel Consumption Minimization Strategy (ECMS);
(4) The Modified ECMS named as ECMS_LL.

Apart from the conventional BEMS and ECMS, modified rule-based MEMS and Modified ECMS-ECMS_LL are considered for studying the effect these energy management strategies on fuel optimality, engine switching frequency and battery parameters. These modified energy management strategies improve the engine's efficiency and thereby increase the fuel optimality by utilizing the electric path favourably over the conventional strategies BEMS and ECMS. However, the increased engine efficiency has a negative impact on battery parameters and this is one of the key focus of this study. The study is performed for two different drive cycles—Worldwide Motorcycle Test Cycle (WMTC) and Indian Drive Cycle (IDC) and three different vehicle configurations for the reliability of the results. Based on the respective simulation results, conclusions are drawn as to which directs a reciprocal relation between the battery parameters and the fuel economy in energy management.

The main contributions of this study are:

1. To design and develop a novel modified ECMS for a plugin hybrid concept two-wheeler, which enhances the fuel optimality further when compared to traditional ECMS.
2. Evaluation and assessment of modified ECMS with traditional ECMS and other rule-based strategies.
3. To perform an all-inclusive study and evaluation of energy management strategies considering battery parameters influencing battery ageing (battery temperature, battery Ah throughput, SOC and charge/discharge C rate), engine switching events and charge sustenance along with fuel economy.

This study helps in laying down a foundation for future improvements in energy management development and it also aids in establishing a basis for comparing energy management controllers.

The paper is organized as follows: In Section 2, the vehicle model of the full parallel two-wheeler HEV is explained. In Section 3, the HEV energy management control problem is formulated. In Section 4, the rule-based and optimal control-based controllers are designed and developed. Along with this, a novel modified ECMS (ECMS_LL) is developed and implemented. In Section 5, the simulation results of the controllers are investigated in detail for two different drive cycle and three different vehicle configurations. The results are analyzed and thereby the conclusions are discussed in Section 6 along with the future direction of the study.

2. Vehicle Model and System Configuration

Two different approaches to HEV modelling can be adopted: backward or forward-facing modelling concerning the physical causality principles [24]. The former assumes that the vehicle meets the target performance so that the vehicle speed is known and the power request is calculated using the kinematical relationships imposed by the drivetrain. Forward-facing modelling takes, as inputs, the driver commands, and simulating the physical behaviours of each component generates the vehicle performance as an output. The powertrain data for the two-wheeler hybrid considered drives this study use a simplified appropriate backwards-facing model for this purpose [24,25]. A backward model developed by sourcing the detailed technical specifications and experimental data of engine, electric machine and battery from the hybrid two-wheeler concept is considered for this study. The power sources, transmission, and control system are developed using MATLAB/Simulink/State flow environment. Since the vehicle is a concept two-wheeler and not production-ready, the actual technical details of the powertrain components are not disclosed. In turn, the normalised values are displayed wherever necessary.

2.1. System Configuration

The vehicle considered for this study is a full parallel plugin hybrid concept two-wheeler and the powertrain architecture for the vehicle is as shown in Figure 1. The system is composed of an engine, a centrifugal clutch, an electric machine, mechanical transmission and an energy storage device (high voltage battery).

Figure 1 shows the mechanical and electrical power flow between the powertrain components. The electric machine is capable of power assist and charging from the engine along with pure electric drive depending on the wheel power requirement and the battery SOC. Hard constraints corresponding to the physical limits of the powertrain components are applied to the control input. The high voltage battery is used for vehicle propulsion and the low voltage battery is used for small dc loads and in starting the engine using a starter motor.

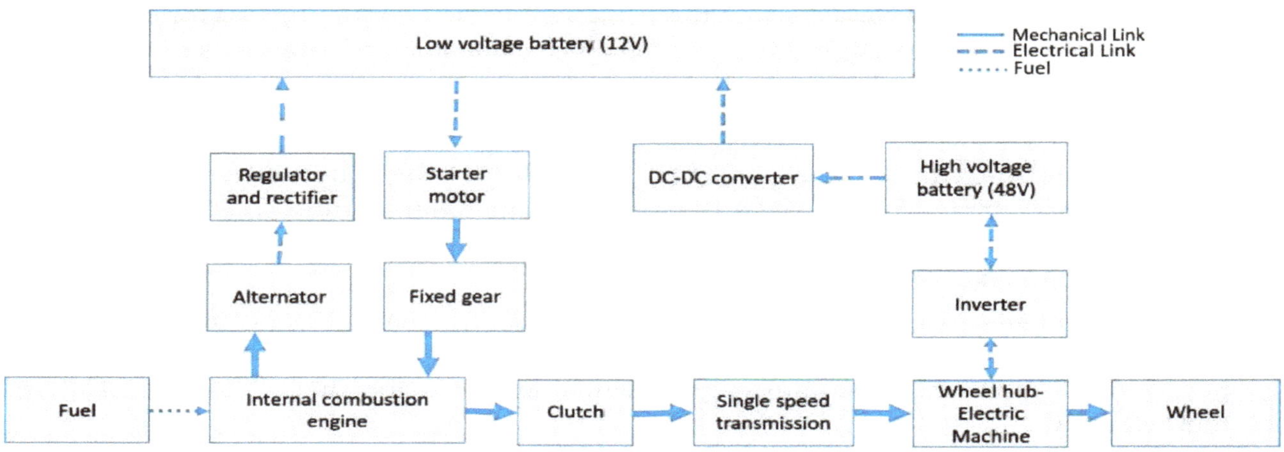

Figure 1. Vehicle powertrain architecture of the hybrid two-wheeler considered.

2.1.1. Vehicle Model

The plant model developed is a backwards-facing vehicle model. The longitudinal dynamics of the vehicle are described by [26]:

$$\frac{d\mathcal{V}(t)}{dt} = \frac{1}{m_v} \cdot \left(\frac{T_w(t)}{r_w} - \frac{1}{2} \rho_{air} C_d A \mathcal{V}^2(t) - C_r m_v g \right) \tag{1}$$

where $\mathcal{V}$ denotes the vehicle speed; T_w denotes the wheel torque; r_w denotes the wheel radius; ρ_{air} denotes the air density; C_d denotes the aerodynamic drag coefficient; A denotes the frontal area of the vehicle; C_r denotes the rolling friction coefficient; m_v denotes the nominal weight; and g denotes the gravitational acceleration. The vehicle is assumed to operate on a flat road. Rolling resistance is considered as one single coefficient. The road material is taken to be hard asphalt (urban Indian roads), the absence of liquid, nominal tyre pressure, nominal tyre temperature and tyre material of rubber.

2.1.2. Engine

The engine model developed is established using steady-state maps based on experimental data from the hybrid two-wheeler concept vehicle. The model of the gasoline engine considered specifies the fuel consumption in (g/kWh) as a function of its operating point, defined by the engine speed (rad/s) and the engine torque (Nm). The engine's fuel energy is used for charging the high and the low voltage batteries along with vehicle propulsion as shown in Figure 1.

The fuel flow rate $\dot{m}_f$ of the engine is given by a steady-state map, which is a function of engine torque T_{ice} and engine speed ω_{ice}, i.e., $\dot{m}_f = f(\omega_{ice}, T_{ice})$.

The power consumption of the engine can be described by the following equation:

$$P_{fuel} = Q_{lhv} \cdot \dot{m}_f \tag{2}$$

where Q_{lhv} is the lower heating value of the fuel.

2.1.3. Electric Machine

The electric machine model is established using static maps derived from the experimental data. The efficiency of the electric machine η_{em} is given by a steady-state map, which is a function of both torque T_{em} and speed ω_{em}. Therefore, the power request of the electric machine is given by:

$$P_{em} = T_{em} \cdot \omega_{em} \cdot \eta_{em}\,(T_{em},\,\omega_{em}) \tag{3}$$

2.1.4. Transmission

The transmission considered here is the single speed with constant efficiency of '0.90'. The clutch used here is centrifugal clutch and is modelled as a simple switch. Regenerative braking is not considered in this study as it can harm the stability of two-wheelers.

2.1.5. Battery Model

The battery is a non-linear system. The battery changes its properties based on many external and internal factors, such as ambient temperature, internal battery temperature, state of charge operation and the depth of discharge [9]. Electrochemical models for a Li-Ion battery as proposed by [27–29] are suitable for battery-level simulation. However, these models are not suitable for optimal control design because they involve a discontinuous mathematical function. Thus, a simple first-order equivalent circuit model is used for this study. Figure 2 shows the typical first-order equivalent circuit model of the battery. The battery can be represented as DC source V_{OC}, RC network (Cp and Rp) and internal resistance Ri.

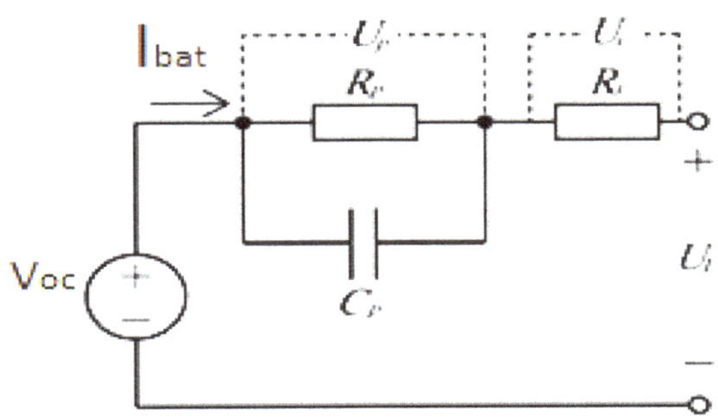

Figure 2. Equivalent circuit model of the battery.

The battery internal resistance is a very important factor to calculate the SOC [30]. The battery internal resistance changes with SOC, battery cell temperature and battery ageing. In the battery model developed, the effect of SOC, C rate and battery cell temperature on internal cell resistance are considered [31].

The equation for calculating the battery current is as follows [32].

$$I_{bat} = \frac{V_{OC} - \sqrt{V_{OC}^2 - 4 \cdot R_{eq} \cdot P_{bat}}}{2 \cdot R_{eq}} \tag{4}$$

where I_{bat} is the battery current, V_{OC} is the open-circuit voltage, R_{eq} is the equivalent battery resistance, and P_{bat} is the power in and out of the battery.

The Joule heating and the entropy effects are the major contributors for heat generation within the battery pack. For automobile applications, where C-rate of operation is high due to varying acceleration from the rider, the Joule heating effect dominates the entropy effect [33,34]. Hence, for the temperature prediction model, only heat generation due to the Joule heating effect is considered. The heat generation through the Joule heating effect is calculated using the relation $Q = I_{bat}{}^2 R_{eq}$, where Q is the heat generated, I_{bat} is the current and R_{eq} is the internal resistance. The battery is assumed to be a homogeneous body with uniform temperature distribution, the temperature difference between the core and the surface of the battery is not considered. The effect of ageing and state of health of the battery is not considered in this model.

A 3D lookup table is created based on experiments results [33], which computes the internal resistance of the battery at every time instant, based on instantaneous SOC, battery temperature, C-rate and the direction of the current flow. The resultant resistance is used in the model for the battery temperature rise calculation and as well for SOC calculation. The rise in the temperature calculated by equating the heat energy in joules against heat capacity equation: $Q = MC\Delta T$, where Q is the heat energy in joules; M is mass of the battery in kg; C is the specific heat capacity of the battery (J/kg °C); ΔT temperature rise in °C [33].

The battery considered in the study is air-cooled. A cooling study has been done on the same type of cells and has been found that cooling constant plays a major role while predicting temperature rise in cells during the vehicle run condition [33]. The average cooling rate °C/hour is calculated from the experiment results [31] conducted on the same cells and implemented in the battery Simulink model.

The powertrain model consisting of power train components are modelled using the experimental data from the hybrid concept two-wheeler. The engine maps, electric machine maps and battery model are based on the experimental data of the concept vehicle. Detailed technical specification of the powertrain components is restricted for publishing as the vehicle is a concept vehicle and is not production ready. However, the normalized data of the powertrain components are shared wherever necessary.

3. HEV Energy Management Control Formulation

For an HEV following a specific driving cycle, the system state equation can be described as [35]:

$$\dot{x}(t) = f(x, u, t) = -\frac{1}{Q_{bat}} I_{bat}(x, u, t) \tag{5}$$

where I_{bat} is denoted as battery current and Q_{bat} is denoted as battery charge capacity.

The control variable $u(t)$ represents the control output, the power split between the engine fuel and battery. The optimal control problem to be solved by energy management is to minimize the following performance index [35]:

$$J(u) = \int_0^{t_f} L(u, t)dt \tag{6}$$

The main optimization objective is to minimize the fuel consumption and therefore the instantaneous cost here is the fuel flow rate or the power equivalent to it [35]:

$$L(u, t) = P_{fuel}(u, t) = Q_{lhv}\dot{m}_f(u, t) \tag{7}$$

where Q_{lhv} is the fuel energy density and is a constant.

This optimization problem is extended for a time interval $\left[0, t_f\right]$ constrained to various local and global constraints, which include:

Local constraints [35]:

$$u_{min}(t) \le u(t) \le u_{max}(t) \quad \forall t \in \left[0, t_f\right] \tag{8}$$

$$-1 \le u(t) \le +1$$

where $u(t)$ is the power split that ranges from '−1' to '+1'. '−1' means full regeneration by engine and vehicle traction power completely supported by engine, and '+1' means full motor mode, the vehicle traction power is fully supported by the motor.

$$x_{min} \le x(t) \le x_{\max} \quad \forall t \in [0, t_f] \tag{9}$$

where $x(t)$ is the SOC, which has a minimum and maximum limit of charge-sustaining mode.

Global constraints [35]:

$$x(t_f) = x_f \tag{10}$$

x_f is the final SOC target set for charge sustenance.

Initial conditions:

$$x(0) = x_0 \tag{11}$$

x_0 is the initial SOC value set for the test.

$$\dot{x}(t) = f(x, u, t) \tag{12}$$

4. Energy Management Strategy Development

4.1. Energy Management Overview

A supervisory energy management controller sets the power split between the electric machine and the internal combustion engine for all the input conditions.

Figure 3 shows the data flow between the energy management controller and the plant model, where P is the power, ω is the rotational speed; τ is the torque; V is the vehicle velocity and B is the brake information. The system constraints considered for the energy management controller are derived from the key powertrain components (engine—torque, power and speed; battery—power and SOC; electric machine—torque, power and speed). The power-split of (P_{ice} and P_{em}) is the key control output.

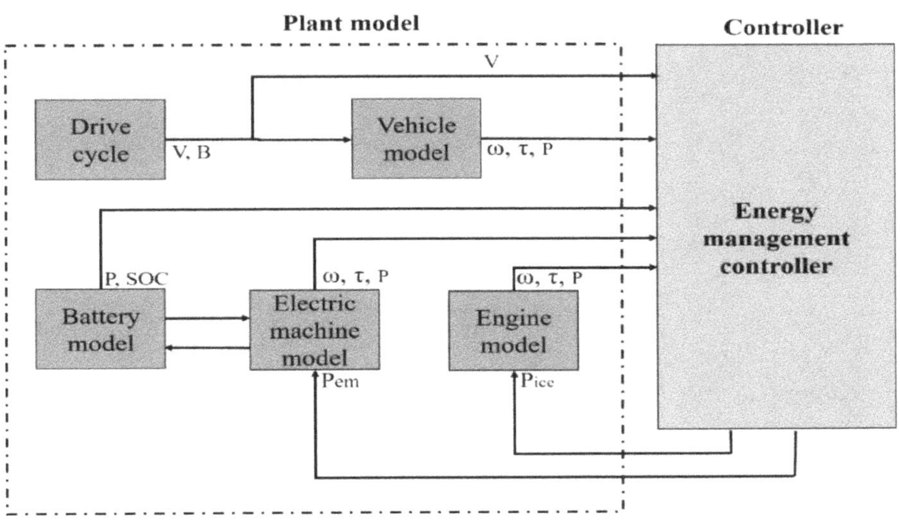

Figure 3. Control data flow block.

4.2. Drive Cycles

For this study, two drive cycles are considered; Worldwide-Motorcycle Test Cycle (WMTC) [36] and an Indian Drive Cycle (IDC) [36], for testing and evaluating the consistency of the energy management control operation.

The WMTC is used as one of the reference drive cycles. This drive cycle is used for the small engine vehicle level tests in many countries (USA, Europe, China, and Japan). It replicates an urban drive cycle, with frequent acceleration, deceleration and start-stop conditions. The second drive cycle considered is IDC, which is a typical Indian urban drive cycle used for testing. According to UN-ECE regulation 168, for L category vehicle (with engine capacity < 150 cm^3) with off-board charge capability, the legislative fuel economy test that considers WMTC is done in the following steps. The CO_2 is calculated as follows.

Condition A: Test with the fully charged electrical system. Condition B: Test with the fully discharged electrical system. Combining conditions A and B results in the following formula:

$$CO_2(\text{g/km}) = \frac{(EVRange \times Condition\ A\ CO_2) + (D_{AV} \times Condition\ B\ CO_2)}{(EVRange + D_{AV})} \tag{13}$$

where D_{AV} is the distance between recharges: 4 km for L category vehicle and EVRange is the distance covered by the vehicle in electric-only mode from full charge to $SOC_{Lowlimit} = 20\%$ SOC.

The condition A test was simulated using the two-wheeler plant model, the simulation result shows the vehicle with fully charged battery takes around 4600 s (24 km), which is the EVRange and D_{AV}: the distance between recharges considered is 4 km (from the UN-ECE regulation). The total drive time considered for test cases is therefore 28 km, which corresponds to 5200 s of the drive cycle. For this study, the WMTC and IDC drive cycles are therefore extended to 1.4 h. (5200 s) with the repetition of the cycle. The velocity and wheel power data of WMTC and IDC for a single test cycle of 600 s and 120 s are shown in Figure 4a,b, respectively.

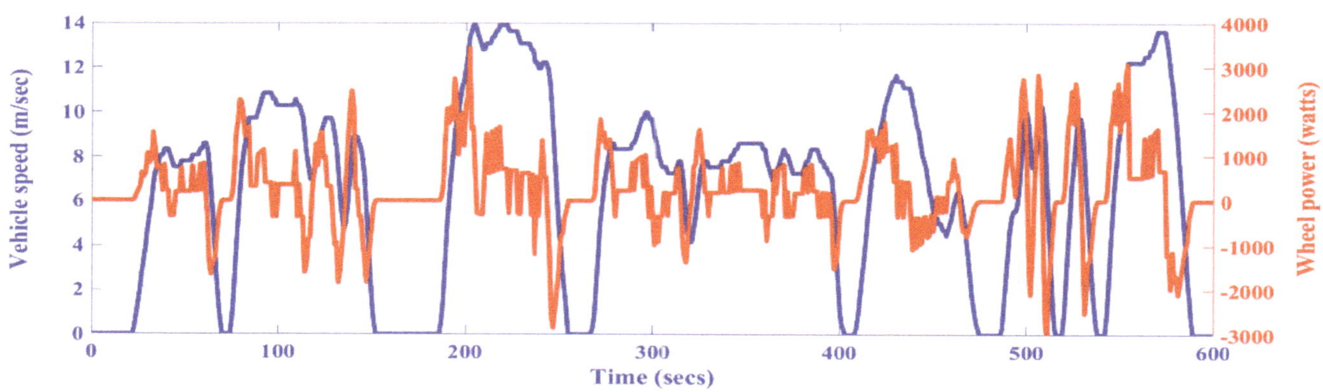

(a) WMTC test cycle

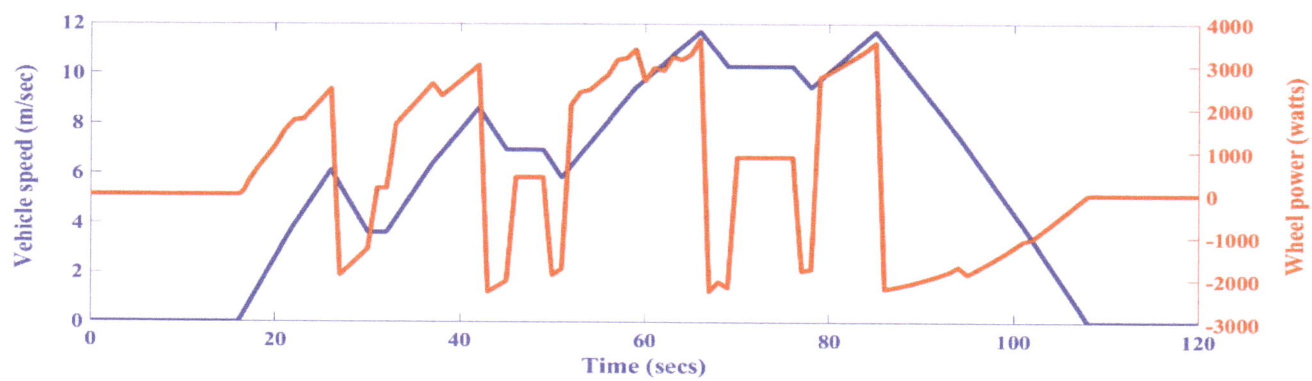

(b) IDC test cycle

Figure 4. Vehicle speed and wheel power for (**a**) Worldwide Motorcycle Test Cycle (WMTC) and (**b**) Indian Drive Cycle (IDC).

4.3. Engine Switching Event

Engine switching is an important criterion to analyze when it comes to two-wheelers. During the engine cranking phase, once the ignition key is turned to initiate an engine start event, electric power from the low voltage battery is applied to the starter motor terminals to activate the starter solenoid and energize the DC starter motor engage the gear to the flywheel and thereby the engine is started. Every time the starter is cycled, it is subjected to wear and tear resulting from metal-on-metal contact on the gears and heat build-up within the electronics. Extreme environmental conditions and extreme usage of starter motor can lead to rapid degradation of life of starting system. According to warranty data gathered from a conventional two-wheeler of similar vehicle specifications considered, starter motors are designed for 50,000 cycles (estimated to correspond 3 years of use). However, the life of the starter motor also depends on the ambient conditions, engine condition (crank time required), and the time between starts (heat build-up) [37]. For this study, it is assumed that the engine starts at the first crank and hence the number of engine start events is equal to the number of starter system cycles.

Every time the engine switches on, it consumes a certain amount of fuel and electric power. From the experimental data, the fuel required for a warm engine-start, including the fuel required to chargeback the low voltage battery, is considered. The number of engine switching events is taken as a metric for comparing the controllers and can be directly related to the starting system life.

4.4. Energy Management Objectives

The problem considered in this paper is the energy management for a full parallel plugin hybrid two-wheeler. The essence of HEV control is the instantaneous management of the power flow from the ICE and the EM. The HEV control strategy aims to minimize the vehicle fuel consumption whilst maintaining the SOC, and meeting the driver demand and respecting the component limitations. The optimization objective considered in this work is to minimize fuel consumption during a trip, and the constraints are:

(1) Charge sustainability: the battery SOC should remain at the defined SOC target reference at the end of the trip.

(2) The satisfaction of driver demand: at each instant, the total torque output of the powertrain should be equal to the driver's demand.

(3) Actuator limitations: at each instant, the output of each powertrain component (engine and electric machine) is within its maximum torque/power rating; similarly, the total battery power is within acceptable limits in both charge and discharge operation.

For the consistency in operation of energy management controllers, different drive cycles and vehicle configurations are considered. Thus, along with different drive cycles, WMTC and IDC, three different system configuration are considered, as shown in Table 1. Table 1 shows the three different vehicle configuration taken into consideration for this study. The engine sizing is taken based on the availability of the engine models available in the production shelf. The three values of uniformly distributed charge C rate values from a minimum limit of 1C to a maximum limit of 5C is considered for the study.

Table 1. Vehicle configuration table.

Vehicle Configuration	Engine Sizing	Battery Charge C Rate
Veh 1	80%	5 C
Veh 2	100%	3 C
Veh 3	70%	1 C

4.5. Basic Rule-Based Energy Management Strategy (BEMS)

The rule-based controllers (BEMS and MEMS) are static controllers and the operating point of the components; engine and electric machine are chosen using rule tables as shown in Table 2 to meet the requirements of the driver and other components (electrical loads and battery) based on prior experience and testing as it is typical for rule-based controllers [6]. This table is formulated based on the rule-based controller used in the hybrid concept two-wheeler. A basic rule-based control strategy (BEMS) implements heuristic knowledge to develop a set of event-triggered rules. This strategy is implemented using Matlab State Flow. The five states are described as launch, electric-only, engine-only, combined (electric and engine) and braking states.

Table 2. Basic rules used in rule-based strategy existing in the hybrid two-wheeler considered.

Mode	State	Battery	Vehicle Conditions	Eng	Mot	Gen	Descriptions
Launch mode	Vehicle launch	SOC > SOCLowlimit	V > 0, DT < Mmt	Off	On	Off	Vehicle start condition with motor alone providing the drive torque
	Vehicle launch	SOC > SOCLowlimit	V > 0, DT > Mmt	On	On	Off	Vehicle start condition with motor and ICE providing the drive torque
	Vehicle launch	SOC < SOCLowlimit	V > 0	On	Off	Off	Vehicle start condition with engine alone providing the drive torque
Pure EV-mode	EV state	SOC > SOCLowlimit	Mmn > V > 0 and DT < Mmt	Off	On	Off	Vehicle run condition with motor providing the drive torque
ICE-mode	Charging state	SOC < SOCLowlimit	0 < V < Mmn, DT < Met	On	Off	On	Drive torque provided by ICE alone and battery charged by engine
	ICE only state	SOC < SOCLowlimit	V > Mmn or DT > Met	On	Off	Off	Drive torque provided by ICE alone
Combined mode	High power demand	SOC > SOCLowlimit	V < Mmn, DT > Mmt	On	On	Off	Drive torque provided by both motor and ICE
Regeneration mode	Braking state	SOC < SOCHighlimit	DT < 0, V > Mgn	Off	Off	Off	No regenerative braking
	Braking state	SOC < SOCHighlimit	DT < 0, V < Mgn	Off	Off	On	Regenerative braking
	Braking state	SOC > SOCHighlimit	DT < 0	Off	Off	Off	No regenerative braking

where V = vehicle velocity; M_{mn} = maximum speed limit of the motor/generator; M_{mt} = maximum motor torque limit; M_{et} = maximum engine torque limit; M_{en} = maximum engine speed limit; M_{cn} = minimum clutch engage speed; D_T = demand torque; $SOC_{Lowlimit}$ = lower SOC limit (20% SOC) and $SOC_{Highlimit}$ = higher limit SOC limit (80% SOC).

4.6. Modified Rule-Based Energy Management Strategy (MEMS)

The second control strategy is MEMS, which uses Load-Levelling Control (LLC). In LLC, the Internal Combustion Engine (ICE) is operated as close as possible to a predetermined value known as the optimal operating point for every instant in time during the vehicle operation [4]. The engine Optimum Operating Line (OOL) is the curve that connects the minimum BSFC for each given power. In actual practice, it is difficult to operate on the optimum operating line as the operating points are based on the wheel torque demand. Therefore, the ICE is operated as much as possible in the optimum operating region near the optimum operating line. The resulting power difference between the ICE and vehicle demand is either used or contributed by an electric machine. The maximum contribution of the electric machine is limited by the state of charge of the battery and its torque and power limitations.

4.7. Optimal Control Solution

Equation (7), which is subjected to Equations (8)–(12), is reduced to the instantaneous minimization problem of the Hamiltonian function H [14]. The global optimal energy management control problem for an HEV is reduced to the instantaneous minimization problem of Hamilton function by Pontryagin's minimum principle, which is explained in detail in the literature [14].

The Hamilton is written as [14]:

$$H(x,u,t) = P_{fuel}(u,t) - \lambda(t)\cdot\frac{1}{Q_{bat}}I_{bat}(x,u,t) \tag{14}$$

The term $\lambda(t)$ can be decoupled as [35]:

$$\lambda(t) = -E_{bat}s(t) = -V_{oc,max}Q_{bat}s(t) \tag{15}$$

where E_{bat} is the battery total energy and $s(t)$ is a dimensionless term. The Hamiltonian function can be further interpreted as equivalent power and the expression is as follows [35]:

$$H(x,u,t) = P_{fuel}(u,t) + s(t)\cdot P_{bat}(x,u,t) \tag{16}$$

where $s(t)$ is called the equivalence factor, which is the weighting factor that transforms the battery power into equivalent fuel power. In charge sustaining, HEV is the ultimate energy is from the fuel. The battery charge and discharge are translated into equivalent fuel consumption. Equation (16) was derived from intuitive considerations on energy balance by Paganelli [26], which resulted in the Equivalent Consumption Minimization Strategy.

4.7.1. Equivalent Fuel-Consumption Minimization Strategy (ECMS)

The third strategy is based on the Equivalent fuel Consumption Minimization Strategy (ECMS), which is an instantaneous approach derived from Pontryagin's Minimum Principle (PMP) [14,15,26]. It is well known that Pontryagin's minimum principle can be applied to the HEV energy management problem to derive an optimal solution in the form of ECMS [38]. While heuristic control strategies such as BEMS and MEMS do not seek to optimize energy consumption or emissions, in contrast, the ECMS explicitly formulates a cost function for the equivalent fuel consumption to be optimized [12,13,35]. In a hybrid vehicle, a global optimization strategy usually sees the minimization of fuel consumption, or emissions, as the main goal. Paganelli [39] used an instantaneous optimization strategy to simplify the global optimization problem into a local optimization problem using minimum instantaneous equivalent fuel for each time cycle as an optimization objective. The instantaneous equivalent fuel consumption is equal to the engine's instantaneous fuel consumption and the electric motor power that can be converted to fuel consumption [39]. ECMS was first proposed by Paganelli and applied to hybrid vehicle energy management strategies; to minimize the instantaneous fuel consumption, the concept of the equivalent factor was introduced [12].

ECMS sees the battery as an energy buffer, and all the energy consumption is equated to the engine's fuel consumption. The battery can be viewed as a reversible energy conversion tank. For a particular parallel hybrid system, there can be two circumstances, which are as follows:

If the power of the battery is positive, as in discharging state, the battery is charged using extra fuel of the engine in future, as shown in Figure 5a.

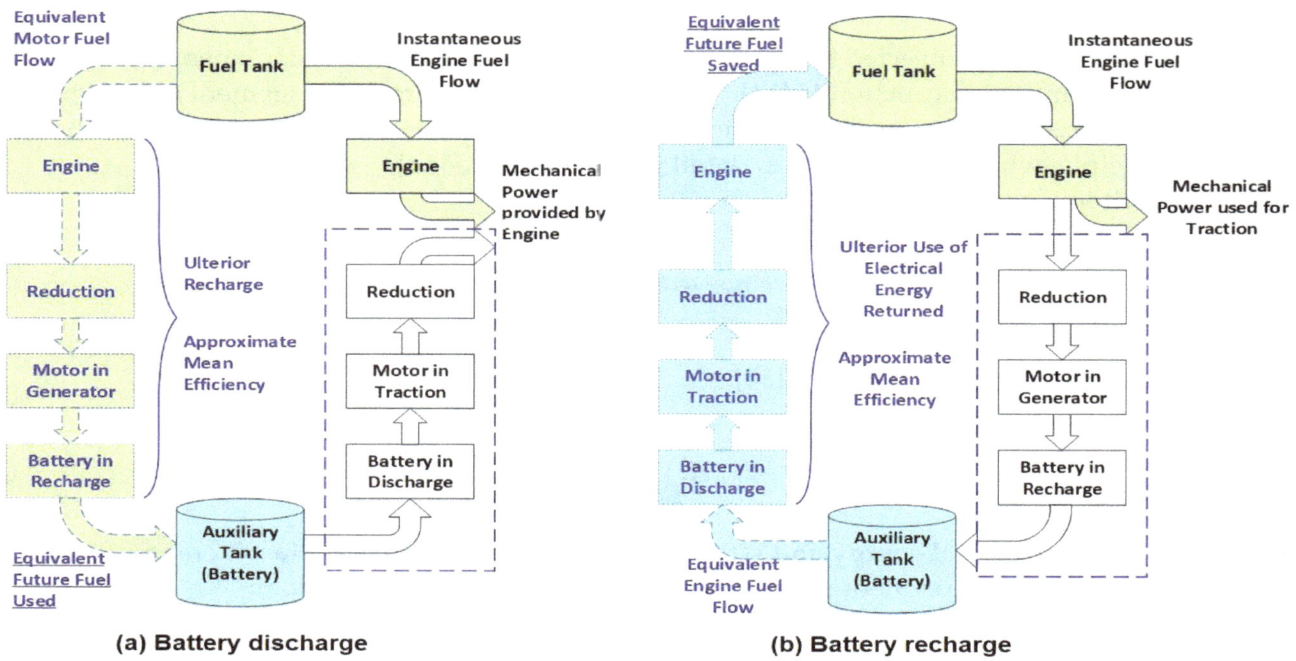

Figure 5. Energy flow diagram of a PHEV during (**a**) Battery charging and (**b**) Battery recharging through engine [26].

If the power of the battery is negative, as in charging state, the battery energy is being stored, which can be utilized to reduce the load on the engine, thereby saving part of the fuel as shown in Figure 5b.

In both cases, the electric power is equated to fuel consumption in the future and the actual fuel consumption by the engine. The instantaneous fuel consumption can be calculated as follows [14]:

$$\dot{m}_{eqv}^{fuel}(t) = \dot{m}_{ICE}^{fuel}(t) + \dot{m}_{BAT}^{fuel}(t) = \dot{m}_{ICE}^{fuel}(t) + \frac{s}{Q_{lhv}} P_{BAT}(t) \tag{17}$$

where $\dot{m}_{eqv}$ instantaneous equivalent fuel consumption $\dot{m}_{ICE}$ is the instantaneous fuel consumption from the engine, $\dot{m}_{BAT}$ is the instantaneous equivalent fuel consumption from the battery power (both in charging and discharging), s is the equivalence factor, which represents the conversion of electric power into fuel consumption and P_{BAT} is the battery power, Q_{lhv} is the low heating value of the fuel.

The cost function for equivalent fuel minimization is:

$$J(x_t, u_t) = \dot{m}_{eqv}^{fuel}(t) = \left(\dot{m}_{ICE}^{fuel}(t) + \dot{m}_{BAT}^{fuel}(t)\right) = \left(\dot{m}_{ICE}^{fuel}(t) + \frac{s}{Q_{lhv}} P_{BAT}(t)\right) \rightarrow min \tag{18}$$

where control input $u_t \in [-1.0 : 0.1 : 1.0]$.

Subject to the following constraints:

$$\begin{aligned} &P_{req}(t) = P_{ice}(t) + P_{em}(t) \\ &SOC_{min} < SOC(t) < SOC_{max} \ \forall\ t \\ &0 \le P_{ice}(t) \le P_{ice,max}(t) \\ &P_{em,min}(t) \le P_{em}(t) \le P_{em,max}(t) \\ &\omega_{ice,min}(t) \le \omega_{ice}(t) \le \omega_{ice,max}(t) \\ &\omega_{em,min}(t) \le \omega_{em}(t) \le \omega_{em,max}(t) \end{aligned} \tag{19}$$

ECMS strongly depends on the definition of the equivalence factor [39]. In the conventional ECMS formulation, the equivalence factor represents the chain of efficiencies through which fuel is transformed into electrical power and vice-versa, and it changes for each operating condition of the powertrain. The equivalence factor is tuned for both charging (s_{chg}) and discharging (s_{dchg}) in charge-sustaining mode for a particular test cycle so that it minimizes the absolute fuel consumption for a given test cycle. The overall fuel consumption can be considered a function of the equivalence factors and a systematic optimization is used to find the equivalence factors that minimize the overall fuel consumption constrained to the *SOC* sustainability, i.e., final *SOC* equals pre-defined SOC_{ref}. The value of SOC_{ref} is taken as 0.30 in this study.

Along with the equivalence factor, to impose the *SOC* constraint, an *SOC* correction factor is used to shift the optimal power split up or down according to the difference between the actual state-of-charge and the target state-of-charge at every time instant. A non-linear "penalty function", p(*SOC*), as depicted in Figure 6 is used to represent the desired characteristics [39]. This function is used to sustain the charge about the *SOC* set point-*SOCref*. This is required avoiding the under or overcharging of the battery regardless of the vehicle demand.

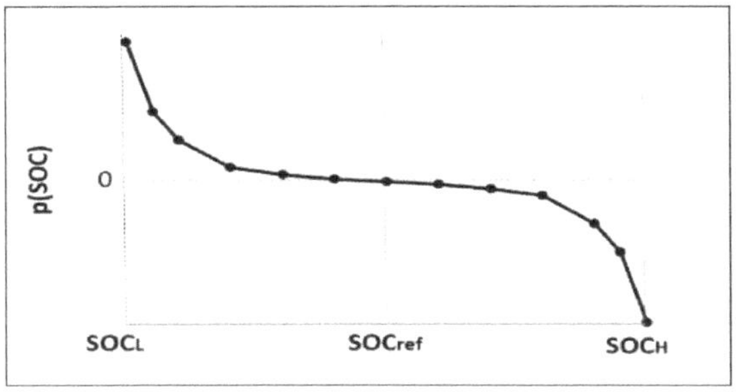

Figure 6. An example of state of charge (*SOC*) penalty weight used in ECMS [39].

The ECMS is only functional in the charge-sustaining mode where both power from the engine and electric machine is optimally split, to provide optimal fuel benefit with [26] minimum *SOC* deviation from the target. The equivalence factors s_{chg} and s_{dchg} can be calculated as follows [14]:

$$S_{chg} = \alpha_1\big(1 - \beta_1\big(SOC - SOC_{ref}\big)\big) \tag{20}$$

$$S_{dchg} = \alpha_2\big(1 + \beta_2\big(SOC - SOC_{ref}\big)\big) \tag{21}$$

$$\alpha_1 > 0, \alpha_2 > 0$$

$$\beta_1 \geq 0,\ \beta_2 \geq 0$$

$$SOC_{ref} = \left(\frac{SOC_H + SOC_L}{2}\right) \tag{22}$$

where α_1 and α_2 are the weighting factors (constant), β_1 and β_2 are penalty factors derived from the battery characteristics, SOC_H and SOC_L are higher and lower *SOC* limits and are calibrated based on the requirement, and SOC_{ref} is the target *SOC*, as shown in Figure 6.

The battery fuel component can be revised as follows [14]:

$$\dot{m}_{BAT}^{fuel}(t) = \frac{S_{chg}}{Q_{lhv}} P_{BAT}(t),\ for\ I_{bat} \leq 0 \tag{23}$$

$$\dot{m}_{BAT}^{fuel}(t) = \frac{S_{dchg}}{Q_{lhv}} P_{BAT}(t),\ for\ I_{bat} > 0 \tag{24}$$

4.7.2. Proposed Equivalent Fuel Consumption Minimization Strategy (ECMS_LL)

The fourth controller, a modified conventional ECMS, is developed in this work and considered for the comparison. In the ECMS_LL, a new penalty function is introduced to the conventional ECMS control in the instantaneous cost function calculation. The original formulation of ECMS derived from impressive engineering intuition and was proved to work well, even without formal proof of optimality [32]. The derivation from the minimum principle includes a method, proposed and implemented in the earlier section, which introduces a penalty weight p(*SOC*) based on the deviation of *SOC* from the target *SOC*. This correction function adjusts the value of the equivalence factor when the *SOC* is close to its maximum or minimum acceptable levels to avoid the over-charging or over-discharging of the battery and thereby helps in the charge-sustaining approach. In fact, in the original ECMS formulation, the equivalence factor represents the chain of efficiencies through which fuel is transformed into electrical power and vice-versa, and it changes for each operating condition of the powertrain. The equivalence factor is considered an optimization variable that acts as a single tuning parameter for the required charge-sustained operation. This method improves the robustness of the strategy; however, it does not adopt the equivalence factor for exploiting engine to operate in the engine optimum region in most of its operational time throughout the drive cycle. The additional penalty function added to the cost function is the cost of engine optimum operating point deviation.

The penalty cost for the deviation from engine optimum operating point improves the optimality and increases the robustness of the strategy for a better fuel economy compared to the conventional ECMS. However, the inclusion of the new penalty factor does affect the equivalence factor tuned for conventional ECMS, which is tested and analyzed further.

The proposed modified ECMS attempts its maximum possibility to bring the engine operating points towards the engine optimum operating region by deriving a penalty function and implementing it into the cost function evaluation. The penalty function depends on engine operating points and its deviation from the engine optimum operating line at every time instant. The addition of this penalty function enhances the engine operating towards the engine optimum operating line, thereby improving the mean engine efficiency for the particular drive cycle considered.

$$J(x_t, u_t) = \dot{m}_{eqv}^{fuel}(t) = \left(\dot{m}_{ICE}^{fuel}(t) + \gamma \frac{S}{Q_{lhv}} P_{BAT}(t) + \beta \left(\frac{ICE_{OP}^{fueleff}}{ICE_{OOP}^{fueleff}}(t) \right) \right) \rightarrow min \tag{25}$$

where $\beta\left(\frac{ICE_{OP}^{fueleff}}{ICE_{OOP}^{fueleff}}(t)\right)$ is the new penalty cost function based on optimum operating point deviation. $ICE_{OP}^{fueleff}$ is the fuel efficiency of the engine operating point at that time instant, and $ICE_{OOP}^{fueleff}$ is the fuel efficiency of engine optimum operating point at that time instant. The addition of $\beta\left(\frac{ICE_{OP}^{fueleff}}{ICE_{OOP}^{fueleff}}(t)\right)$ affects the equivalence factor of conventional ECMS for the charge-sustained requirement and therefore a multiplier term γ is introduced in the function for the electric path. The value of the gain β is based on systematic testing for a particular drive cycle for charge-sustained *SOC* values. The main objective of the charge-sustained approach is to keep the *SOC* to the required reference value at the end of the trip. A range of β values is used and varied until the fuel benefit saturates under charge-sustained operation. For different β, the change in the equivalence factor 'γs' for the charge-sustained operation SOC_{ref} is calculated. Figure 7 shows the relationship between β and γs for WMTC and IDC for three different vehicle configurations considered.

Figure 7 shows the characteristic plot of the equivalence factor and the gain β of the new penalty function added to cost function for the minimization of fuel consumption. The characteristic plot is similar and consistent for different drive cycles and system configurations considered. By adding this penalty function, the mean engine efficiency for a particular drive cycle and system configuration can increase compared to the conventional ECMS. However, the percentage increase in the engine's efficiency depends on the vehicle torque requirement (drive cycle dependent), different component sizing and charge C rate limits considered while formulating the control law.

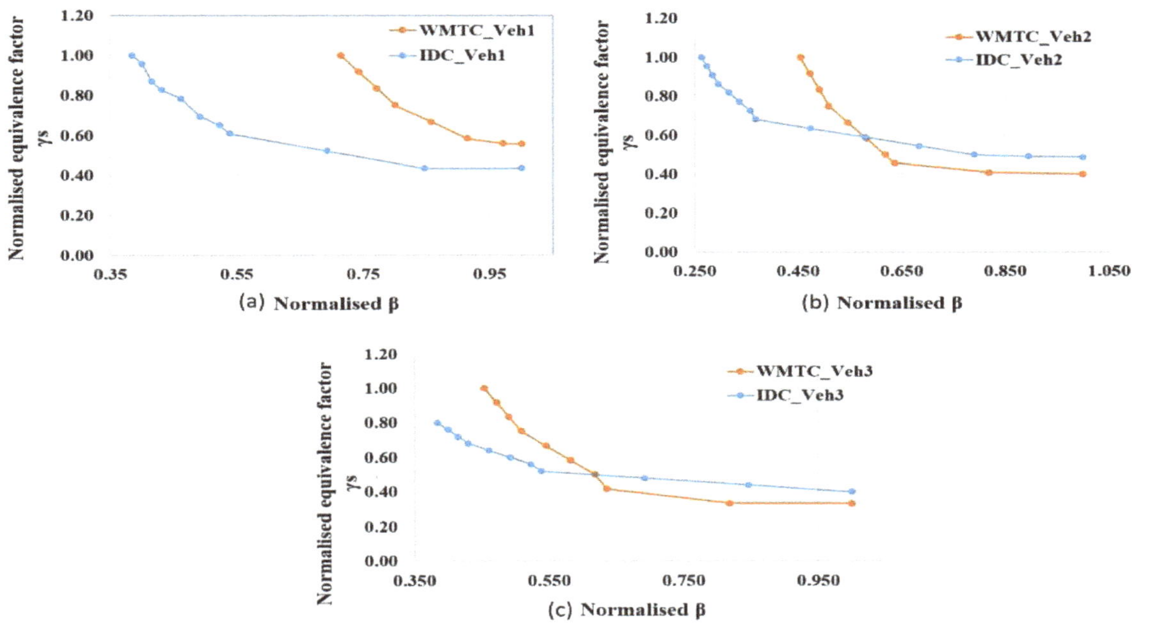

Figure 7. Relationship between the new equivalence factor 'γs' and gain 'β' for 3 different system configurations: (**a**) Veh1, (**b**) Veh2 and (**c**) Veh3 tested for WMTC and IDC test cycles

5. Simulation Results and Discussion

An all-inclusive evaluation of rule-based and optimal based energy management strategies concerning battery ageing, engine switching, fuel economy along with charge sustenance is accomplished. To further check the reliability and consistency of the results, two different drive cycles (WMTC and IDC) and three different vehicle configuration (Veh 1, Veh 2 and Veh 3) are studied. All the cycles replicated to fulfil the distance of 28 km, which is defined as one-day driving. The simulation results of the optimization-based controllers (ECMS and ECMS_LL) are presented and compared with rule-based controllers (BEMS and MEMS).

5.1. Engine Operating Points and Fuel Economy

Figures 8 and 9 show the distribution of engine operating points for Veh 1 with BEMS, MEMS, ECMS and ECMS_LL controllers for the WMTC and IDC test cycles, respectively. From the figures, it is clear that irrespective of the drive cycles, the engine operating points clearly show that the engine efficiency is highest with ECMS_LL followed by ECMS, MEMS and BEMS. A similar trend is seen with Veh 2 and Veh 3 configurations. This is because the cost function calculated with optimal control ECMS and ECMS_LL is designed to minimize the fuel consumption at every time instant by choosing the best optimum power split. As discussed in the earlier session, ECMS_LL is designed by adding an extra penalty (based on the deviation of engine operating point from the engine optimum line) function to the cost function of ECMS, which improves the engine efficiency further and thereby fuel economy. When it comes to rule-based strategies and optimal control-based strategies, the latter is the clear winner when it comes to fuel economy. The optimal control strategies evaluate the best power split by prioritizing the ICE operating points near to OOL. This is done by using the resultant power difference between the ICE and vehicle demand being used or contributed by an electric machine and battery. This gives the optimal control strategies an advantage of better fuel economy, over the rule-based strategies. The rigid rules of the rule-based controllers constrain the possibility of the use of load levelling in certain conditions.

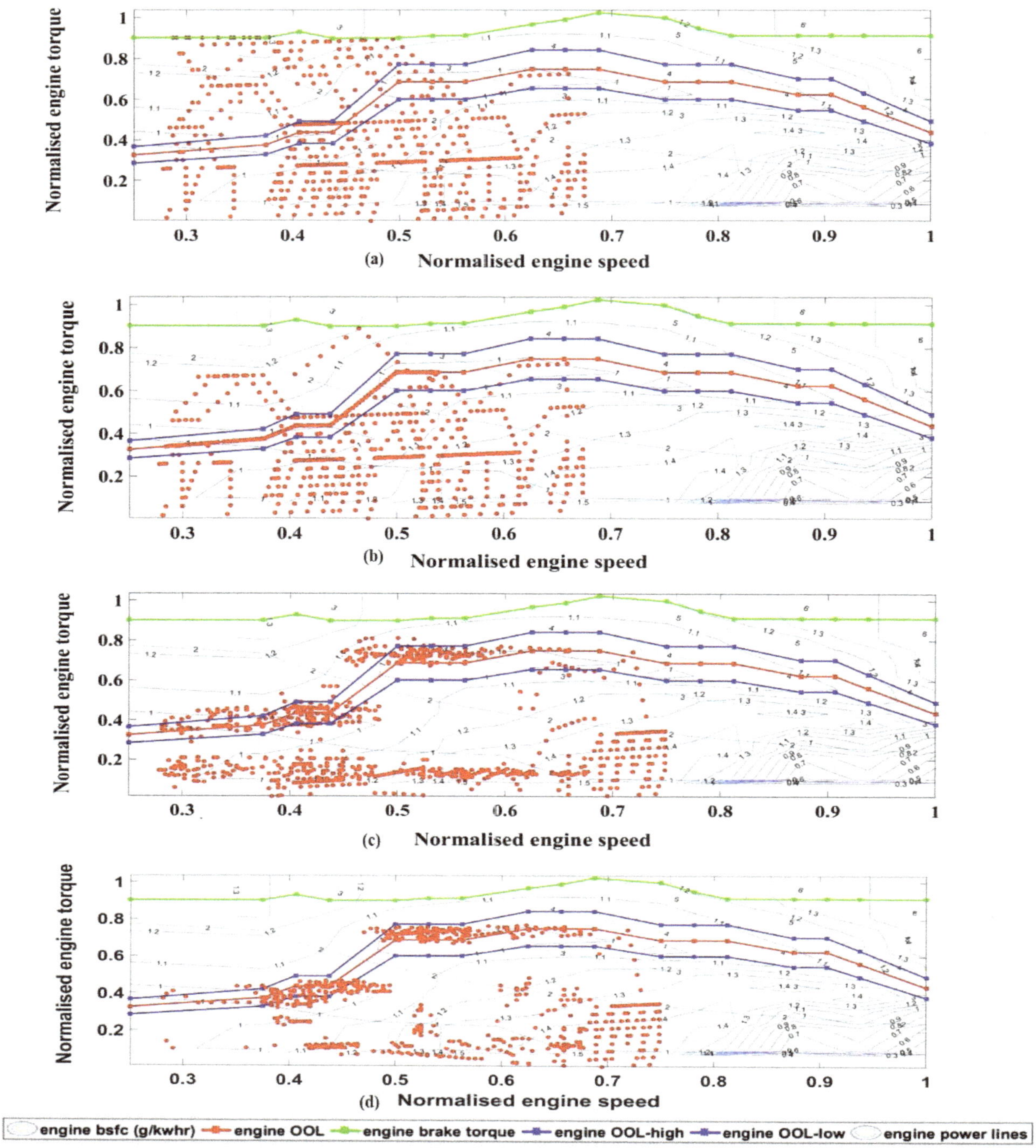

Figure 8. Veh1 engine operating points for different controllers tested for the WMTC test cycle: (**a**) (BEMS), (**b**) MEMS, (**c**) ECMS and (**d**) ECMS_LL.

When implemented with optimal control, the ECMS_LL tries to choose the operating point that minimizes the total equivalent consumption by placing the engine operating point in and around the region of engine OOL. However, the ECMS has to select the engine operating point that has the minimum fuel consumption and yet satisfies the best SOC correction to display the necessary charge sustainability. Similar trends are observed with other vehicle configurations for WMTC and IDC drive cycles.

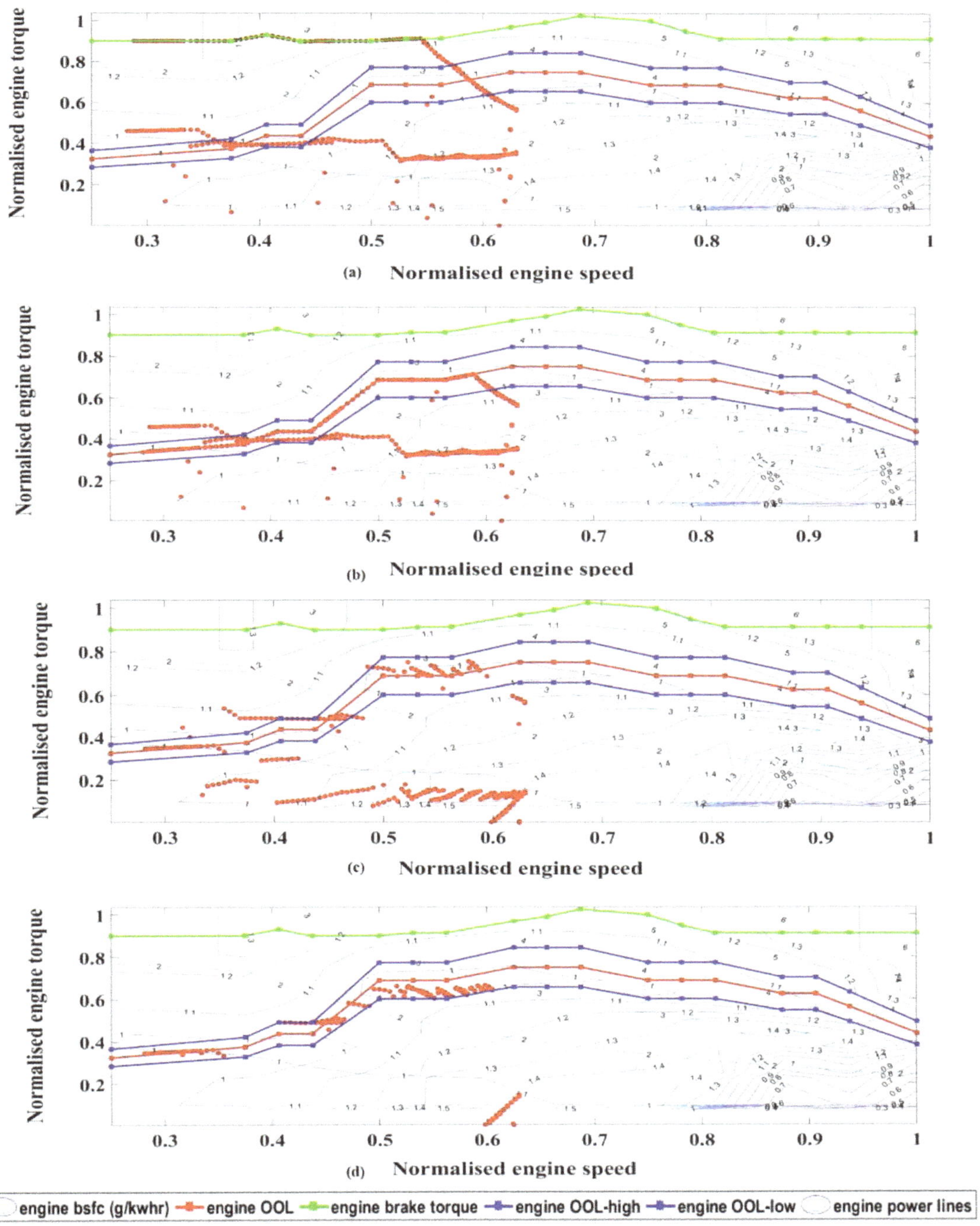

Figure 9. Veh1 engine operating points for different controllers tested for the IDC test cycle: (**a**) BEMS, (**b**) MEMS, (**c**) ECMS and (**d**) ECMS_LL.

Table 3 shows the absolute fuel consumption (FC) and percentage FC reduction of the considered controllers. The fuel consumption improvement of optimal control ECMS and ECMS_LL over BEMS and MEMS is quite significant because of the engine operating in the best efficient regions. It shows consistent positive results with both drive cycle for all the three-vehicle configurations considered. A maximum fuel benefit of 21.18% is seen with ECMS_LL controller for vehicle configuration Veh 1 when tested for IDC drive cycle. A fuel benefit of 11.36% is seen with ECMS_LL controller for vehicle configuration Veh 1 when tested for WMTC drive cycle. The percentage of fuel benefit varies with drive cycle and vehicle configuration and the same is seen in Table 3. This is because of the engine sizing considered, charge C rate limits, wheel torque requirement from the drive cycles and other system constraints. However, they all show similar trends in fuel consumption reduction, with ECMS_LL displaying the highest benefit. The engine is made to operate near the optimum operating zone in most of the times by utilizing the electric path wherever applicable. Thus, here, the battery and electric machine are operated conveniently for increasing the engine's efficiency.

Table 3. Absolute fuel consumption and fuel consumption reduction percentage for different controllers tested with the WMTC and IDC test cycle for Veh 1, Veh 2 and Veh 3.

Controllers	Veh 1				Veh 2				Veh 3			
	WMTC		IDC		WMTC		IDC		WMTC		IDC	
	Abs FC (gms)	FC Red (%)	Abs FC (gms)	FC Red (%)	Abs FC (gms)	FC Red (%)	Abs FC (gms)	FC Red (%)	Abs FC (gms)	FC Red (%)	Abs FC (gms)	FC Red (%)
BEMS	146.7	0.0	142.8	0.0	148.1	0.0	148.0	0.0	145.7	0.0	141.1	0.0
MEMS	138.1	5.9	139.1	2.6	146.3	1.3	145.3	1.8	144.1	1.1	138.1	2.1
ECMS	137.1	6.5	129.3	9.4	144.0	2.8	139.0	6.0	143.2	1.7	133.0	5.7
ECMS_LL	130.0	11.4	112.6	21.2	136.5	7.9	121.8	17.7	140.0	3.9	128.2	9.1

5.2. Charge Sustainability

One of the control objective considered during the formulation of energy management strategy is charge sustainability along with the minimization of fuel consumption. For a fair trade-off, the absolute fuel consumption is calculated by considering the equivalent fuel value of the difference of final SOC and final target SOC_{ref} (0.30). Table 4 shows the final SOC values for the controllers with WMTC and IDC test cycles for different vehicle configurations. The final SOC for the optimal controllers is within the stringent bandwidth of ±1% of the SOC target (SOC = 0.30) when compared to rule-based controllers. The optimal controllers ECMS and ECMS_LL satisfy the control objective of charge sustainability. In the case of rule-based strategies, the final SOC depends on the rules defined and varies with the vehicle power demand and torque requirement demand from test cycles considered. Nevertheless, the final SOC values with rule-based controllers achieve the SOC target like optimal controllers.

Table 4. Final SOC values for different vehicle configuration and drive cycles.

Controllers	Veh 1		Veh 2		Veh 3	
	WMTC	IDC	WMTC	IDC	WMTC	IDC
BEMS	0.30	0.31	0.31	0.31	0.31	0.32
MEMS	0.33	0.31	0.32	0.31	0.31	0.31
ECMS	0.30	0.30	0.30	0.30	0.30	0.30
ECMS_LL	0.30	0.30	0.30	0.30	0.30	0.30

5.3. Engine Switching Events

Table 5 shows the engine switching events (ESE), for the considered controllers for different vehicle configurations and test cycles. With the rule-based controllers, the engine switching depends on the rules that are based on SOC limits considered and the drive cycle requirements for the wheel power and torque. It is seen that, with optimal control, the engine switching events consistently show lower numbers when compared to rule-based strategies. The engine-switching event is reduced by 23% and 55% for WMTC and IDC, respectively, using optimal control when compared to rule-based controllers. One of the important objectives of energy management control is to keep the rider comfort, safety and reliability of starting system in mind while evaluating the control strategies for a vehicle.

The starting system life of a typical conventional vehicle of a similar vehicle is 50,000 cycles/3 years. The one-day travel is defined as 28 Kms. If 50,000 cycles are distributed equally across 365 days/year, the per-day switching limit would be = (50,000/(3 × 365)) ~ approx. 46 switching events per day. Assuming that the engine switches in the first very cranking event, the number of engine switching events can be equated to the number of starting system cycles.

The optimal controllers fare better in reducing the engine switching events when compared to rule-based strategies. The rule-based controllers show almost double the switching events of those allowed for a normal life span of the starting system. Irrespective of the vehicle configurations and the test cycles considered, the rule-based strategies reduce the life span of the starting system by approximately 50%. Since the engine switching depends on the rules, which are based on defined SOC limits, the engine is switched on/off accordingly, which increases the number of ESE.

Table 5. Engine switching event for different controllers tested for different vehicle configurations (Veh1, Veh2 and Veh3) and test cycles (WMTC and IDC).

Controllers	Veh 1		Veh 2		Veh 3	
	WMTC	IDC	WMTC	IDC	WMTC	IDC
BEMS	89	96	89	96	89	96
MEMS	89	96	89	96	89	96
ECMS	69	43	69	43	69	43
ECMS_LL	69	43	69	43	69	43

5.4. Battery Temperature

Table 6 shows the absolute battery temperature and battery temperature rise for the respective controllers considered for WMTC and IDC test cycles for the considered vehicle configuration. From Table 6, it is seen that the rise in battery temperature is found higher in the optimal controllers compared to rule-based strategies. The lithium-ion batteries are capable of operating from −20 to 60 °C [40]. Lithium-ion batteries have maximum efficiency when operated between 10 and 35 °C; this is called the optimum temperature range [40].

Table 6. Battery temperature and battery temperature rise for different vehicle configurations tested for WMTC (W) and IDC (I) test cycles.

Controllers	Absolute Battery Temperature (°C)						Battery Temperature Rise (°C)					
	Veh 1		Veh 2		Veh 3		Veh 1		Veh 2		Veh 3	
	W	I	W	I	W	I	W	I	W	I	W	I
BEMS	31.4	32.0	31.4	31.9	31.7	31.8	1.4	2.0	1.4	1.9	1.7	1.8
MEMS	33.9	32.3	33.2	32.1	33.3	32.3	3.9	2.3	3.2	2.1	3.3	2.3
ECMS	41.7	41.2	41.0	41.4	39.3	38.4	11.7	11.2	11.0	11.4	9.3	8.4
ECMS_LL	43.8	41.9	44.3	42.0	42.7	38.4	13.8	11.9	14.3	12.0	12.7	8.4

Initial battery temperature considered for the study = 30 °C.

Figure 10 shows the variation of battery internal resistance along with the battery temperature for all the controllers tested for IDC test cycle with vehicle configuration Veh 1. With ECMS_LL and ECMS optimal controllers, the battery temperatures for all the test cases (irrespective of vehicle configuration and drive cycle) rise above 35 °C, which is the upper limit of the optimum temperature range. Increase in battery temperatures increases the degradation rates of the battery components, which includes maximum charge/discharge capacity, the effectiveness of the electrode in storing Li-ions, charge transfer rate constant, the total resistance of electrode resistance and electrode/electrolyte resistance [41]. The battery temperature rise with optimal controllers increases approximately up to five times when compared to rule-based strategies. It is seen that the optimal controllers provide a significant increase in the battery temperatures and it is known that higher battery temperature results in a faster battery degradation process [9,33]. Optimal controllers are designed to prioritize fuel economy and in the process, they exploit the battery operation, which leads to faster battery degradation when compared to rule-based controllers. The battery temperature is one of the key factors influencing the battery ageing. Using optimal control, the fuel economy benefit is achieved at the cost of increased battery temperature, which would accelerate the process of battery ageing.

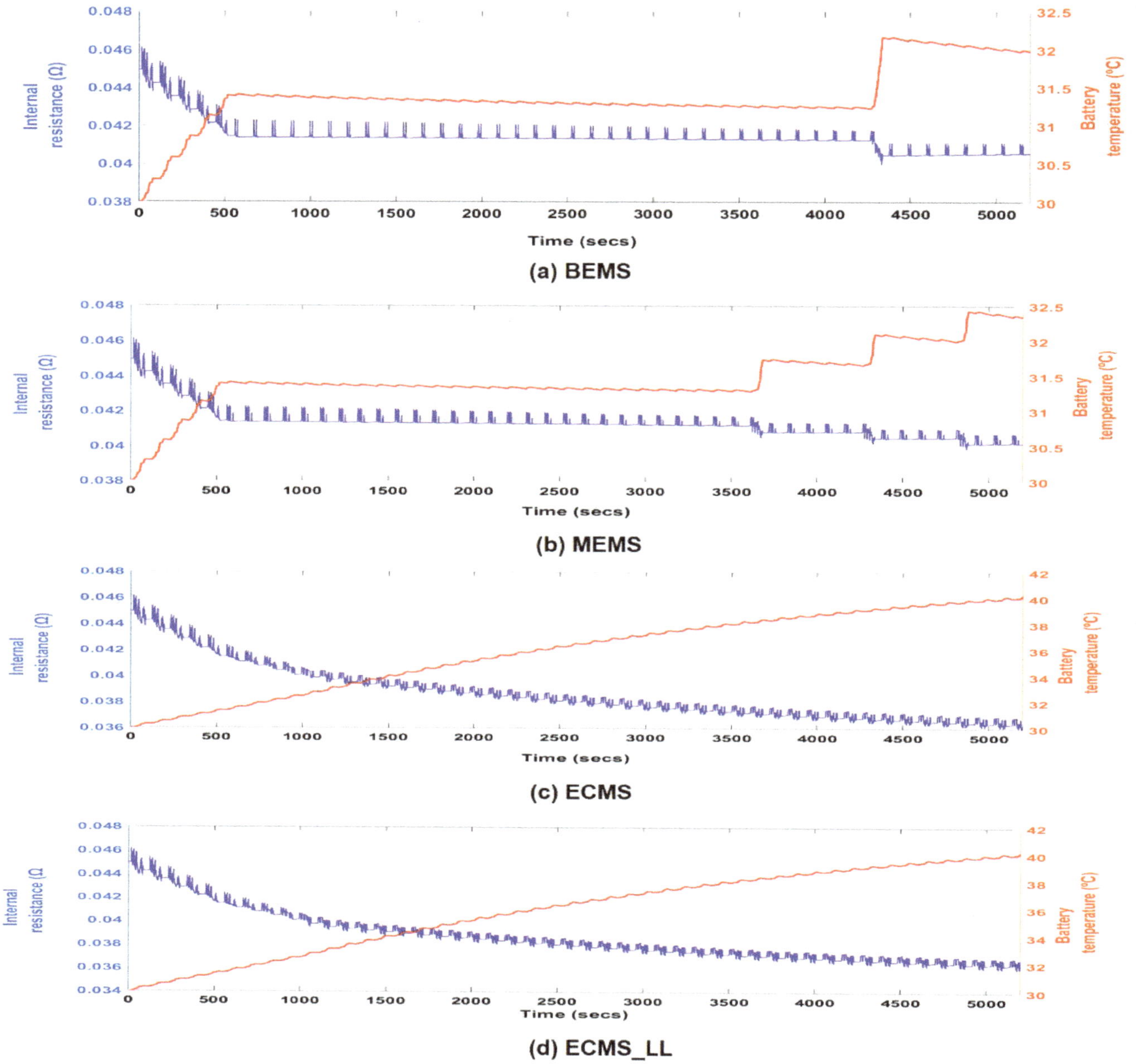

Figure 10. Battery internal resistance and temperature profiles for Veh 1 configuration tested for the IDC test cycle: (**a**) BEMS, (**b**) MEMS, (**c**) ECMS and (**d**) ECMS_LL.

5.5. Battery Throughput

Figure 11 and Table 7 show the trends of battery throughput, for optimal based and rule-based controllers for different drive cycle and vehicle configuration considered. It is seen that optimal controllers ECMS and ECMS_LL show increased battery throughput when compared to rule-based BEMS and MEMS. Increased battery throughput is one of the potential factors affecting battery ageing [9,33,41]. Higher effective battery throughput increases the heat generation and thereby accelerating the battery ageing process [34,41]. The increased levels of battery throughput indicate faster battery ageing process with optimal controllers when compared to rule-based controllers.

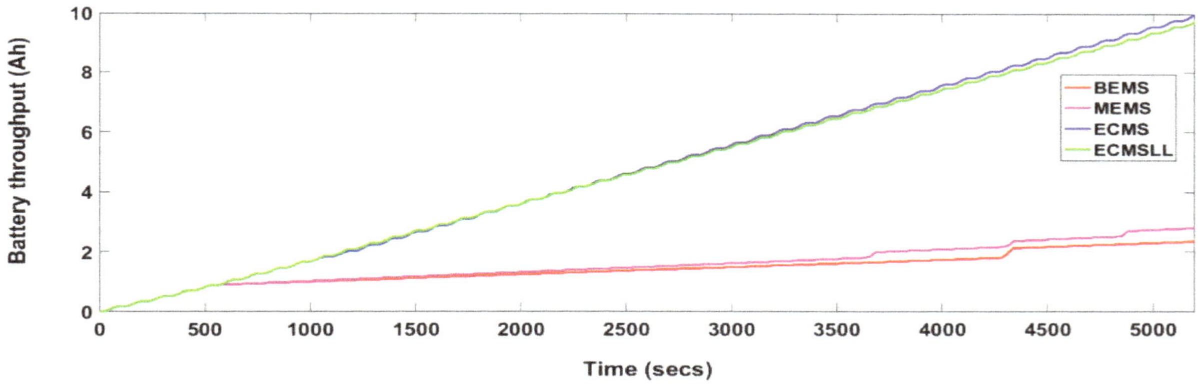

Figure 11. Battery throughput profile for the IDC test cycle.

Table 7. Absolute battery throughput and battery throughput rise in percentage for different vehicle configurations tested for WMTC (W) and IDC (I) test cycles.

Controllers	Battery throughput in Ah						Battery throughput Rise in %					
	Veh 1		Veh 2		Veh 3		Veh 1		Veh 2		Veh 3	
	W	I	W	I	W	I	W	I	W	I	W	I
BEMS	1.8	2.3	1.7	2.3	2.0	2.4	0	0	0	0	0	0
MEMS	3.7	2.7	3.0	2.5	3.7	3.1	52	13	74	6	84	29
ECMS	9.5	10.5	9.4	10.7	9.0	9.5	437	350	446	357	349	290
ECMS_LL	10.9	11.2	11.5	11.2	11.0	9.5	516	378	566	381	451	291

The main objective of the cost function design was limited to improvement in fuel economy through improvement in the engine's efficiency and appropriate power split during the vehicle run. To increase the engine's efficiency, the battery is exploited to its fullest operation. In this process, the battery throughput increases to a maximum of five times when compared to the rule-based strategies. Frequent charging and discharging to accommodate the engine running in its optimum region increases the overall battery throughput for the journey. Battery throughput is a critical factor influencing the process of battery ageing [40].

5.6. Battery Charge and Dis-Charge C Rates

The distribution of charge and discharge C rate for rule-based and optimal control-based controllers tested for IDC test cycles and vehicle configuration Veh 1 is shown in Figure 12. The maximum charge rate limit for vehicle configuration Veh 1 is fixed to 5 C.

The figure clearly shows that the optimal controllers exhibit higher frequency and peaks of charge/discharge C rates when compared to rule-based strategies. A similar trend is found with the other vehicle configurations. Higher peaks of C rates, contribute to higher Joule heating, which in turn increases the internal temperature of the battery [34]. The higher C rates and higher frequency of charge and discharge cycles increase the battery temperature as well as battery throughput, thereby affecting the battery's health and accelerating the ageing process [41]. The cost function designed in optimal controllers improves fuel economy by running the engine around the optimum region by employing the electric path wherever applicable. This, in turn, exploits the battery parameters which accelerates the battery ageing.

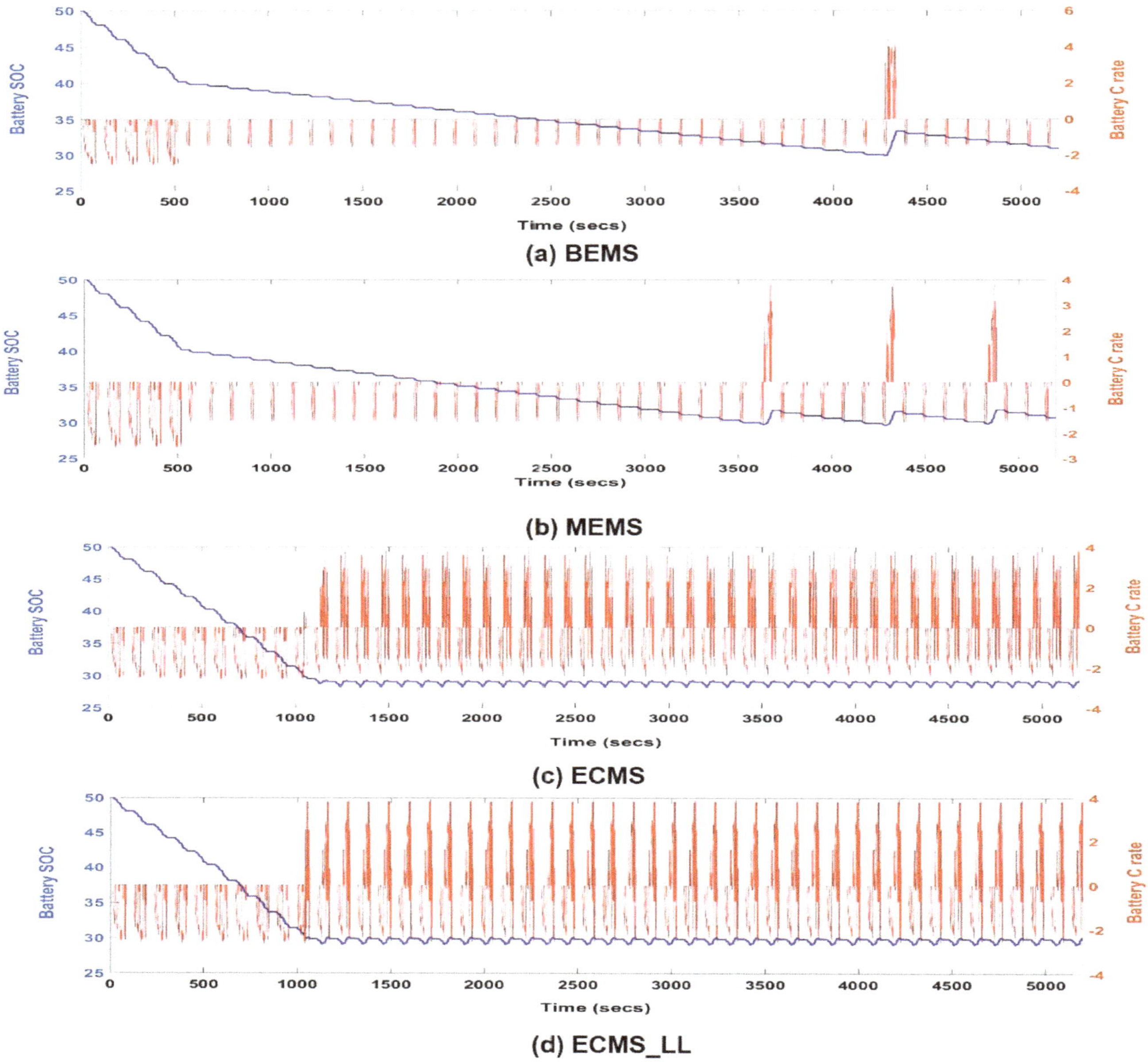

Figure 12. Battery SOC and charge C rate for Veh 1 configuration for different controllers tested for IDC test cycle: (**a**) BEMS, (**b**) MEMS, (**c**) ECMS and (**d**) ECMS_LL.

5.7. Discussion

Very limited work has been done on energy management control for two-wheeler HEVs. Most of the studies of energy management strategies on PHEVs were focussed on fuel economy, tailpipe emissions and charge sustenance. This study focuses on the aspect that optimal control-based energy management strategies with a desire to minimize fuel consumption display potential negative effects on the battery parameters influencing battery ageing. To study this an optimal based and a rule-based strategy was developed for parallel plugin hybrid two-wheeler. The study analysed all the critical battery parameters (battery temperature, charge/discharge C rate and battery throughput) along with engine switching and fuel economy and thereby evaluate the rule-based control strategies and optimal based control strategies. To display consistency and reliability of the results, the controllers were tested for two drive cycles and three different vehicle configurations. Two basic rule-based controllers,

(a) BEMS and (b) MEMS, and two optimal control-based strategies, (c) ECMS and (d) ECMS_LL, were developed and considered for the evaluation.

The simulation results highlight that the optimal controllers considerably reduces the fuel consumption and engine-switching events when compared to rule-based controllers, irrespective of the drive cycles and vehicle configurations considered. The IDC test cycle results show a significant reduction in fuel consumption of up to 21.18% and a reduction in engine switching events of up to 55% with ECMS_LL when compared with rule-based strategy BEMS. However, the benefit of fuel consumption and engine switching showed significant adverse effects on the battery parameters influencing battery ageing. The battery temperature and battery throughput increased by 31% and 378% respectively.

Similarly, with the WMTC drive cycle, fuel consumption is reduced by 11.4% and ESE by 22.5%. However, there was a significant increase in battery temperature and battery throughput by 39% and 515%, respectively. Similar results are observed with other vehicle configurations. The optimal controller utilizes the electric path favourably to operate the engine in the most optimal region wherever possible. Thus, it reduces overall fuel consumption. Using the electric path for keeping the engine operation in the optimum region exploits battery usage, thus increasing the battery temperature, battery throughput and high peaks of charge/discharge C rates. These battery parameters, in turn, affect battery ageing.

The simulation results show that optimal controllers fare better when it comes to reducing the engine switching events. Unlike rule-based energy management strategies, ECMS cost function is designed to minimize fuel consumption at every instant of vehicle running. In ECMS cost function, the energy required for the operation of the starting system and extra fuel required to crank and start the engine is taken into account in the cost function. Thus, it inhibits unnecessary engine switching on/off. A benefit of 55% and a 22% reduction in engine switching events with IDC and WMTC drive cycles are noticed, respectively, for Veh 1 configuration. A similar trend of engine switching events has been observed with other vehicle configurations. This reduction in engine switching events helps in improving the starting system life and also the reliability of engine operation.

Figure 13 shows consolidated simulation results in the form of a radar chart for WMTC and IDC test cycles for vehicle 1 configuration. The normalized results of controllers (BEMS, MEMS, ECMS and ECMS_LL) for fuel consumption reduction, engine switching events, battery temperature rise, final SOC and battery throughput for WMTC and IDC test cycles are shown in Figure 13. The maximum and minimum are represented as '1' and '0', respectively, while comparing the controllers. The '1.0' is defined as the maximum battery throughput, the maximum SOC deviation from target SOC, the maximum number of engine switching events, the maximum reduction of fuel consumption and the maximum rise in battery temperature and '0' is defined as the minimum most.

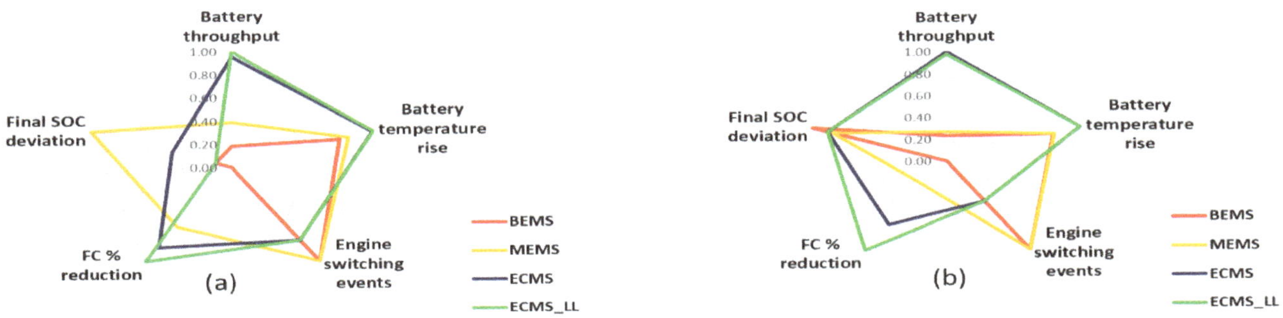

Figure 13. Simulation results for the Veh 1 configuration considered: (**a**) WMTC test cycle and (**b**) IDC test cycle.

It is evident from the simulation results that the optimal controllers have been able to provide the best fuel economy, minimal engine switching events and better SOC sustainability. However, the improvement in fuel consumption and engine switching has significant adverse effects on the battery health and the battery ageing factors (effective battery throughput, charge/discharge C rates and battery temperature). The battery temperature and battery throughput increased by a maximum of five times and the frequency and peaks of charge/discharge C rate have increased substantially. Similar results are observed with different vehicle configurations, as discussed in the results section. The benefit of fuel economy with energy management strategy is partially at the cost of battery health and ageing. Thus, consideration of battery parameters is a must in the evaluation of energy management strategies.

It can thus be suggested that optimal controllers, which are primarily designed for reducing fuel consumption, engine switching and charge sustenance for the particular test cycles display promising results for the control objectives considered. However, the simulation results show with optimal control-based strategies there is an increased negative effect on the battery ageing factors. This study emphasizes that the assessment of energy management strategy for HEV would be incomplete without considering the study of battery parameters influencing ageing and engine switching along with fuel economy. The comparison of controllers presented in this paper provides insights into the interdependency between fuel consumption, engine switching and battery parameters influencing the battery ageing.

6. Conclusions and Future Work

For an appropriate evaluation of energy management strategies, it is necessary to capture all the important factors related to the critical powertrain components. Battery replacement cost is one of the major hurdles in a PHEV that limit its prevalence in the automotive industry. Therefore, without consideration of the battery parameters influencing the battery ageing, the evaluation of energy management strategies remains incomplete.

This paper has analyzed an optimal control-based ECMS for a parallel plugin HEV two-wheeler. The paper presents an evaluation of rule-based and optimal control-based controllers for the HEV two-wheeler. A modified ECMS strategy, ECMS_LL, was proposed and developed. The ECMS_LL further improves fuel economy when compared to conventional ECMS by maximizing the convergence of the engine operating points to the optimum working region of the engine. This is done by exploiting the electric path favourably wherever necessary. From the simulation results, it is evident that the improvement in fuel economy is partly on the cost of battery health degradation.

This study emphasizes the need for an all-inclusive evaluation for the HEV energy management strategy. The work presented in this paper forms the basis for the future study of the optimal behaviour of energy management strategies, which would include battery ageing and engine switching events along with fuel economy and charge sustainability. The work emphasizes the requirement of an appropriate battery ageing-conscious energy management strategy to enhance the overall performance of PHEVs.

Designing an optimal control energy management with a consideration of battery parameters in the cost function formulation is taken as future work. The further development of a simple battery-ageing model to evaluate the battery capacity loss is under study. As a next step, the simulation results in this study are planned to be validated using HIL testing and then on the actual vehicle.

Author Contributions: Conceptualization, Methodology, Validation, and Writing—Original Draft Preparation: N.K.K.; Supervision: A.M. and A.A., Writing—Review & Editing: A.M., A.A. and D.Q.T. All authors have read and agreed to the published version of the manuscript.

References

1. Chan, C.C. *The State of the Art of Electric and Hybrid Vehicles*; IEEE: Piscataway, NJ, USA, 2002; Volume 90, pp. 247–275.
2. Markel, T.; Simpson, A. Cost-benefit analysis of plug-in hybrid electric vehicle technology. *World Electr. Veh. J.* **2007**, *1*, 294–301. [CrossRef]

3. Wirasingha, S.G.; Emadi, A. Classification and review of control strategies for plug-in hybrid electric vehicles. *Trans. Veh. Technol.* **2010**, *60*, 111–122. [CrossRef]
4. Salmasi, F.R. Control strategies for hybrid electric vehicles: Evolution, classification, comparison, and future trends. *Trans. Veh. Technol.* **2007**, *56*, 2393–2404. [CrossRef]
5. Zeng, Y.; Cai, Y.; Kou, G.; Gao, W.; Qin, D. Energy management for plug-in hybrid electric vehicle based on adaptive simplified-ECMS. *Sustainability* **2018**, *10*, 2060. [CrossRef]
6. Huang, Y.J.; Yin, C.L.; Zhang, J.W. Design of an energy management strategy for parallel hybrid electric vehicles using a logic threshold and instantaneous optimization method. *Int. J. Automot. Technol.* **2009**, *10*, 513–521. [CrossRef]
7. Škugor, B.; Deur, J.; Cipek, M.; Pavković, D. Design of a power-split hybrid electric vehicle control system utilizing a rule-based controller and an equivalent consumption minimization strategy. *Proc. Inst. Mech. Eng. Part J. Automob. Eng.* **2014**, *228*, 631–648.
8. Chen, B.C.; Wu, Y.Y.; Wu, Y.L.; Lin, C.C. Adaptive power split control for a hybrid electric two-wheeler. *IEEE Trans. Veh. Technol.* **2011**, *60*, 1430–1437. [CrossRef]
9. Tang, L.; Rizzoni, G.; Onori, S. Energy management strategy for HEVs including battery life optimization. *Trans. Transp. Electrif.* **2015**, *1*, 211–222. [CrossRef]
10. Sun, C.; He, H.; Sun, F. The role of velocity forecasting in adaptive-ECMS for hybrid electric vehicles. *Energy Procedia* **2015**, *75*, 1907–1912. [CrossRef]
11. Han, J.; Kum, D.; Park, Y. Synthesis of predictive equivalent consumption minimization strategy for hybrid electric vehicles based on closed-form solution of optimal equivalence factor. *Trans. Veh. Technol.* **2017**, *66*, 5604–5616. [CrossRef]
12. Paganelli, G.; Delprat, S.; Guerra, T.M.; Rimaux, J.; Santin, J.J. Equivalent consumption minimization strategy for parallel hybrid powertrains. In Proceedings of the Vehicular Technology Conference. IEEE 55th Vehicular Technology Conference, Birmingham, AL, USA, 6–9 May 2002; Volume 4, pp. 2076–2081.
13. Musardo, C.; Rizzoni, G.; Guezennec, Y.; Staccia, B. A-ECMS: An adaptive algorithm for hybrid electric vehicle energy management. *Eur. J. Control* **2005**, *11*, 509–524. [CrossRef]
14. Serrao, L.; Onori, S.; Rizzoni, G. ECMS as a realization of Pontryagin's minimum principle for HEV control. In Proceedings of the 2009 American Control Conference, St. Louis, MO, USA, 10–12 June 2009; pp. 3964–3969.
15. Kim, N.; Cha, S.; Peng, H. Optimal control of hybrid electric vehicles based on Pontryagin's minimum principle. *Trans. Control Syst. Technol.* **2010**, *19*, 1279–1287.
16. Choi, S.S.; Lim, H.S. Factors that affect cycle-life and possible degradation mechanisms of a Li-ion cell based on $LiCoO_2$. *J. Power Sources* **2002**, *111*, 130–136. [CrossRef]
17. Vetter, J.; Novák, P.; Wagner, M.R.; Veit, C.; Möller, K.C.; Besenhard, J.O.; Hammouche, A. Ageing mechanisms in lithium-ion batteries. *J. Power Sources* **2005**, *147*, 269–281. [CrossRef]
18. Broussely, M.; Biensan, P.; Bonhomme, F.; Blanchard, P.; Herreyre, S.; Nechev, K.; Staniewicz, R.J. Main aging mechanisms in Li ion batteries. *J. Power Sources* **2005**, *146*, 90–96. [CrossRef]
19. Belt, J.R.; Ho, C.D.; Miller, T.J.; Habib, M.A.; Duong, T.Q. The effect of temperature on capacity and power in cycled lithium ion batteries. *J. Power Sources* **2005**, *142*, 354–360. [CrossRef]
20. Serrao, L.; Onori, S.; Sciarretta, A.; Guezennec, Y.; Rizzoni, G. Optimal energy management of hybrid electric vehicles including battery aging. In Proceedings of the 2011 American Control Conference, San Francisco, CA, USA, 29 June–1 July 2011; pp. 2125–2130.
21. Kum, D.; Peng, H.; Bucknor, N.K. Control of engine-starts for optimal drivability of parallel hybrid electric vehicles. *J. Dyn. Syst. Meas. Control* **2013**, *2*, 135. [CrossRef]
22. Kim, H.; Kim, J.; Lee, H. Mode transition control using disturbance compensation for a parallel hybrid electric vehicle. *Proc. Inst. Mech. Eng. Part J. Automob. Eng.* **2011**, *225*, 150–166. [CrossRef]
23. Babu, D.S.; Raj, S.S.; Malsoor, B. Stability Enhancement of A Powered Two Wheeler Vehicle Under Curve Negotiation. *Int. Refereed. J. Eng. Sci.* **2016**, *5*, 62–71.
24. Delavaux, M.; Lhomme, W.; McGordon, A. Comparison between forward and backward approaches for the simulation of an electric vehicle. In Proceedings of the IEEE Vehicle Power and Propulsion Conference (VPPC2010), Lille, France, 3–5 September 2010.
25. Dixon, G.; Stobart, R.; Steffen, T. Unified Backwards Facing and Forwards Facing Simulation of a Hybrid Electric Vehicle using MATLAB Simscape. *SAE Tech. Paper* **2015**. [CrossRef]

26. Paganelli, G.; Guerra, T.M.; Delprat, S.; Santin, J.J.; Delhom, M.; Combes, E. Simulation and assessment of power control strategies for a parallel hybrid car. *Proc. Inst. Mech. Eng. Part J. Automob. Eng.* **2000**, *214*, 705–717. [CrossRef]
27. Tremblay, O.; Dessaint, L.A. Experimental validation of a battery dynamic model for EV applications. *World Electr. Veh. J.* **2009**, *3*, 289–298. [CrossRef]
28. Tremblay, O.; Dessaint, L.A.; Dekkiche, A.I. A generic battery model for the dynamic simulation of hybrid electric vehicles. In Proceedings of the 2007 IEEE Vehicle Power and Propulsion Conference, Arlington, TX, USA, 9–12 September 2007; pp. 284–289.
29. Hussein, A.A.H.; Batarseh, I. An overview of generic battery models. In Proceedings of the 2011 IEEE Power and Energy Society General Meeting, At Detroit, Michigan, USA, 24–29 July 2011; pp. 1–6.
30. Wang, D.; Bao, Y.; Shi, J. Online lithium-ion battery internal resistance measurement application in state-of-charge estimation using the extended Kalman filter. *Energies* **2017**, *10*, 1284. [CrossRef]
31. Vignesh, B.R.; Grandjean, T.; McGordon, A.; Greenwood, D. Thermal modeling of lithium ion batteries for temperature rise predictions in hybrid vehicle application. In Proceedings of the 2018 Thirteenth International Conference on Ecological Vehicles and Renewable Energies (EVER), Monte Carlo, Monacco, 10–12 April 2018; pp. 1–7.
32. Onori, S.; Serrao, L.; Rizzoni, G. Adaptive equivalent consumption minimization strategy for hybrid electric vehicles. In Proceedings of the ASME 2010 dynamic systems and control conference, Cambridge, MA, USA, 13–15 September 2010; pp. 499–505.
33. Vignesh, B.R. Parametric Modelling, Simulation and Validation of Lithium Ion Batteries for Temperature Rise Prediction in Hybrid Two-Wheelers. Ph.D. Thesis, University of Warwick, Coventry, UK, 2016.
34. Srinivasan, R.; Baisden, A.C.; Carkhuff, B.G.; Butler, M.H. The five modes of heat generation in a Li-ion cell under discharge. *J. Power Sources* **2014**, *262*, 93–103. [CrossRef]
35. Onori, S.; Serrao, L. On Adaptive-ECMS strategies for hybrid electric vehicles. In Proceedings of the International Scientific Conference on Hybrid and Electric Vehicles, RHEVE 2011, Malmaison, France, 6–7 December 2011; Volume 67.
36. Urdhwaresh, R. Limits, Regulations, Measurement of Exhaust Emissions and Calculation of Fuel Consumption. In *Indian Emissions Regulations*; Automotive Research Association of India: Pune, India, 2017.
37. Windover, P.R.; Owens, R.J.; Levinson, T.M.; Laughlin, M.; Gaines, L. *Stop and Restart Effects on Modern Vehicle Starting System Components*; Argonne National Lab. (ANL): Argonne, IL, USA, 2015.
38. Sciarretta, A.; Guzzella, L. Control of hybrid electric vehicles. *Ieee Control Syst. Mag.* **2007**, *27*, 60–70.
39. Paganelli, G.; Ercole, G.; Brahma, A.; Guezennec, Y.; Rizzoni, G. General supervisory control policy for the energy optimization of charge-sustaining hybrid electric vehicles. *JSAE Rev.* **2010**, *22*, 511–518. [CrossRef]
40. Warner, J.T. *The Handbook of Lithium-ion Battery Pack Design: Chemistry, Components, Types and Terminology*; Elsevier: Amsterdam, The Netherlands, 2005.
41. Leng, F.; Tan, C.M.; Pecht, M. Effect of temperature on the aging rate of Li ion battery operating above room temperature. *Sci. Rep.* **2015**, *5*, 12967. [CrossRef]

11

The Integrated Energy Consumption Index for Energy Biomass Grinding Technology Assessment

Weronika Kruszelnicka *, Robert Kasner, Patrycja Bałdowska-Witos, Józef Flizikowski and Andrzej Tomporowski

Faculty of Mechanical Engineering, University of Science and Technology in Bydgoszcz, Al. Prof. S. Kaliskiego 7, 85-796 Bydgoszcz, Poland; robert.kasner@utp.edu.pl (R.K.); patrycja.baldowska-witos@utp.edu.pl (P.B.-W.); fliz@utp.edu.pl (J.F.); a.tomporowski@utp.edu.pl (A.T.)

* Correspondence: weronika.kruszelnicka@utp.edu.pl

Abstract: The assessment of engineering objects in terms of energy consumption is an important part of sustainable development. Many materials, including those from the energy sector, need to undergo earlier processing, e.g., grinding. Grinding processes still demand a significant amount of energy, whereas current energy assessment methods do not take into account important parameters of the process, which makes it difficult to choose their optimal values. The study presents the analysis, testing, and assessment of mechanical engineering systems in terms of the energy consumption involved in the grinding of biomass intended for energy production purposes. A testing methodology was developed to improve the parameters of multi-disc grinding, including the reduction of energy consumption, power input, product quality improvement, and process efficiency. An original model of integrated energy consumption was developed. Tests were carried out on a five-disc grinder for five programs to assess the programmable angular speeds of the grinder discs. Output values, including specific energy demand, fragmentation degree, and integrated energy consumption, were assigned to each testing program. The test results were subjected to statistical analysis. Based on the authors' own research, it was found that the angular speed of the discs and, consequently, the linear speed of the grinding blades, have a significant influence on the values of the integrated energy consumption of the preliminary process.

Keywords: energy consumption; comminution; grinding; multi-disc mill; energy consumption index; energy assessment

---

1. Introduction

The simplest and the most common method of biomass utilization is its combustion, e.g., in the boilers of a power station [1]. There are also other forms of energy production that are increasing in popularity, including: gasification, biomethane production, esterification, the acquisition of bioethanol, etc. [2,3]. Depending on the energy purpose and type of biomass, the comminution process will vary—in particular, in terms of the desired final dimension of the comminution product [4,5]. In the case of anaerobic digestion, biomass is comminuted to several millimeter fractions [4]. Pellets are usually made of fibrous biomass ground to 2–4 mm or smaller, depending on the briquetting machine design, while briquettes, from particles smaller than 10 mm [6]. In the production of biofuels, in turn, biomass particles should be in the range of 100–500 μm, and for gasification and direct combustion processes, below 100 μm [4]. Previous research has shown that the smaller the biomass particles are used for energy purposes, the greater the efficiency and easier control of biomass thermochemical transformation processes [5].

Acquiring energy from biomass requires its earlier preparation: drying, grinding, palletizing, etc. [7]. These procedures require certain energy inputs (costs) which significantly decrease the energy and environmental balance in the entire lifecycle of the energy material (wood, grain biomass, etc.) [8–13]. According to the idea of sustainable development concerning the design of systems, devices, and processes, in three areas—the environment, society, and economy—the biomass processing lines should be characterized by the lowest possible energy consumption and environmental impact [14,15]. It also aims to improve the competitiveness of biomass fuels in relation to conventional fossil fuels [16,17].

Before being used for energy purposes, energy media need to be properly prepared, e.g., their dimensions need to be reduced by means of grinding [2]. Roller, drum, ball, hammer, and disc mills are used to grind granular biomaterials (biomass grains) [7,18]. Tests have shown that the best relations between product quality, yield, and grinding energy consumption occur for hammer and disc mills [2,7]. The quality of the grinding product is most often determined by the grinding degree, which is defined as the ratio of the characteristic dimensions of the particles before and after size reduction [19,20]. Other indicators, e.g., bulk density, grain size distribution, specific surface area, energy value, and their relationship with the comminution process, are also important from the point of view of the final product quality [4,21–24]. Previous studies conducted by other researchers have shown that the reduction of biomass particles' sizes result in improved flowability and movement properties; the material moves better, e.g., on feeders [25]. Grinding processes also increase the bulk density of the product, which allows, among others, to reduce storage space [5]. In the case of biomass, there was also a trend of increasing energy demand along with an increase in the particle size of the grinding input material [26]. It has also been shown that the energy demand increases as the particle size of the comminution product decreases; i.e., the smaller the particles we want to achieve, the greater the energy input [5,27]. The conducted analyzes have shown that the smaller the particles after comminution, the easier they are conducted; e.g., anaerobic digestion processes and the efficiency of biomass conversion into biofuels increases, due to the increase in the specific surface area of the particles in grinding processes [4,28,29].

Current legislation imposes requirements which are supposed to reduce the energy consumption of technological lines, including energy material grinding. Unfortunately, currently used grinding machines and devices consume much energy and are inefficient [2,30–32], which motivates researchers to take up research in order to reduce the energy consumption of grinders and mills. Thus, upon implementing innovative structural solutions, a multi-criteria analysis needs to be used. It should take into consideration such parameters as efficiency, energy consumption, fragmentation degree, and power consumption. The design of structural solutions should meet the criteria of sustainable development. Criteria to be used for the structure assessment should be matched in such a way that the choice of an optimal solution is possible.

Biomaterial grinding (e.g., granular biomass) is accompanied by many phenomena caused by the impact of the material, the machine elements, the internal structure of biomaterials, and their strength parameters [33]. Strength properties vary depending on the material and depend on, among others, internal structure, humidity, and biological features [34], and in the case of biomass, a significant diversification of properties is observed even within one species [35–37]. The increases in strength and hardness increase the energy demand for fragmentation [38–40]. The increase in the humidity of the input material also contributes to the increase in energy demand during the size reduction process. The internal structure of biomass grains is not without significance—materials with greater glassiness are characterized by increased energy demands in size reduction processes; however, the fragmentation efficiency is greater, resulting in smaller particles [36,38,41,42]. Wiercioch et al. [35] showed proportionality of the grinding energy demand and mass of material. Other studies, conducted for single grains, have shown that the amount of grinding energy demand increases as grain thickness increases [37]. Thus, the criteria for the assessment of the structure and process need to be formulated. They must be adjusted both to the material and the grinding machine, as well as to the specificity of the company. Hence, the identification of variables and factors affecting the process assessment in terms of energy consumption, efficiency, and product quality seems to be of key importance [7,31].

Research on energy-related aspects involved in grinding has been ongoing since the end of the 20th century. Since that time, many theories link energy consumption with product quality and the efficiency of the grinding process. These include the hypotheses of Rittinger, Kick, Holmes, Charles, Rebinder, Mielnikow, and Bond, as well as the statistical hypothesis. They, however, are based on certain simplifications regarding the similarity of particles and the failure to consider the forces of grinding. In addition, there are restrictions involved in their applicability [30,43–46]. Polish experiences in research on energy consumption include experiments conducted by: J. Flizikowski, Malewski, Tomporowski, Zawada, Brach, and Sidor [47–55], who have been involved in modeling the grinding process [56].

The literature provides but only a few studies, including a comprehensive energy assessment of grinding machines and devices with a focus on the process parameters. A number of studies indicate energy consumption and the degree of grinding as basic assessment criteria [57,58].

The criteria for energy assessment provided in the literature (Table 1) represent a machine-oriented approach to the grinding process. They can be considered to belong to the group of efficiency criteria proposed by M. Bielski [59], presenting the ratio of profits to expenditures [60]. Dependences (1)–(8) indicate the energy "productivity" of the process; that is, they define the amount of energy needed to grind the material unit (unit of mass, volume). The function of adaptation (Equation (9)) proposed by J. Flizikowski [49] views many grinding factors as a machine process and is designed for global assessment of the process in terms of biomass-processing for the needs of feeding animals. It combines the effects of material grinding—an increase in the processed grain assimilability by living organisms and the amount of energy used by the machine elements and power transmission system. Effectiveness understood in terms of the energy exchange efficiency (Equation (10)) has been discussed in paper [30].

A criterion of a unit demand for energy is often used as an energy measure (dependence (6)). Depending on the material properties and the required grain dimensions, energy demands can be: from 20 $kWh{\cdot}Mg^{-1}$ to 130 $kWh{\cdot}Mg^{-1}$ for rough and medium grinding and up to 800 $kWh{\cdot}Mg^{-1}$ for micro and nanogrinding.

For brittle materials such as glass, cement, and clinker, the energy required for very fine grinding is much lower than for polymer and biological fibrous materials. The energy demand for these materials is nearly two times higher, whereas the final particles are from 10 to 100 times larger. Hence, the grinding efficiency of fibrous materials is lower than that of brittle materials [61].

An insufficient number and quality of grinding-process descriptions, ineffective technological solutions, and unsatisfying attempts to improve grinding system efficiency means major problems remain unsolved, such as the high energy costs of grinding, poor yield, and insufficient quality of the product represented by granulometric parameters. These problems should be solved by monitoring and supervising grinding parameters such as linear speed, the angular speed of discs, energy consumption, power input, etc. In view of the above, taking up research on this subject can bring numerous benefits.

There is an urgent need to provide an explanation of the essence of engineering and the control of granular biomass quasi-shearing by means of a multi-hole grinding unit with a focus on the process of energy consumption.

Hence, the following research goals were established:

The development of a mathematical model of energy consumption and granular biomass grinding for the needs of the integrated design of multi-hole (disc) grinding units. Experimental determination of the impact of selected parameters of the multi-disc grinding process on the sustainable energy consumption of granular biomass.

Table 1. Energy assessment criteria of the grinding process presented in the literature.

Criterion	Mathematical Model
Overall net energy for a grinding unit of dry mass [2]	$E_T = \int_0^T (P_T - P_0)\, dt/m_{DM} = \int_0^T \Delta P_t\, dt/m_{DM}$ (1) E_T—Overall net energy for grinding unit of a dry biomass, kJ kg_{DM}^{-1}; P_T—power consumed by the machine in time, t; W,—average power input of the grinder neutral gear, W; ΔP_T—net power input for biomass grinding in time t, W; and m_{DM}—mass of the ground dry feed mass, kg.
Specific grinding Energy [7]	$E_{spec} = 1/m \int_0^T (P_T - P_1)\, dt$ (2) E_{spec}—Specific energy of grinding, J·kg^{-1}; P_T—total power consumption, W; and P_l—power consumed by the grinder without being loaded by the feed, W.
Specific energy consumption [20]	$SEC = E_{tot}/Q$ (3) SEC—Specific energy consumption, kJ·kg^{-1}; E_{tot}—overall energy consumption, kJ; and Q—productivity, kg.
Energy density [20]	$Ed = SEC \cdot \rho$ (4) Ed—Energy density, kJ·m^{-3}; ρ—bulk energy density of the ground material, kg·m^{-3}; and SEC—specific energy consumption, kJ·kg^{-1}.
Specific energy of grinding [62]	$E_r = (E_c - E_s)/m$ (5) E_r—specific energy of grinding, J·kg^{-1}; m—mass of the ground sample of material, kg; E_c—overall energy consumption, J; and E_s—neutral gear energy consumption, J.
Unit energy demand [63–65]	$E_j = P/Q$ (6) E_j—unit energy demand, kWh·kg^{-1}; P—power on the grinder shaft, kW; and Q—mass yield, kg·h^{-1}.
Energy efficiency [58]	$EE = E_m(SP)/E_m(Mill) \cdot 100$ (7) EE—energy efficiency, $E_m(SP)$—specific energy used for grinding a single particle, and $E_m(Mill)$ – specific energy net energy used for production of a given product.
Target energy efficiency [58]	$EE = E_m(SP)/E_m(Mill)X_{50} \cdot 100$ (8) EE—energy efficiency, $E_m(SP)$—specific energy used to generate a product with a given average dimension X_{50}, and $E_m(Mill)X_{50}$—net specific energy used to generate a product with average dimension X_{50}.
Adjustment function quasi-shearing [50]	$e_{RP} = ((\eta_{q-s} - \eta_o) \cdot E_{brutto} \cdot \eta_s \cdot \eta_p)/((k_j \cdot v_r + \tau_{q-s} \cdot F_{q-s} + \varepsilon \cdot F_R{}' \cdot v_r{}^2) \cdot M_k \cdot v_r \cdot t)$ (9) e_{RP}—adjustment function for quasi-shearing; η_{q-s}—combustion efficiency of ground energy carriers; η_o—efficiency of whole grain combustion of selected energy carriers; E_{brutto}—energy of processed grains, kJ·kg^{-1}; $\eta_s \cdot \eta_p$—efficiency of the grinder drive; η_s—engine efficiency; η_p—transmission efficiency; k_j—idle motion coefficient, kJ·s^{-1}; v_r—speed of quasi-shear, m·s^{-1}; τ_{q-s}—stress during quasi-shearing, N·m^{-2}; F_{q-s}—temporary cross-section of quasi-shearing, m^2; $F_R{}'$—cross-section of secondary impacts during quasi-shearing, m^2; ε—proportionality coefficient, N·s^{-2}·m^{-4}; and M_k—times indicator.
Energy change efficiency [30]	$Ef_{ze} = E_{roz}/E_w \cdot 100$ (10) Ef_{ze}—energy change efficiency, %; E_{roz}—grinding energy, J; and E_w—energy input, J.

Research problems were formulated in the form of questions in relation to the research goals established herein:

1. Will the development of an adequate mathematical model of energy consumption considering basic parameters of the process such as the yield, fineness degree, unit energy consumption, and power input enable the integrated design of the structure, disc parameters, and hole movement in shearing culvert discs?
2. What is the impact of deliberately changed characteristics and relations of the unit motion on the yield, fineness degree, power input, unit energy consumption, and integrated index of sustainable granular biomass grinding energy consumption in a multi-hole grinding unit?

2. Materials and Methods

2.1. Integrated Energy Consumption Index

The introduction of energy and pro-environment-based assessment into the construction–optimization process of the technology of energy biomass grinding leads to the creation of new, better structural and process solutions. In a sense, it is a driving force for biomass-processing competitiveness and innovativeness [66,67]. Paper [68] indicates that energy efficiency assessments make it possible to improve the grinder process motion parameters and indicates technologies which are the best for a given product. It has also been proven that the developed efficiency models can be used not only for process verification but also for verification of the whole structure of grinding, particularly the working unit.

Developing an adequate mathematical model and finding functional dependences between the variables are important elements of each assessment (structural-optimization efficiency) [69]. The first step should involve the identification of the assessed processes and subprocesses and the determination of criteria [68].

The assumption of the model of integrated energy consumption for grinding takes into account the most important grinding indexes, which are indirectly or directly connected with power and energy demand. First of all, the criteria for power consumption assessment, unambiguously characterizing the grinding device, were established. The following criteria were accepted as components of the model of integrated energy consumption, in accordance with the desired states of the grinding process [70,71]:

- Yield of grinding K_{Qr}:
$$K_{Qr} = Q_r, \tag{11}$$
- Power demand K_P:
$$K_P = 1/P_r, \tag{12}$$
- Unit energy demand K_{Ej}:
$$K_{Ej} = 1/E_j, \tag{13}$$
- Fragmentation degree K_i:
$$K_i = i_{80}, \tag{14}$$

where Q_r—yield of grinding, kg·h^{-1}; P_r—average power consumption of grinding, kW; E_j—unit of energy consumption, kWh·kg^{-1}; and i_{80}—80% fragmentation degree.

The yield of grinding is indirectly related to the power and energy consumption of grinding. Increasing the yield is usually associated with an increase in the grinder dimensions and, subsequently, an increase in the power of the drive unit, which involves a change of energy parameters. In the machine conditions, grinding yield Q_r can be determined on the basis of a dependence of grinding product mass change Δm in the intake hopper during the time of observation Δt [19]:

$$Q_r = \Delta m/\Delta t, \tag{15}$$

which can further be written:

$$Q_r = dm/dt, \tag{16}$$

$$Q_r = \int_0^t m\,dt. \tag{17}$$

Power consumption for grinding changes in time, which is caused by an irregularity in the operation of the machine [72]. In the simplest way, it can be defined as a periodical occurrence of rises and falls in power consumption as a result of variable amounts of materials between the working elements, which causes momentary periods of the machine running idle during grinding (with no feed). In connection with this, the overall power consumption in time t P_{toti} can be defined as [7,73]:

$$P_{toti} = P_{ui} + P_{ii}, \tag{18}$$

where P_{ui}—power consumption for the medium grinding in a given moment t, kW and P_{ii}—power consumption of the idle gear in a given moment t, kW.

Then, the total grinding power consumption P_r in a certain time interval T is an average of the momentary power consumption, according to the equation:

$$Qr = \sum_{i=0}^{T} P_{toti}/T. \tag{19}$$

Grinding processes are characterized by the parameters of the unit energy demand; therefore, they were accepted to be a component of the integrated energy consumption index model. Generally, a unit energy demand is defined as the amount of energy per mass unit of the material (volume) [2,7,20,63,64]. For the grinding machine unit, demand E_j is expressed as a ratio of average power consumed by the grinder P_r to yield Q_r [19,63,73]:

$$E_j = P_r/Q_r \tag{20}$$

The last criterion of the model of integrated energy consumption was accepted to be the fragmentation degree, being a determiner of the product quality and effectiveness of the grinding process [20,74]. Most generally, the fragmentation degree is defined as a ratio of the material mean dimensions before grinding to its mean dimensions after grinding [21,75]. So far, it has been proven that increasing the frequency of the machine element contact (by increasing the rotational, angular, and linear speeds of the grinding elements) and repetitions of grinding provides an output product with smaller dimensions—that is, with a higher fragmentation degree [72,76]. Unfortunately, it causes higher energy consumption [27,77,78]. The energy consumption model assumes a fragmentation degree of 80% i_{80}, defined as [19]:

$$i_{80} = D_{80}/d_{80} \tag{21}$$

where D_{80}—dimension of the sieve hole through which 80% of the feed material passes and d_{80}—dimension of the sieve hole through which 80% of the grinding product passes.

The model of integrated energy consumption was constructed in the same way as the model of efficiency described in paper [68]. This made it possible to establish a quantitative relation between the most important grinding indexes in terms of energy consumption. Integrated energy consumption was described in the criteria-based approach proposed by J. Zawada in the following way [79]:

$$E_{zint} = K_{Qr} \cdot K_P \cdot K_{Ej} \cdot K_i \tag{22}$$

Bearing in mind dependences (11)–(14) and (22), integrated energy consumption depends, in direct proportion, on the yield and fragmentation degree and, in an inverse proportion, to the power consumption and unit energy demand. Assuming that the state to be desired is that in which $Q \rightarrow \infty$, $i_{80} \rightarrow \infty$, $P_r \rightarrow 0$, and $E_j \rightarrow 0$, then $E_{zint} \rightarrow \infty$. In the sustainable approach, a better process will be the one with a higher integrated energy consumption index value (22).

2.2. *Raw Materials*

Rice and corn grains that are unsuitable for food purposes due to having been stored in inappropriate conditions that were used for the production of pellets underwent grinding in the experimental tests, verifying the correctness of the integrated energy consumption model. Both crops are extensively grown, and both of them have found applications in the energy sector due to their good energy characteristics, which make them suitable for the production of pellets whose calorific value is similar to that of a wooden biomass [80]. Grain moisture was determined on a moisture balance MAC 210/NP (RADWAG, Radom, Poland) by means of the weight method. Rice grains were characterized by moisture at the level of 13.47% ± 0.02%, whereas corn was at the level of 12.68% ± 0.02%. The characteristic dimension of grain D_{80} for corn was 8.15 mm and, for rice, 2.14 mm (Figure 1). These values were obtained on the basis of a granulometric analysis carried out on a Camsizer (Retsch Technology GmbH, Haan, Germany) design according to the ISO norm 13322-2:2006 [81].

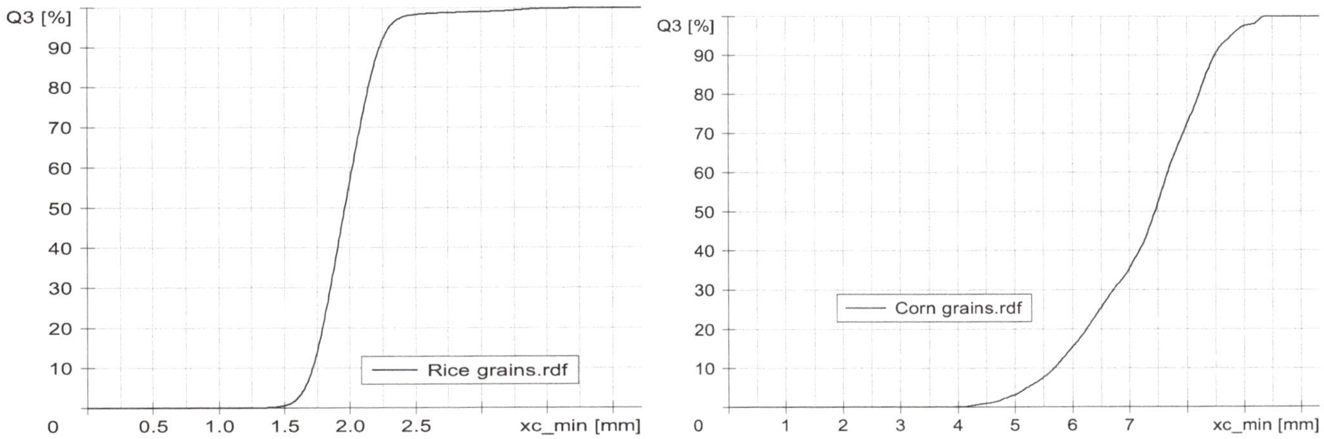

(**a**) (**b**)

Figure 1. Granulometric distribution of (**a**) rice grains (**b**) corn grains.

2.3. Test Stand

Verification of the model of integrated energy consumption was performed on a test stand for the intelligent monitoring of performance characteristics of a multi-disc grinder. The test stand was composed of: a five-disc mill, a control panel with a performance monitoring system, and an integrated system for analysis of the grinding product's particle size (Figure 2).

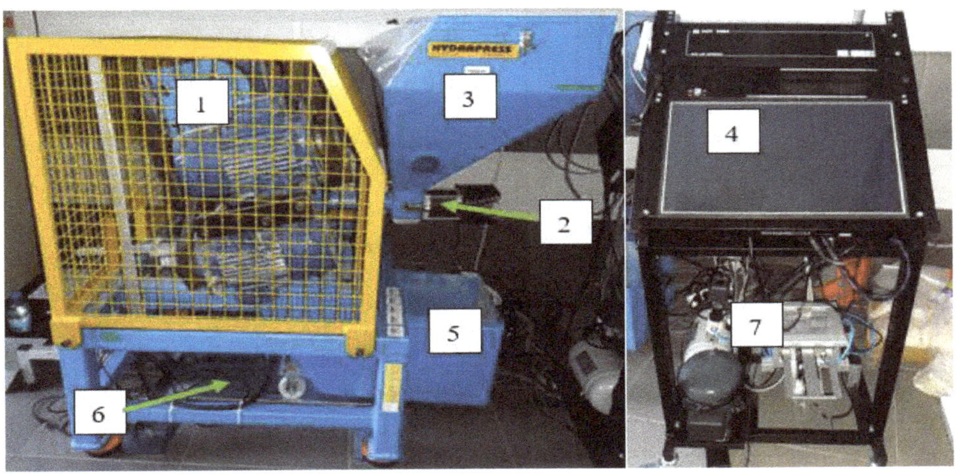

Figure 2. Test stand for the intelligent monitoring of granular mass characteristics: 1—five-disc grinder, 2—worm feeder, 3—hopper, 4—control unit, 5—reception basket of the product of grinding, 6—weight of the product of grinding, and 7—integrated system for the analysis of the grinding product's particle size [28].

A five-disc mill was equipped with a special integrated system for feed-dosing with a screw conveyor feeder, which makes it possible to control the feeding of the grain material into the grinding chamber. The grinding unit consists of five multi-hole discs fixed on the grinder shaft (Table 2). Each disc is powered individually by means of a dedicated motor which enables precise control of the angular speed of each disc. The final product is collected in the intake hopper with a scale fixed underneath, by means of which, the mass of the fragmented material is weighed. This makes it possible to determine the actual yield of the process. The integrated system of the particle size analysis makes it possible to assess the grinding product's granulometric composition and to determine the degree of fragmentation. The whole system is equipped with a set of sensors and archive modules. Control of the grinder and data reading is included in the control panel by means of the authors' own application (MŁYN, 2019).

Table 2. Structural parameters of the five-disc grinder.

Parameter	Unit	Disc 1	Disc 2	Disc 3	Disc 4	Disc 5
Disc diameter D_n	mm	274	274	274	274	274
Number of holes l_n	pcs.	14	22	27	33	39
Holes diameter d_n	mm	30	23	21	17,5	17,5

2.4. Research Methodology

Verification of the integrated energy consumption index model was carried out consistently with the plan presented in Figure 3.

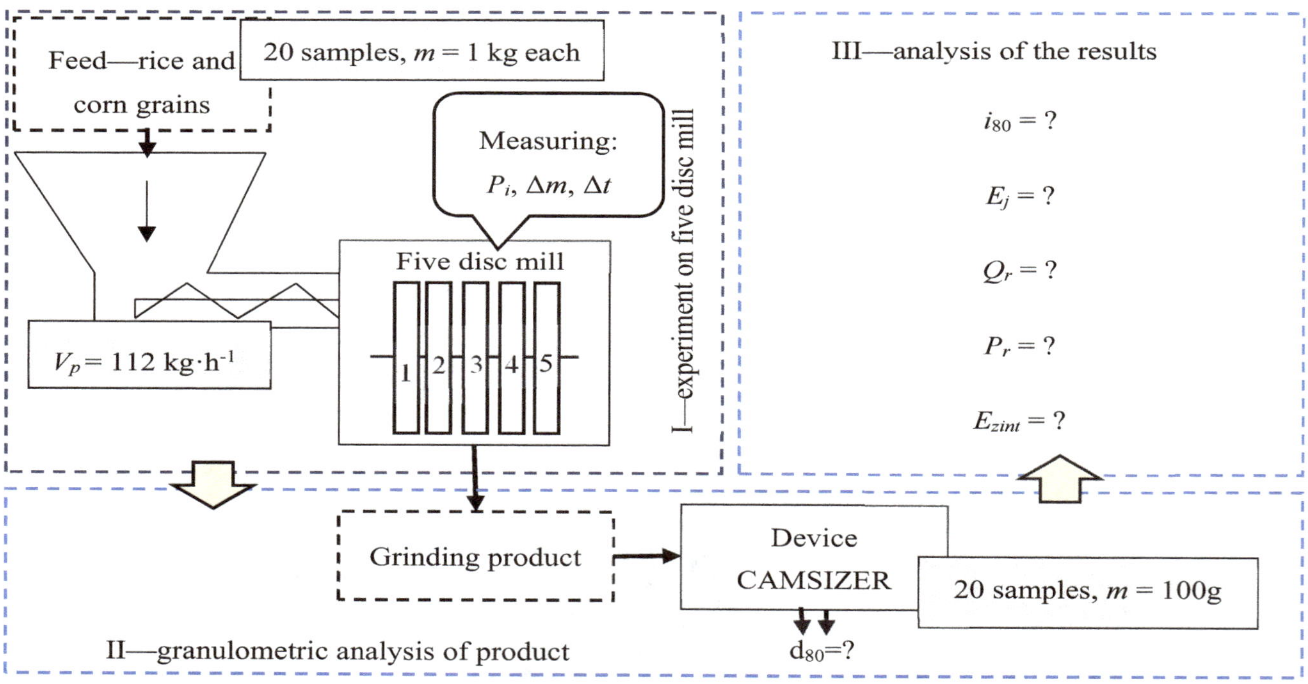

Figure 3. Research plan. V_p—feed dosing capacity, P_i—power consumption on the i-th disc, Δm—mass increase in the intake hopper, d_{80}—dimension of the sieve hole through which 80% of the grinding product passes, i_{80}—80% fragmentation degree, E_j—unit energy consumption, Q_r—yield of grinding, P_r—power consumed by the grinder, and E_{zint}—integrated energy consumption.

First, 20 one-kilogram samples of rice and corn were prepared. The moisture level was determined for each of them on a moisture balance MAC 210/NP (RADWAG, Radom, Poland) in accordance with ISO norm 1446 [82], and, additionally, average grain size before grinding on a CAMSIZER (Retsch Technology GmbH, Haan, Germany) device was also determined in accordance with ISO norm 13322-2:2006 [81]. Next, the prepared material was subjected to grinding on a five-disc grinder for 20 different grinding disc angular speed settings in accordance with the accepted test program [28,79] in order to investigate the dependence of the integrated energy consumption index on the total increase in the grinder disc angular speed $S\Delta\omega$ (Table 3). The minimal speed of the grinder discs accepted to be the point of reference was 20 rad·s^{-1}. The accepted settings of the disc angular speed were different in terms of the angular speed gradient $\Delta\omega$. The feed was delivered by means of a screw conveyor feeder with the speed of 112 kg·h^{-1}. Changes in the time of parameters such as power consumption, product mass in the intake hopper, angular speeds of the grinder discs, torques, and size of the product particles were recorded (with a sampling frequency every 0.5 s) in order to determine the values of the integrated energy consumption model criteria (components)—that is, power consumption, yield, fragmentation degree, and unit energy consumption.

Table 3. Settings of the five-disc mill control parameters [28].

Test Program	No. of Configuration	$S\Delta\omega$	$\Delta\omega$	ω_1	ω_2	ω_3	ω_4	ω_5
		rad·s^{-1}	rad·s^{-1}	rad·s^{-1}	rad·s^{-1}	rad·s^{-1}	rad·s^{-1}	rad·s^{-1}
I	1	50	5	20	25	30	35	40
	2	100	10	20	30	40	50	60
	3	150	15	20	35	50	65	80
	4	200	20	20	40	60	80	100
II	1	200	20	100	80	60	40	20
	2	150	15	80	65	50	35	20
	3	100	10	60	50	40	30	20
	4	50	5	40	35	30	25	20
III	1	40	20	20	40	20	40	20
	2	85	20	45	25	45	25	45
	3	225	25	75	50	75	50	75
	4	360	20	100	80	100	80	100
IV	1	40	20	20	40	20	40	20
	2	80	40	20	60	20	60	20
	3	120	60	20	80	20	80	20
	4	160	80	20	100	20	100	20
V	1	240	80	100	20	100	20	100
	2	280	60	100	40	100	40	100
	3	320	40	100	60	100	60	100
	4	360	20	100	80	100	80	100

RP—research program; ω_1, ω_2, ω_3, ω_4, and ω_5—angular speeds of discs, rad·s^{-1}; $\Delta\omega$—increase in angular speeds, rad·s^{-1}; and $S\Delta\omega$—total increase in angular speeds, rad·s^{-1}.

The next step involved both statistical and substantive analyses. The values of the integrated energy consumption were determined for each setting of the grinder disc according to dependence (22). Statistical analysis of the results was carried out by means of MS EXCEL (Microsoft, Redmond, WA, USA) and Statistica software (TIBCO Software Inc., Palo Alto, CA, USA). A general statistical analysis of the results, including the determination of the characteristics of the integrated energy consumption distribution and its variable components, was performed. Tests of the distribution of the normality of the variables were conducted by means of a Shapiro-Wilk test. This test is used to assess the distribution of the results. The null hypothesis of this test is that the population is normally distributed [83]. Thus, if the p-value is less than the chosen alpha level (in this study, $p > 0.05$), then the null hypothesis is rejected, and it can be assumed that the data are not characterized by normal distribution. When the p-value is greater than the chosen alpha level, it cannot be rejected that the data are characterized by normal distribution [83]. A correlation analysis using the Pearson method was performed for variables whose distribution was close to normal, while a Spearman correlation analysis was applied for the remaining ones. Next, a regression analysis was used. The results of the statistical analyses were found to be significant for $p < 0.05$.

3. Results and Discussion

In this part of the study, the test results of the integrated energy consumption index variables are described along with an analysis of their relations with the independent variable: the increase in the grinder disc angular speed $S\Delta\omega$. The results of the tests performed for the settings of the accepted grinder disc angular speed configurations for rice grinding are presented in Table S1 and, for corn, in Table S2.

3.1. *Test Results of the Integrated Energy Consumption Model Variables and Their Analysis*

The analysis included general descriptive characteristics of the resultant variables of power consumption, fragmentation degree, unit energy consumption, yield, the Shapiro-Wilk test for the normality of distribution, and correlation analysis with the Pearson method.

3.1.1. Power Consumption

Figure 4 shows the test results of the power consumption for rice and corn grinding on a five-disc grinder. The results indicate that the disc angular speed changes do affect power consumption. An increase in power consumption along with an increase in the grinder disc angular speed was observed. It can also be noticed that the power consumption for each disc setting was different depending on the material used (Figure 4). The power consumption for rice grinding was lower than the power consumption for corn grinding for all cases, which is mostly connected with the differing properties and internal structures of these materials [34].

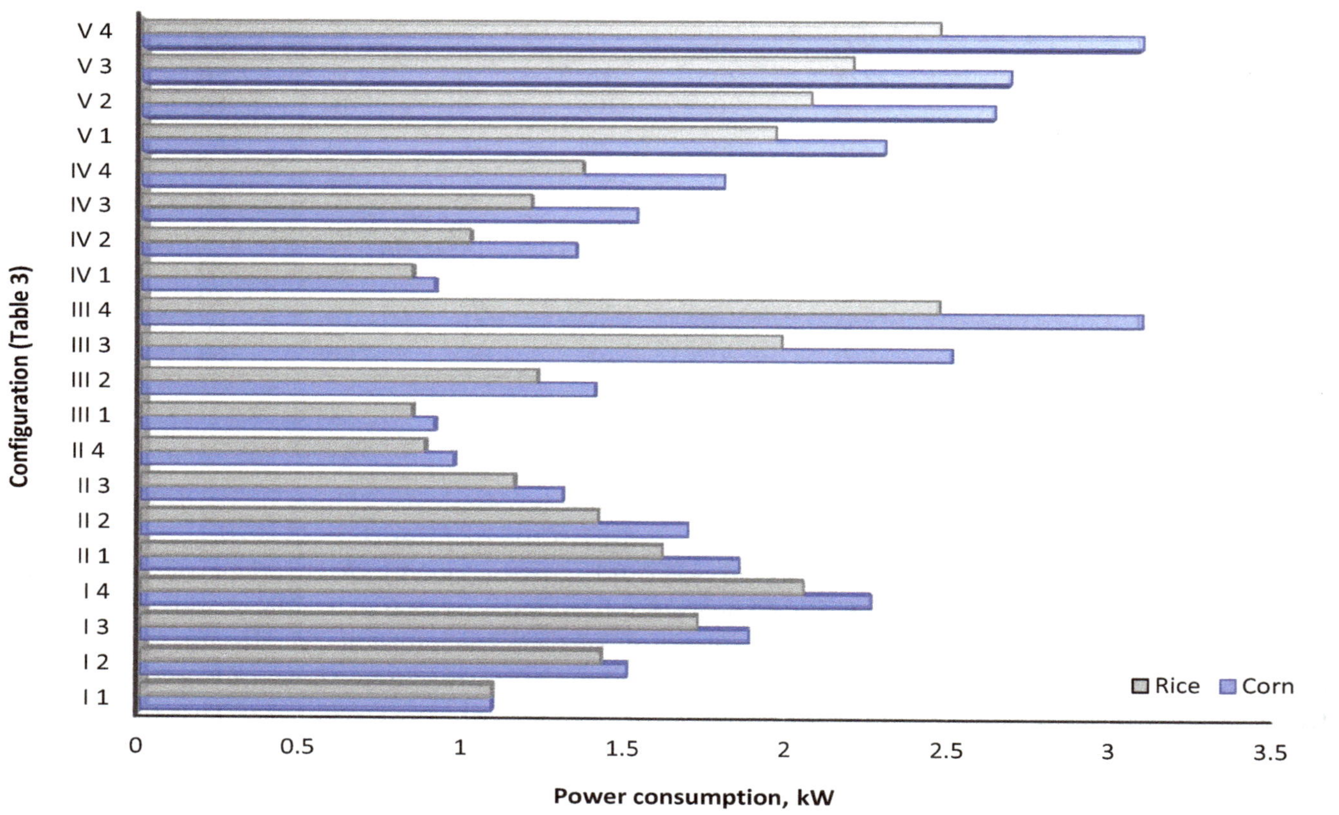

Figure 4. Test results of the power consumption for rice and corn grinding.

Earlier analyses of rice and corn strength indicated that higher forces are needed to grind corn grains, which is reflected by the power demand of grinding machines, due to the necessity of overcoming the higher resistance of moving elements [37]. An increased demand for power and energy in the case of a higher-strength grinding material was confirmed by [38–40], among others. An increased demand for power for corn grinding is also the effect of differences in the internal structure thereof [38–40]. Rice grains deprived of husks were used in the tests, whereas corn grains had the seed coat. Moreover, corn grains differ from rice grains in terms of endosperm structure. Grains with higher endosperm glassiness have been observed to need more energy and power in the process of grinding [36,41,42].

On the basis of the Shapiro-Wilk test (Appendix A, Tables A1 and A2) and the values of skewness and kurtosis (Appendix A, Tables A3 and A4) of the distribution of the power consumption test results, it was found that this distribution is approximately normal. Subsequently, the Pearson method was used for the analysis of correlation between the power consumption of grinding P_R and total speed increase $S\Delta\omega$. Correlation analysis revealed that there was a strong relation between the power consumption in grinding P_R and a total speed increase $S\Delta\omega$ during both rice grinding (R = 0.9658, Appendix A, Table A5) and corn grinding (R = 0.9847, Appendix A, Table A6).

On the basis of the linear regression analysis, equations describing the dependence of power consumption as a function of total speed increase were determined. It was found that power consumption for grinding can be described by means of a linear model with a high match (Figure 5). Both the coefficient in the equations and the regression models themselves were statistically significant (Tables 4 and 5).

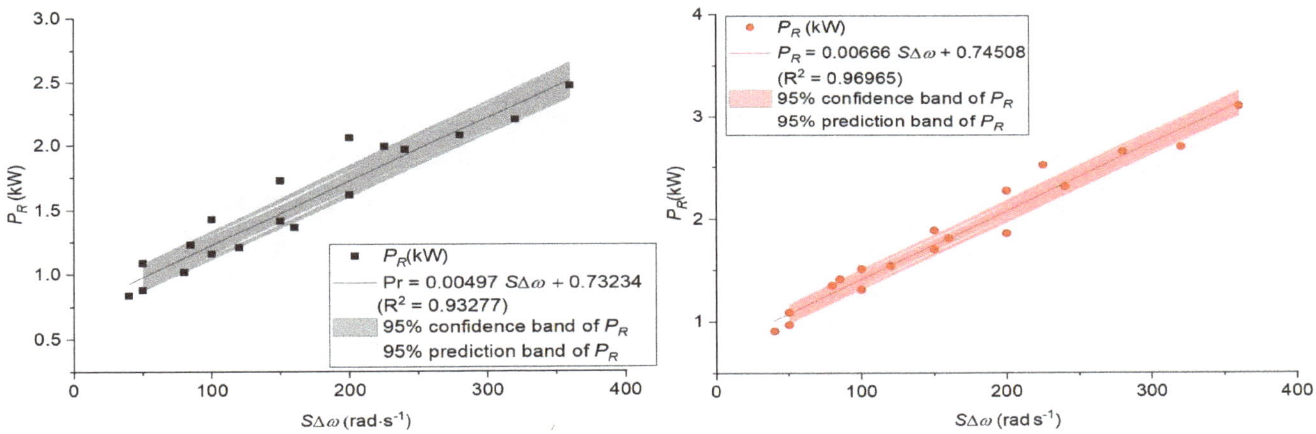

(a) (b)

Figure 5. Power consumption for grinding as a function of total speed increase; marking the confidence interval and values predicted for (**a**) rice grinding and (**b**) corn grinding.

Table 4. Test results of significance of the coefficients of the linear regression model of power consumption.

-	-	Value	Standard Error	*t*-Value	Prob > |*t*|
Rice	Intercept	0.73234	0.06115	11.97547	5.21551×10^{-10}
	Slope	0.00497	3.14735×10^{-4}	15.80313	5.36891×10^{-12}
Corn	Intercept	0.74508	0.05392	13.8175	5.05108×10^{-11}
	Slope	0.00666	2.77525×10^{-4}	23.98018	4.11081×10^{-15}

Table 5. Test results of the significance of the power consumption linear regression equation.

	Sum of Squares	Mean Square	F-Value	Prob > F
Rice	5.12697	5.12697	249.7388	5.36891×10^{-12}
Corn	9.17894	9.17894	575.04915	4.11081×10^{-15}

3.1.2. Yield

Figure 6 shows the results of rice and corn grinding yield tests performed on a five-disc grinder. The results indicate that the disc angular speed changes have an impact on this parameter change. As in the case of power consumption, it was observed that grinding yields increased along with an increase in the disc angular speeds. It could also be observed that, for the settings of the grinder disc angular speeds at which the first disc speed was the lowest, the rice grinding yield was significantly higher than that of the corn (Figure 6). For the remaining settings, the yield was similar though slightly higher for corn. This might result from the relations between the structural characteristics of working discs, speed, and the size of the grain used. The average size of rice grains was definitely lower than that of corn; hence, more of them can be introduced into the open space of the grinding discs.

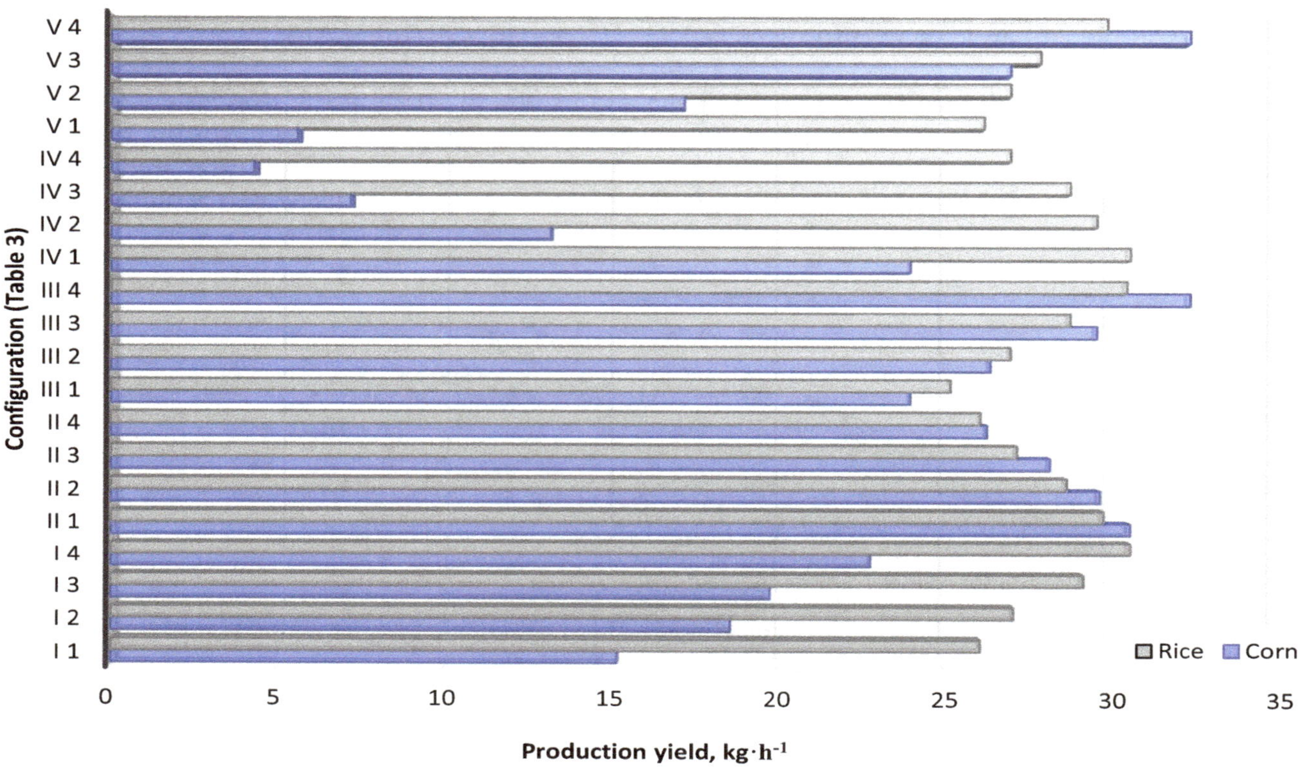

Figure 6. Test results of rice and corn grinding.

On the basis of the Shapiro-Wilk test (Appendix A, Tables A1 and A2) and the values of skewness and kurtosis (Appendix A, Tables A3 and A4) of the distribution of the performance test results, it was found that this distribution is approximately normal. Subsequently, the Pearson method was used in the analysis of correlation between grinding yield Q_r and the total speed increase. A correlation analysis showed that the correlation between the yield of grinding Q_r and the total speed increase $S\Delta\omega$ during both rice grinding (Appendix A, Table A5) and corn grinding was poor (Appendix A, Table A6).

Poor correlations between the disc total angular speed increased, and the yield might be the result of the grain strength properties and their relations and interactions with the grinding unit. The phenomena accompanying the flow of granular materials through the grinding chamber, the inter-hole space, and inter-disc space of the quasi-shearing unit are not without importance. There can occur, among other things, periods when the material lingers in the grinding chamber due to the adhesion of wet grains, which can be observed, e.g., in the case of corn and the poor results in the fourth test program.

3.1.3. Unit Energy Consumption

Figure 7 shows the test results of the unit energy consumptions for rice and corn grain grinding. For all the considered settings of the five-disc grinder angular speeds, the unit energy demand was higher for corn grinding (Figure 7), which was mostly connected with a higher power consumption during corn grinding, which, in turn, results from the grain strength properties which are discussed in Section 3.1.1. Earlier research also indicated a higher unit energy consumption for corn grinding [19]. The results indicate that an increase in the angular speed of discs caused an increase in energy consumption (Figure 7).

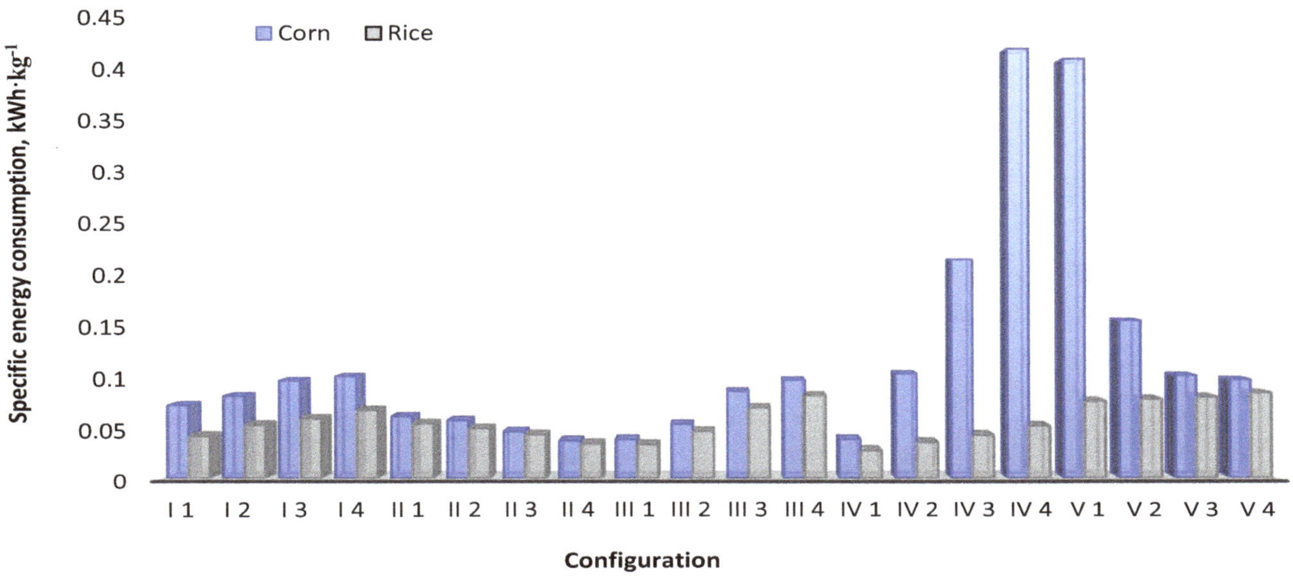

Figure 7. Test results of the unit energy consumption for rice and corn grinding.

On the basis of the Shapiro-Wilk test (Appendix A, Tables A1 and A2) and the values of skewness and kurtosis (Appendix A, Tables A3 and A4) of the distribution of the results of the unit energy consumption tests, it was found that this distribution is approximately normal for the grinding of rice, while for the grinding of corn, it deviated from normal. Subsequently, the Pearson method was used in an analysis of correlation between unit energy consumption E_j for rice grinding and total speed increase $S\Delta\omega$; whereas, in the case of corn, the Spearman method was used to assess these relations. Correlation analysis showed that there was a very strong positive correlation between the unit energy consumption E_j and total speed increase $S\Delta\omega$ for rice grinding (R = 0.95332, Appendix A, Table A5). In the case of corn grinding, the correlation of these two variables was moderate (R = 0.5963, Appendix A, Table A8).

An equation describing the dependence of the unit energy consumption for rice grinding as a function of the total speed increase was determined based on a linear regression analysis. It was found that the unit energy demand can be described by means of a linear model with a high match (Figure 8a).

In the case of the unit energy consumption involved in corn grinding, after the removal of outstanding values (results for settings IV3, IV4, and V1), it was the exponential model that provided the best match (Figure 8b) when compared to the others. This model was chosen on the basis of a ranking of the different nonlinear models. As a selection criterion, the value of the model match coefficient R^2 (the higher the value, the better the model) and the value of the standard error of the model coefficients (the lower the error, the better the model) were used. Unfortunately, the chosen model explained the variability of the unit energy consumption as a function of the total speed increase merely partially—slightly over 46% of the variability was explained by means of the chosen function.

Both the coefficients in the equations and the regression models were statistically significant (Tables 6 and 7). The standard error of the coefficients in the equations did not exceed 12%.

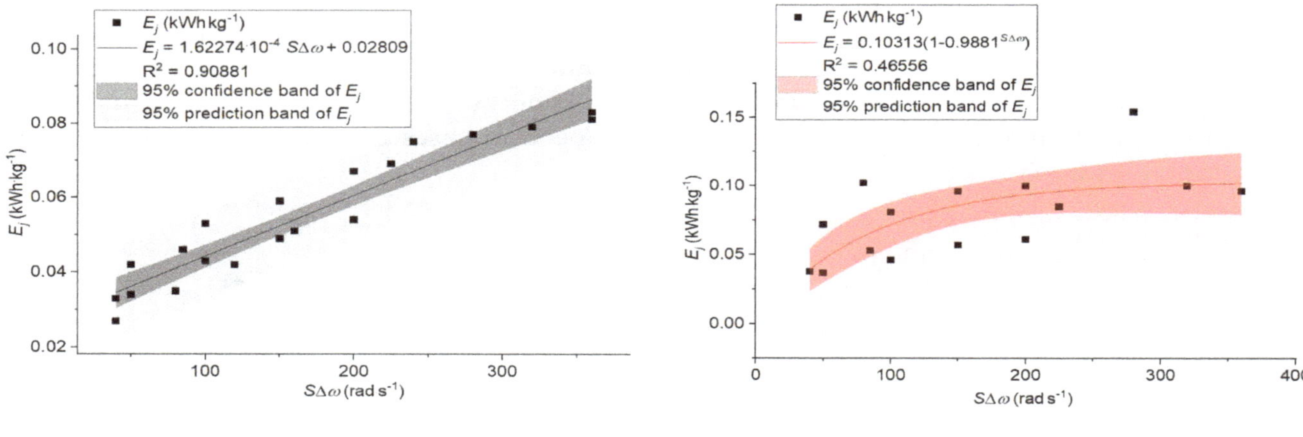

Figure 8. Unit energy consumption for grinding as a function of the total speed increase with indication of the confidence interval and predicted values for: (**a**) rice grain grinding and (**b**) corn grinding.

Table 6. Test results of the significance of the coefficients of the unit energy consumption regression model.

-	-	Value	Standard Error	*t*-Value	Prob > |*t*|
Rice	Intercept	0.02809	0.00235	11.93399	5.5149×10^{-10}
	Slope	1.62274×10^{-4}	1.21157×10^{-5}	13.39367	8.44112×10^{-11}
Corn	a	0.10313	0.01227	8.40195	4.68401×10^{-7}
	b	0.9881	0.00401	246.4428	0

a, b—coefficients of regression model (Figure 8b).

Table 7. Test results of the significance of the equations of the unit energy consumption regression model.

-	Sum of Squares	Mean Square	F-Value	Prob > F
Rice	0.00546	0.00546	179.39049	8.44112×10^{-11}
Corn	0.10853	0.05427	97.41927	2.55003×10^{-9}

The unit (specific) energy consumption depend on two factors: the energy consumed for the grinding process and the mass of grinded material, and it described the amount of energy needed for grinding one kg of material. According to [19], the unit energy consumption can be expressed as the ratio of power demand and grinding yield. For grinded materials, the dependence between the yield and discs' total speed increase was not specified, and as it is one of the unit energy components, it influences the dependence between the unit energy consumption and the discs' total speed increase. For corn, the yield values spread was significant; that is why the fitting is very low. The problems with the yield in the case of corn grinding was presented in Section 3.1.2 and was also reported in the previous study [28].

3.1.4. Fragmentation Degree

Figure 9 shows the test results of rice and corn grain fragmentations at 80% after grinding on a five-disc mill. A higher fragmentation degree was found for corn grain grinding for all analyzed disc angular speed settings (Figure 9), which was connected with the ratio of the grain dimension and the inter-disc gap. The grains of rice and corn had similar average dimensions in all tests, and this value depended on the value of the inter-disc gap—the size of a particle after grinding will always be smaller than the gap. Due to the fact that the characteristic size of corn grains was bigger before grinding than the size of rice grains (Table S1 and Table S2 in the Supplementary Materials) and their dimensions decreased down to those comparable to rice grains, the fragmentation degree of the corn grains was higher. It was observed, like in other studies [5,26], that for grinding bigger particles (corn has bigger particles than rice), the energy consumption is higher. It was also reported that the higher the disc speed (so the contacts between materials and cutting edges are more intensive), the grinding product has smaller particles [4,84]. On the basis of the Shapiro-Wilk test (Appendix A, Tables A1 and A2) and the values of the skewness and kurtosis (Appendix A, Tables A3 and A4) of the distribution of the grinding results, it was found that this distribution is approximately normal for both rice and corn grinding. Subsequently, the Pearson method was used in an analysis of the correlation between the grain fragmentation degree i_{80} and summary speed increase $S\Delta\omega$. Correlation analysis showed that there was a very strong positive relation between the rice fragmentation degree i_{80} and the total speed increase $S\Delta\omega$ (R = 0.93237, Appendix A, Table A5). In the case of corn grinding, the positive correlation between these two variables was moderate (R = 0.6893, Appendix A, Table A6).

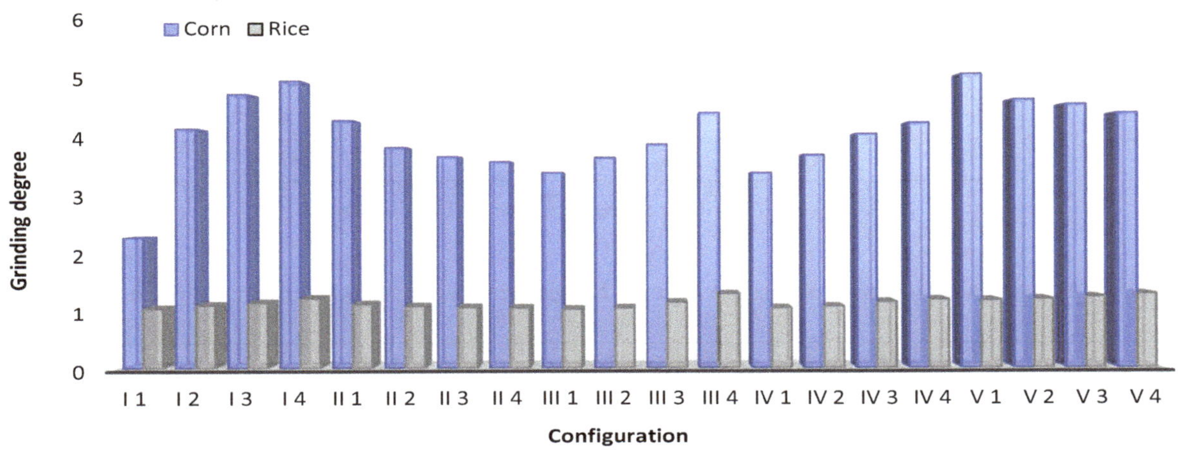

Figure 9. Test results of an 80% fragmentation degree for rice and corn grinding.

The equation describing the dependence of an 80% degree of rice fragmentation as a function of the total speed increase was determined based on an analysis of linear regression. It was found that an 80% fragmentation degree can be described with a high match by means of a linear model (Figure 10a). The obtained linear model was able to provide a 90% explanation of the fragmentation degree as a function of the total angular speed increase $S\Delta\omega$.

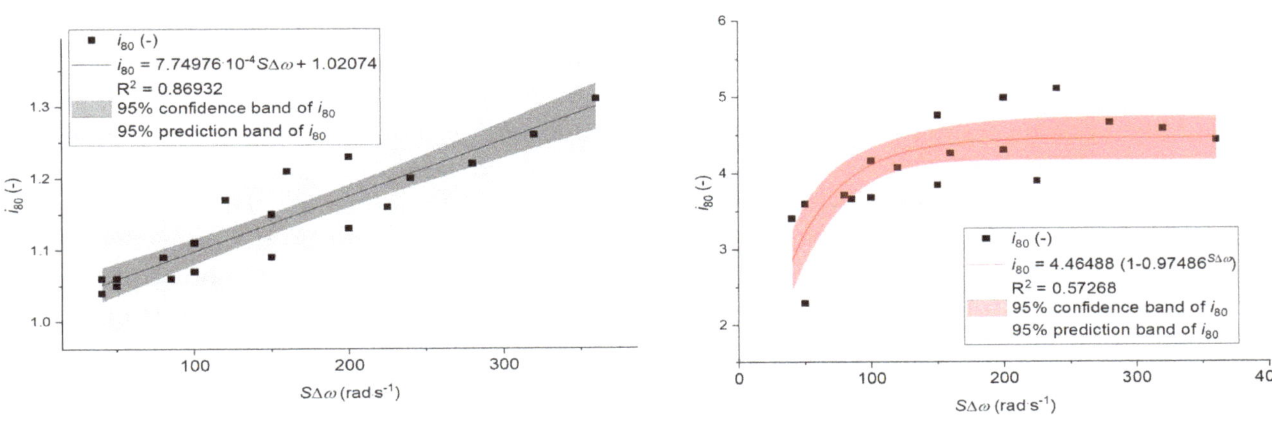

(a) (b)

Figure 10. An 80% fragmentation degree as a function of the total speed increase with the markings of the confidence interval and values predicted for: (**a**) rice grain comminution and (**b**) corn grain comminution.

For the case of the 80% fragmentation degree of corn, the linear model explained the fragmentation degree changes as a function of the total speed increase $S\Delta\omega$ only to a small extent ($R^2 = 0.47513$). Thus, the decision was made to check nonlinear models. As in the case of the unit energy demand, a ranking of models which took into account the value of the match coefficient R^2 and the value of the standard error of the model coefficients was used. It was the exponential model which best explained the highest percent of the 80% corn fragmentation degree variability as compared to the other models (Figure 10b). Unfortunately, the chosen model did not sufficiently explain the variability of the 80% fragmentation degree as a function of the total speed increase. Both the coefficients in the equations and the regression models themselves were statistically significant (Tables 8 and 9). The standard error of the coefficients in the equations did not exceed 10%.

Table 8. Test results of the regression model coefficient significance for an 80% fragmentation degree.

-	-	Value	Standard Error	*t*-Value	Prob > |*t*|
Rice	Intercept	1.02074	0.01376	74.17752	7.73924×10^{-24}
	Slope	7.74976×10^{-4}	7.08224×10^{-5}	10.94254	2.19489×10^{-9}
Corn	a	4.46488	0.13804	32.34576	0
	b	0.97486	0.00357	273.12834	0

a, b—coefficients of regression model (Figure 10b).

Table 9. Test results of the regression equation significance for an 80% fragmentation degree.

-	Sum of Squares	Mean Square	F-Value	Prob > F
Rice	0.12447	0.12447	119.73909	2.19489×10^{-9}
Corn	335.73877	167.86938	864.38272	1.31021×10^{-18}

3.2. *Analysis of the Integrated Energy Consumption Results*

The next step of the tests involved an analysis of the integrated energy consumption for rice and corn grain comminution. Figure 11 shows the results of the integrated energy consumption tests for both grains. In the case of corn, the highest values of the integrated energy consumption were found for setting no. 4 with RP II and, next, for setting no. 1 with RP III and setting no. 1 with RP IV. For rice grinding, the highest values of the integrated energy consumption were found for setting no. 1 in RP IV and, next, for setting no. 1 in RP III, setting no. 4 in RP II, and setting no. 2 in RP IV.

An analysis of the relations between the integrated energy consumption made it possible to determine the yield ranges, power consumption, unit energy consumption, and fragmentation degree for which the integrated efficiency is the highest (Figures 12 and 13, red color). The integrated energy consumption for rice grinding on a five-disc grinder assumes values within the interval (928–1420) when the grinding yield is included in the range (25.2–30.6) $kg \cdot h^{-1}$, unit energy consumption in the range (0.027–0.049) $kWh \cdot kg^{-1}$, power consumption in the range (0.88–1.02) kW, and fragmentation degree in the range (1.04–1.13) (Figure 12). The integrated energy consumption of corn grinding on a five-disc grinder assumes values in the range (1592–2640) when the grinding yield is within (19.8–32.4) $kg \cdot h^{-1}$, unit energy consumption within (0.037–0.061) $kWh \cdot kg^{-1}$, power consumption within (0.91–1.31) kW, and fragmentation degree in the range (2.75–4.21) (Figure 13).

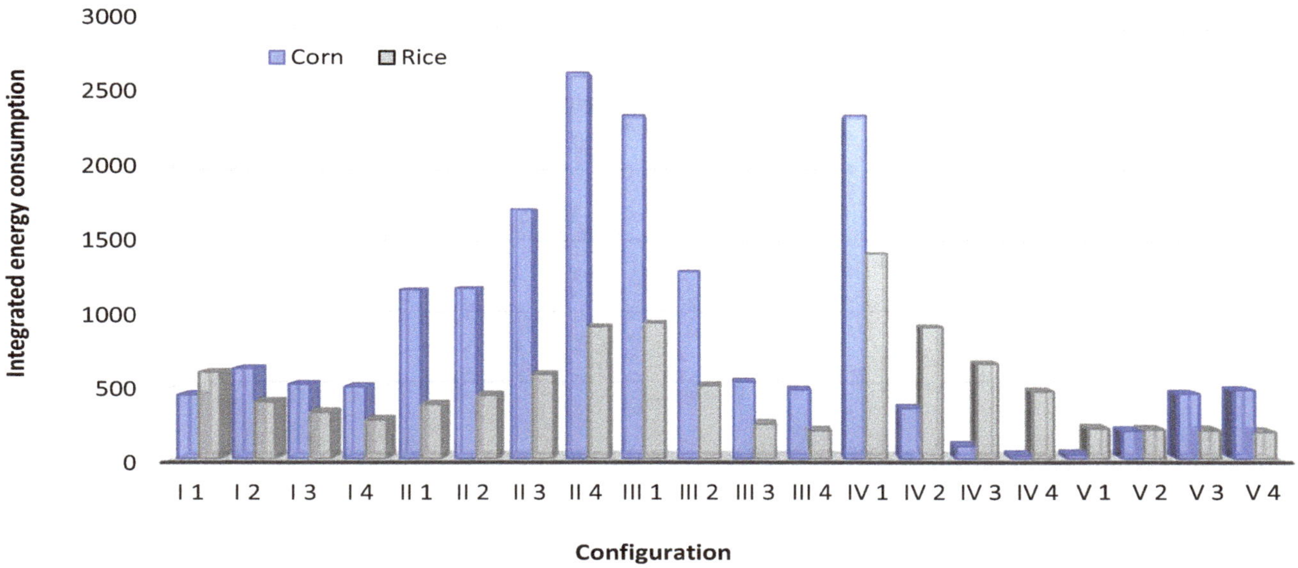

Figure 11. Test results of the integrated energy consumption for rice and corn grinding.

The results of the integrated energy consumption tests are characterized by a higher value scatter for corn grinding, and the scatter of results for power consumption, yield, fragmentation degree, and unit energy consumption were also higher for corn grinding than for rice grinding (Figures 12 and 13). Thus, by knowing the components of the integrated energy consumption, its values can be predicted with a higher accuracy for rice grinding than for corn grinding.

Next, a basic statistical analysis of the variable was performed. On the basis of the Shapiro-Wilk test (Appendix A, Tables A1 and A2) and the values of the skewness and kurtosis (Appendix A, Tables A3 and A4) of the distribution of the results, it was found that the distribution of the integrated energy consumption deviated from normal; therefore, in the correlation analysis, the Spearman coefficient was used to describe the monotonic relationships between the variables. Correlation analysis showed that there is a very strong negative correlation between the integrated energy consumption of rice grinding and total speed increase $S\Delta\omega$ (R = −0.95554, Appendix A, Table A7). In the case of corn grinding, the negative correlation of these two variables was moderate (R = −0.51131, Appendix A, Table A8).

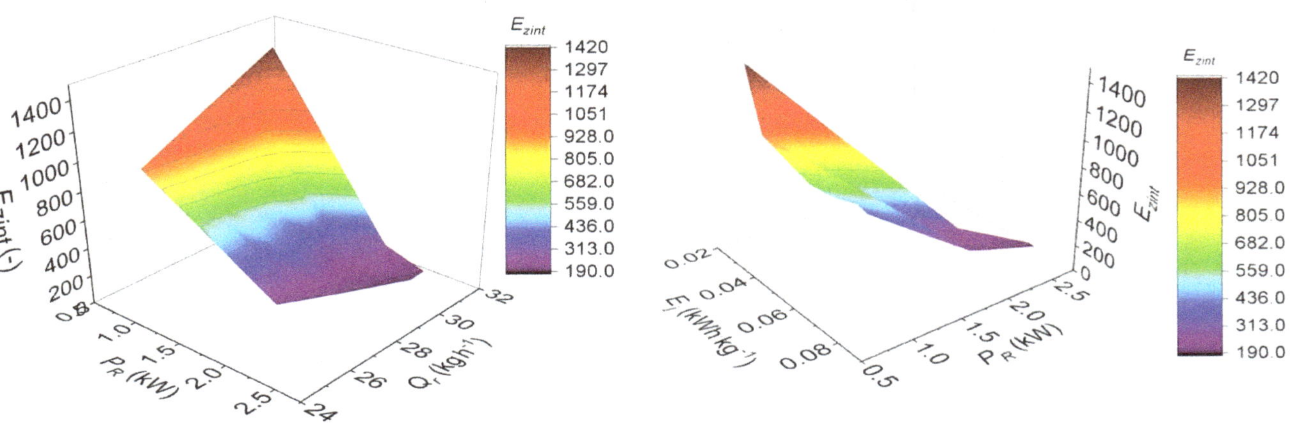

Figure 12. *Cont.*

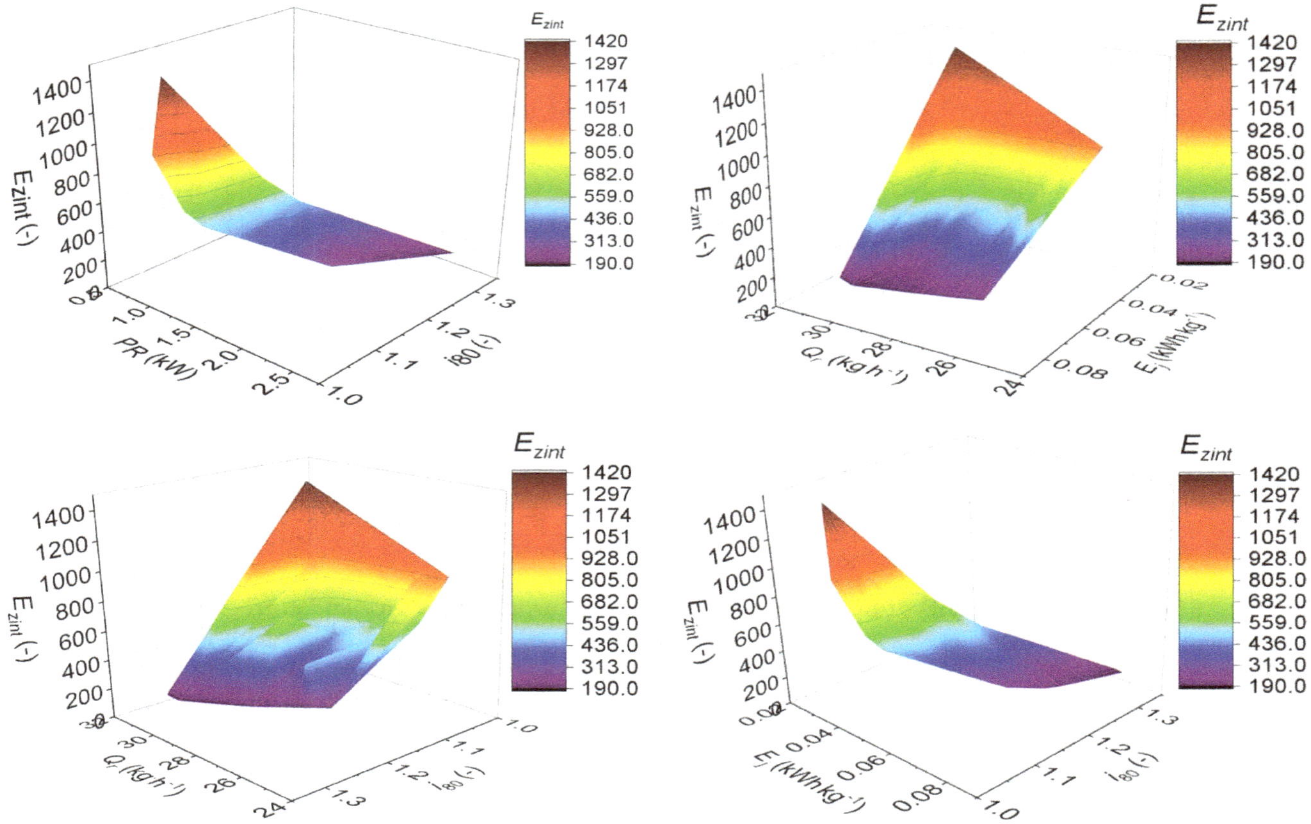

Figure 12. The integrated energy consumption of rice grinding as a function of the model components.

Nonlinear regression analysis was used to determine the equations describing the dependence of the rice and corn grinding's integrated energy consumption as a function of the total speed increase. In this case, the best-matched model was determined on the basis of the value of matching coefficient R^2 and the error values of the standard coefficients of the determined dependencies. It was found that the integrated energy consumption of rice grinding is best explained, nearly 77%, by an exponential model (Figure 14a). As regards corn, the best match was an exponential model, though it accounted for only 46% of the integrated energy consumption variability (Figure 14b). Both coefficients in the equations and the regression models themselves were statistically significant (Tables 10 and 11). The standard error of coefficients did not exceed 11% in the rice grinding's integrated energy consumption equations, and, in the case of corn, it did not exceed 26%.

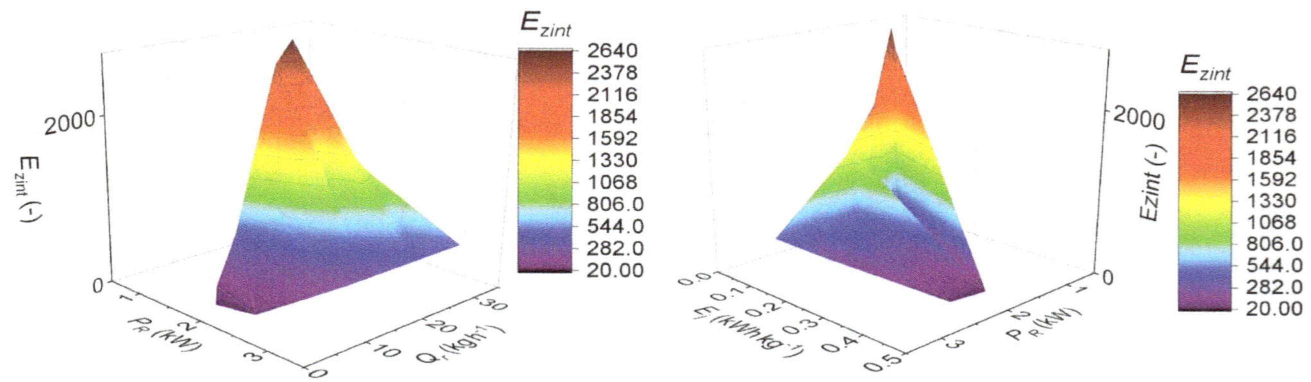

Figure 13. *Cont.*

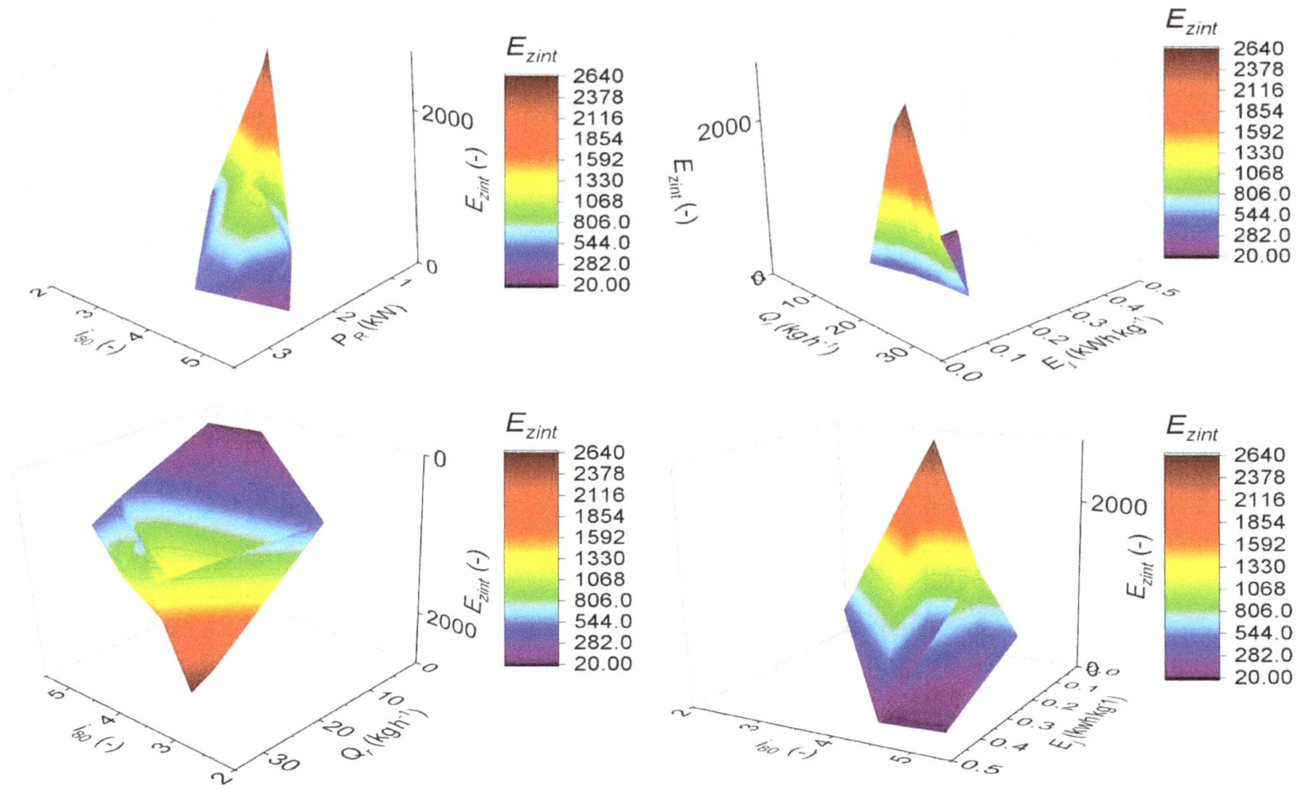

Figure 13. The integrated energy consumption for corn grinding as a function of the model variable components.

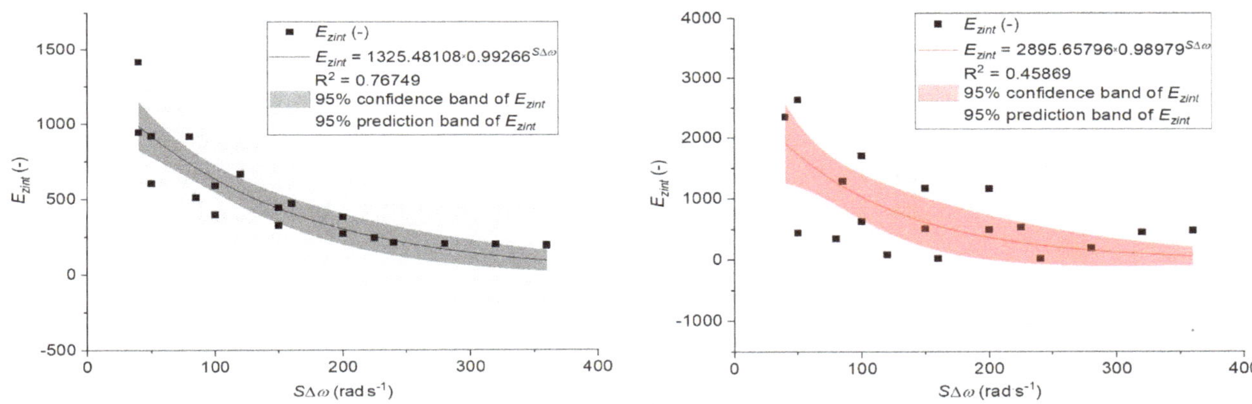

(**a**) (**b**)

Figure 14. The integrated energy consumption as a function of the total speed increase with the indication of the confidence interval and predicted values for: (**a**) rice grain comminution and (**b**) corn grain comminution.

The results of the analyses indicate that the rice and corn grinding processes differ in terms of yield, power consumption, fragmentation degree, unit energy consumption, and integrated energy consumption. In the case of corn, the process of grinding has more of a random character, which is mostly the result of the large differentiation of the grains of one species, which was also observed in other tests [20,28,85]. Hence, the match of models of unit energy consumption, yield, and fragmentation degree as a function of corn grinding speed increases, and subsequently, the models of the integrated energy consumption, which include the above-listed components, is rather poor.

Table 10. Results of significance tests for the coefficients of the integrated energy consumption regression model.

-	-	Value	Standard Error	*t*-Value	Prob > |*t*|
Rice	a	1325.48108	155.50266	8.52385	9.81711×10^{-8}
	b	0.99266	0.00124	802.86997	0
Corn	a	2895.65796	776.51222	3.72906	0.00154
	b	0.98979	0.00339	292.34877	0

a, b—coefficients of regression model (Figure 14a,b).

Table 11. Results of the significance tests for the equations of the integrated energy consumption regression.

	Sum of Squares	Mean Square	F-Value	Prob > F
Rice	6,691,362.97097	3,345,681.48549	125.99749	2.60167×10^{-11}
Corn	2.09017×10^{7}	1.04509×10^{7}	28.02063	2.96612×10^{-6}

Analyzing the values of the integrated energy consumption for the two grain species considered herein, it can be said that, from the point of view of the accepted criterion, a smaller increase in the angular speeds in relation to the reference speed on each disc (20 rad·s^{-1}) provides higher values of integrated energy consumption; e.g., the process of grinding is characterized by a low power consumption while maintaining the remaining parameters of grinding at a satisfactory level.

Verification of the integrated energy consumption model was performed for granular materials. It should be expected that, for a different type of biomass, e.g., wooden biomass, the value of the energy consumption index and its relations with the grinder disc total angular speeds would be different from the materials used in the study.

The proposed index of the integrated energy consumption can successfully be used to assess the grinding process and to select the grinder angular speeds to provide a low energy consumption, high yield, and a high fragmentation degree of the product. This index can find applications in automatic and control systems for industrial lines grinding energy materials, e.g., biomass, coal, and post-use polymer materials.

4. Summary and Conclusions

The aim of the study was achieved by developing and verifying the model of integrated energy consumption for grinding a biomass, which takes into account the fundamental parameters of the process: power consumption, unit energy consumption, yield, and the determination of the impact of changes in the grinding unit motion parameters on the values of the index developed herein.

The analyses of this study have proven that the index developed herein can be used successfully to assess a biomass grinding's energy consumption in an integrated approach—that is, one which includes not only energy consumption but also other parameters important from the point of view of the grinding process (yield, fragmentation degree, power consumption, and unit energy consumption). The verification of the model on a real object—a five-disc grinder—made it possible to determine the dependencies of the integrated energy consumption index and the motion parameters of the grinder discs for the total increase in disc annular speeds. Based on the analyses, it was found that the total speed increased, and thus, the increase in the disc angular speeds causes an increase in power consumption, in the fragmentation degree, and the unit energy consumption for both rice and corn grinding. In the case of rice grinding, these dependencies can be described with a high match by means of linear models. In the case of corn, a linear dependence was found only for power consumption. Nonlinear models were used for the unit energy consumption and fragmentation degree, though they accounted for only 60% of the studied parameter variability. The yield did not correlate with the total increase in angular disc speeds for either rice or corn grinding. This results mainly from physical mechanical properties of the grains and grain-grinding unit relations and confirms the variety and variability of grains observed so far within one type of grain and, in addition, confirms the randomness of the grinding process.

The test results prove that integrated energy consumption is negatively correlated with a total increase in the grinder angular speeds. The models which best represent the variability of the integrated energy consumption were exponential models; though, in the case of corn, this model was characterized by a lower matching coefficient ($R^2 = 0.45896$) than the energy consumption model for rice grinding ($R^2 = 0.76749$). Upon analyzing the values of the integrated energy consumption for two grain species, it can be observed that, in terms of the criterion accepted herein, a smaller angular speed increase in relation to the reference speed on each disc of the grinder (20 rad·s^{-1}) causes higher values of integrated energy consumption; that is, the process of grinding is characterized by a low power consumption while sustaining the remaining parameters of grinding at a satisfactory level.

The model of integrated energy consumption considered in this study can be a useful tool for integrated design and to select structural and process grinding parameters in preimplementation tests for grinders. Moreover, it can find an application in the automation and control systems of industrial lines' grinding energy materials, such as biomass, coal, and post-use polymers.

Further research on the structures and operations of biomass grinders in terms of their energy consumptions should be focused on finding the relations between selected structural characteristics of the grinding unit and negative environmental factors such as CO_2 emissions and natural resource depletion, as well as operation potentials.

Author Contributions: Conceptualization: W.K.; methodology, W.K. and J.F.; software, W.K.; validation, A.T. and J.F.; formal analysis, R.K. and P.B.-W.; investigation, W.K. and A.T.; resources, W.K.; data curation, W.K.; writing—original draft preparation, W.K.; writing—review and editing, W.K., A.T., J.F., R.K., and P.B.-W.; visualization, W.K.; supervision, J.F. and A.T.; project administration, W.K.; and funding acquisition, W.K. All authors have read and agreed to the published version of the manuscript.

Appendix A

Table A1. Tests results of the normality of the Shapiro-Wilk distribution of variables for the process of rice comminution.

-	DF	Statistic	*p*-Value	Decision at Level (5%)
P_R	20	0.93151	0.1651	Cannot reject normality
Q_r	20	0.93426	0.18646	Cannot reject normality
E_j	20	0.93621	0.20316	Cannot reject normality
i_{80}	20	0.92006	0.09935	Cannot reject normality
E_{zint}	20	0.85728	0.00708	Reject normality
$S\Delta\omega$	20	0.91736	0.08817	Cannot reject normality

P_r—grinding power consumption, kW; Q_r—grinding yield, kg·h^{-1}; E_j—unit energy consumption, kWh·h^{-1}; i_{80}—80% fragmentation degree; E_{zint} —integrated energy consumption; and $S\Delta\omega$—total increase in angular speeds, rad·s^{-1}.

Table A2. Test results of the normality of the Shapiro-Wilk distribution of variables for the process of corn grinding.

	DF	Statistic	*p*-Value	Decision at Level (5%)
P_R	20	0.93699	0.21026	Cannot reject normality
Q_r	20	0.91059	0.06541	Cannot reject normality
E_j	20	0.65731	1.22879×10^{-5}	Reject normality
i_{80}	20	0.95111	0.38428	Cannot reject normality
E_{zint}	20	0.83526	0.00305	Reject normality
$S\Delta\omega$	20	0.91736	0.08817	Cannot reject normality

P_r—power consumption for grinding, kW; Q_r—grinding yield, kg·h^{-1}; E_j—unit energy consumption, kWh·h^{-1}; i_{80}—80% fragmentation degree; E_{zint}—integrated energy consumption; and $S\Delta\omega$—total angular speed increase, rad·s^{-1}.

Table A3. Results of the descriptive statistical analysis of variables for the process of rice grinding.

-	Mean	Standard Deviation	Sum	Skewness	Kurtosis	Minimum	Median	Maximum	Range (Max.–Min.)
P_R	1.5555	0.53786	31.11	0.31892	−1.15303	0.84	1.425	2.48	1.64
Q_r	28.165	1.69528	563.3	−0.02646	−1.28024	25.2	28.3	30.6	5.4
E_j	0.05495	0.01778	1.099	0.2127	−1.26079	0.027	0.052	0.083	0.056
i_{80}	1.149	0.08681	22.98	0.50272	−0.89842	1.04	1.14	1.31	0.27
E_{zint}	505.65	328.92877	10113	1.32281	1.62855	190	422	1417	1227
$S\Delta\omega$	165.5	104.43961	3310	0.60399	−0.72265	40	150	360	320

P_r—power consumption for grinding, kW; Q_r—yield of grinding, kg·h^{-1}; E_j—unit energy consumption, kWh·h^{-1}; i_{80}—80% fragmentation degree; E_{zint}—integrated energy consumption; and $S\Delta\omega$ – total increase in angular speeds, rad·s^{-1}.

Table A4. Results of the descriptive statistical analysis of variables for the process of corn grinding.

-	Mean	Standard Deviation	Sum	Skewness	Kurtosis	Minimum	Median	Maximum	Range (Max.–Min.)
P_R	1.8465	0.70585	36.93	0.41186	−0.92384	0.91	1.755	3.11	2.2
Q_r	21.725	8.82627	434.5	−0.74904	−0.4979	4.3	24	32.4	28.1
E_j	0.118	0.11051	2.36	2.23553	4.35658	0.037	0.0905	0.422	0.385
i_{80}	4.0685	0.65617	81.37	−0.77544	1.47486	2.29	4.125	5.11	2.82
E_{zint}	872.15	807.93116	17443	1.12497	0.13387	24	511.5	2639	2615
$S\Delta\omega$	165.5	104.43961	3310	0.60399	−0.72265	40	150	360	320

P_r—power consumption for grinding, kW; Q_r—yield of grinding, kg·h^{-1}; E_j—unit energy consumption, kWh·h^{-1}; i_{80}—80% fragmentation degree; E_{zint}—integrated energy consumption; and $S\Delta\omega$—total increase in angular speeds, rad·s^{-1}.

Table A5. Results of the analysis of variable correlation using the Pearson method for the process of rice grinding.

		P_R	Q_r	E_j	i_{80}
$S\Delta\omega$	r-Pearson	0.9658*	0.37153	0.95332*	0.93237*
	p-value	5.36891×10^{-12}	0.10676	8.44112×10^{-11}	2.19489×10^{-9}

Two-tailed test of significance is used and * correlation is significant at the 0.05 level. P_r—power consumption for grinding, kW; Q_r—yield of grinding, kg·h^{-1}; E_j—unit energy consumption, kWh·h^{-1}; i_{80}—80% fragmentation degree; and $S\Delta\omega$—total angular speed increase, rad·s^{-1}.

Table A6. Results of the analysis of variable correlation using the Pearson method for the process of corn grinding.

		P_R	Q_r	i_{80}
$S\Delta\omega$	r-Pearson	0.98471*	0.25495	0.6893*
	p-value	4.11081×10^{-15}	0.27801	7.741×10^{-4}

Two-tailed test of significance is used and * correlation is significant at the 0.05 level. P_r—power consumption for grinding, kW; Q_r—yield of grinding, kg·h^{-1}; i_{80}—80% fragmentation degree; and $S\Delta\omega$—total angular speed increase, rad·s^{-1}.

Table A7. Results of the correlation analysis with the Spearman method for the process of rice grinding.

-	-	E_{zint}
$S\Delta\omega$	Spearman Corr.	−0.95554*
	p-value	5.48274×10^{-11}

Two-tailed test of significance is used and * correlation is significant at the 0.05 level. E_{zint}—integrated energy consumption and $S\Delta\omega$—total angular speed increase, rad·s^{-1}.

Table A8. Results of the analysis of variable correlation with the Spearman method for corn grinding.

-	-	E_j	E_{zint}
$S\Delta\omega$	Spearman Corr.	0.5963*	−0.51131*
	p-value	0.00552	0.02121

Two-tailed test of significance is used and * correlation is significant at the 0.05 level. E_j—unit energy consumption, kWh·h^{-1}; E_{zint}—integrated energy consumption; and $S\Delta\omega$—total angular speed increase, rad·s^{-1}.

References

1. Dzikuć, M.; Piwowar, A. Ecological and economic aspects of electric energy production using the biomass co-firing method: The case of Poland. *Renew. Sustain. Energy Rev.* **2016**, *55*, 856–862. [CrossRef]
2. Miao, Z.; Grift, T.E.; Hansen, A.C.; Ting, K.C. Energy requirement for comminution of biomass in relation to particle physical properties. *Ind. Crops Prod.* **2011**, *33*, 504–513. [CrossRef]
3. Wasiak, A. Technologies of Biofuel Production. In *Modeling Energetic Efficiency of Biofuels Production*; Green Energy and Technology; Springer: New York, NY, USA, 2019; pp. 29–34. ISBN 978-3-319-98431-5.
4. Mayer-Laigle, C.; Blanc, N.; Rajaonarivony, R.K.; Rouau, X. Comminution of dry lignocellulosic biomass, a review: Part I. from fundamental mechanisms to milling behaviour. *Bioengineering* **2018**, *5*, 41. [CrossRef] [PubMed]
5. Tumuluru, J.S.; Heikkila, D.J. Biomass grinding process optimization using response surface methodology and a hybrid genetic algorithm. *Bioengineering* **2019**, *6*, 12. [CrossRef]
6. Wróbel, M.; Mudryk, K.; Gąsiorski, A.; Posyłek, Z.; Dróżdż, T. Energy consumption during grinding of biomass for pelletisation [Nakłady energetyczne podczas mielenia biomasy przygotowywanej do procesu peletowania]. *Prz. Elektrotech.* **2017**, *93*, 229–232.
7. Eisenlauer, M.; Teipel, U. Comminution of biogenic materials. *Can. J. Chem. Eng.* **2017**, *95*, 1236–1244. [CrossRef]
8. Yılmaz, S.; Selim, H. A review on the methods for biomass to energy conversion systems design. *Renew. Sustain. Energy Rev.* **2013**, *25*, 420–430. [CrossRef]
9. Nnaemeka, N.; MacManus, N. Measurement of energy requirements for size reduction of palm kernel and groundnut shells for downstream bioenergy generation. *J. Eng. Technol. Res.* **2016**, *8*, 47–57. [CrossRef]
10. Hess, J.R.; Wright, C.T.; Kenney, K.L.; Searcy, E.M. *Uniform-Format Solid Feedstock Supply System: A Commodity-Scale Design to Produce an Infrastructure-Compatible Bulk Solid from Lignocellulosic Biomass–Executive Summary*; Idaho National Laboratory (INL): Idaho Falls, ID, USA, 2009.
11. Cundiff, J.S.; Grisso, R.D. Containerized handling to minimize hauling cost of herbaceous biomass. *Biomass Bioenergy* **2008**, *32*, 308–313. [CrossRef]

12. Kruszelnicka, W.; Bałdowska, P.; Tomporowski, A.; Piasecka, I.; Mroziński, A. Analiza korzyści energetycznych rozdrabniania nośników energii. *Inż. Apar. Chem.* **2018**, *57*, 7–8.
13. Wasiak, A.L. Effect of Biofuel Production on Sustainability of Agriculture. In *Proceedings of the 7TH International Conference on Engineering, Project, and Production Management*; Halicka, K., Nazarko, L., Eds.; Elsevier Science Bv: Amsterdam, The Netherlands, 2017; Volume 182, pp. 739–746.
14. Al-Hamamre, Z.; Saidan, M.; Hararah, M.; Rawajfeh, K.; Alkhasawneh, H.E.; Al-Shannag, M. Wastes and biomass materials as sustainable-renewable energy resources for Jordan. *Renew. Sustain. Energy Rev.* **2017**, *67*, 295–314. [CrossRef]
15. Buytaert, V.; Muys, B.; Devriendt, N.; Pelkmans, L.; Kretzschmar, J.G.; Samson, R. Towards integrated sustainability assessment for energetic use of biomass: A state of the art evaluation of assessment tools. *Renew. Sustain. Energy Rev.* **2011**, *15*, 3918–3933. [CrossRef]
16. Zhu, J.Y.; Pan, X.; Zalesny, R.S. Pretreatment of woody biomass for biofuel production: Energy efficiency, technologies, and recalcitrance. *Appl. Microbiol. Biotechnol.* **2010**, *87*, 847–857. [CrossRef] [PubMed]
17. Adapa, P.; Tabil, L.; Schoenau, G. Compaction characteristics of barley, canola, oat and wheat straw. *Biosyst. Eng.* **2009**, *104*, 335–344. [CrossRef]
18. Tomporowski, A. *Rozdrabnianie Alternatywnych Nośników Energii*; Wydawnictwa Uczelniane Uniwersytetu Technologiczno-Przyrodniczego w Bydgoszczy: Bydgoszcz, Poland, 2018.
19. Tomporowski, A.; Flizikowski, J.; Kruszelnicka, W. A new concept of roller-plate mills. *Przem. Chem.* **2017**, *96*, 1750–1755.
20. Mi, D.; Bahnasawy, A.; Ali, S.; El-Haddad, Z. Grinding Parameters and their Effects on the Quality of Corn for Feed Processing. *J. Food Process. Technol.* **2015**, *6*, 1.
21. Schwechten, D.; Milburn, G.H. Experiences in dry grinding with high compression roller mills for end product quality below 20 microns. *Miner. Eng.* **1990**, *3*, 23–34. [CrossRef]
22. Lee, H.-W.; Lee, J.-W.; Gong, S.-H.; Song, Y.-S. Study on the size reduction characteristics of miscanthus sacchariflorus via image processing. *J. Korean Wood Sci. Technol.* **2018**, *46*, 309–314. [CrossRef]
23. Lee, H.-W. Size reduction characteristics of yellow poplar in a laboratory knife mill. *J. Korean Wood Sci. Technol.* **2016**, *44*, 166–171. [CrossRef]
24. Williams, O.; Lester, E.; Kingman, S.; Giddings, D.; Lormor, S.; Eastwick, C. Benefits of dry comminution of biomass pellets in a knife mill. *Biosyst. Eng.* **2017**, *160*, 42–54. [CrossRef]
25. Tannous, K.; Lam, P.S.; Sokhansanj, S.; Grace, J.R. Physical Properties for Flow Characterization of Ground Biomass from Douglas Fir Wood. *Part. Sci. Technol.* **2013**, *31*, 291–300. [CrossRef]
26. Lopo, P. The right grinding solution for you: Roll, horizontal or vertical. *Food Manag.* **2002**, *53*, 23–26.
27. Mani, S.; Tabil, L.G.; Sokhansanj, S. Grinding performance and physical properties of wheat and barley straws, corn stover and switchgrass. *Biomass Bioenergy* **2004**, *27*, 339–352. [CrossRef]
28. Kruszelnicka, W. A New Model for Environmental Assessment of the Comminution Process in the Chain of Biomass Energy Processing. *Energies* **2020**, *13*, 330. [CrossRef]
29. Silva, G.G.D.; Couturier, M.; Berrin, J.-G.; Buléon, A.; Rouau, X. Effects of grinding processes on enzymatic degradation of wheat straw. *Bioresour. Technol.* **2012**, *103*, 192–200. [CrossRef]
30. Salman, A.D.; Ghadiri, M.; Hounslow, M. *Particle Breakage*; Elsevier: Amsterdam, The Netherlands, 2007; ISBN 978-0-08-055346-7.
31. Wołosiewicz-Głąb, M.; Foszcz, D.; Saramak, D.; Gawenda, T.; Krawczykowski, D. Analysis of a grinding efficiency in the electromagnetic mill for variable process and feed parameters. In Proceedings of the E3S Web of Conferences; EDP Sciences: Les Ulis, Francja, 2017; Volume 18, p. 01012.
32. Wang, Y.; Forsberg, E. *International Overview and Outlook on Comminution Technology*; Lulea Tekniska Universitet: Lulea, Sweden, 2014.
33. Pandiselvam, R.; Thirupathi, V.; Mohan, S. Engineering properties of rice. *Agric. Eng.* **2015**, *XL*, 69–78.
34. Tumuluru, J.S.; Tabil, L.G.; Song, Y.; Iroba, K.L.; Meda, V. Grinding energy and physical properties of chopped and hammer-milled barley, wheat, oat, and canola straws. *Biomass Bioenergy* **2014**, *60*, 58–67. [CrossRef]
35. Wiercioch, M.; Niemiec, A.; Roma, L. The Impact Of Wheat Seeds Size On Energy Consumption Of Their Grinding Process. *Inz. Rol.* **2008**, *103*, 367–372.
36. Warechowska, M. Some physical properties of cereal grain and energy consumption of grinding. *Agric. Eng.* **2014**, 239–249.

37. Dziki, D.; Laskowski, J. Influence of wheat kernel geometrical properties on the mechanical properties and grinding ability. *Acta Agrophys.* **2003**, *2*, 735–742.
38. Warechowska, M.; Warechowski, J.; Skibniewska, K.A.; Siemianowska, E.; Tyburski, J.; Aljewicz, M.A. Environmental Factors Influence Milling And Physical Properties And Flour Size Distribution Of Organic Spelt Wheat. *Tech. Sci.* **2016**, *19*, 387–399.
39. Dziki, D. Ocena energochłonności rozdrabniania ziarna pszenicy poddanego uprzednio zgniataniu. *Inz. Rol.* **2007**, *11*, 51–58.
40. Dziki, D.; Cacak-Pietrzak, G.; Miś, A.; Jończyk, K.; Gawlik-Dziki, U. Influence of wheat kernel physical properties on the pulverizing process. *J. Food Sci. Technol.* **2014**, *51*, 2648–2655. [CrossRef] [PubMed]
41. Greffeuille, V.; Mabille, F.; Rousset, M.; Oury, F.-X.; Abecassis, J.; Lullien-Pellerin, V. Mechanical properties of outer layers from near-isogenic lines of common wheat differing in hardness. *J. Cereal Sci.* **2007**, *45*, 227–235. [CrossRef]
42. Greffeuille, V.; Abecassis, J.; Barouh, N.; Villeneuve, P.; Mabille, F.; Bar L'Helgouac'h, C.; Lullien-Pellerin, V. Analysis of the milling reduction of bread wheat farina: Physical and biochemical characterisation. *J. Cereal Sci.* **2007**, *45*, 97–105. [CrossRef]
43. Shashidhar, M.G.; Murthy, T.P.K.; Girish, K.G.; Manohar, B. Grinding of Coriander Seeds: Modeling of Particle Size Distribution and Energy Studies. *Part. Sci. Technol.* **2013**, *31*, 449–457. [CrossRef]
44. Haffez, G.S.A. Correlation between Bond work index and mechanical properties of some Saudi ores. *J. Eng. Sci.* **2012**, *40*, 271–280.
45. Tavares, L.M. Optimum routes for particle breakage by impact. *Powder Technol.* **2004**, *142*, 81–91. [CrossRef]
46. Comminution and Laws of Comminution. Available online: http://www.thecementgrindingoffice.com/lawsofcomminution.html (accessed on 6 January 2018).
47. Sidor, J.; Froszcz, D.; Tomach, P.; Krawczykowski, D. Młyny wysokoenergetyczne do mielenia rud i surowców mineralnych. *CUPRUM Czas. Nauk. Tech. Gór. Rud* **2015**, *75*, 71–85.
48. Flizikowski, J.; Sadkiewicz, J. Inżynieria innowacji rozdrabniaczy ziarnistej biomasy. *Inż. Apar. Chem.* **2013**, *52*, 36–37.
49. Flizikowski, J. Test methods for the intelligent development of grinding processes. *Inż. Apar. Chem.* **2016**, *55*, 90–92.
50. Flizikowski, J.; Bieliński, K.; Bieliński, M. *Podwyższanie Energetycznej Efektywności Wielotarczowego Rozdrabniania Nasion Zbóż na Paszę*; Wydawnictwa Akademii Techniczno-Rolniczej: Bydgoszcz, Poland, 1994.
51. Zawada, J.; Chochoł, K. Doświadczalne badania energii kruszenia (na przykładzie modelowej kruszarki dźwigniowej Blake'a). *Gór. Geoinż.* **2009**, *33*, 317–326.
52. Zawada, J. *Wstęp do Mechaniki Procesów Kruszenia*; Wydawnictwo Technologii Eksploatacji: Radom, Poland, 1998.
53. Malewski, J. Analiza układu kruszenia rudy w ZWR Lubin. *Min. Sci.* **2015**, *22*, 111–118.
54. Krzemińska, M.; Malewski, J. Energochłonność operacji przygotowania rud do wzbogacania w kopalniach LGOM. *Prz. Gór.* **2011**, *67*, 143–147.
55. Brach, I. *Podstawowe Problemy w Procesach Rozdrabniania Minerałów*; Ośrodek Wydawnictw IOMB: Warszawa, Poland, 1963.
56. Sidor, J. Modelowanie procesu rozdrabniania materiału w młynach wibracyjnych. *Inż. Apar. Chem.* **2006**, *Nr 4s*, 129–130.
57. Kawatra, S.K. *Advances in Comminution*; Society for Mining, Metallurgy, and Exploration: Littleton, Colorado, 2006; ISBN 978-0-87335-246-8.
58. Fuerstenau, D.W.; Abouzeid, A.-Z.M. The energy efficiency of ball milling in comminution. *Int. J. Miner. Process.* **2002**, *67*, 161–185. [CrossRef]
59. Bielski, M. *Podstawy Teorii Organizacji i Zarządzania*; Wydawnictwo C. H. Beck: Warszawa, Poland, 2002.
60. Kruszelnicka, W. Efektywność wielotarczowego rozdrabniania biomasy ziarnistej. *Ekol. Tech.* **2018**, *28*, 24–26.
61. Macko, M. Size reduction by grinding as an important stage in recycling. In *Post-Consumer Waste Recycling and Optimal Production*; InTechOpen: London, UK, 2012; pp. 273–294.
62. Dziki, D.; Laskowski, J. Influence of selected factors on wheat grinding energy requirements. *TEKA Kom. Mot. Energ. Roln* **2005**, *5*, 56–64.
63. Steyn, C.W. Optimisation of a Fully Autogenous Comminution Circuit. Master's Thesis, University of Pretoria, Pretoria, South Africa, 2013.

64. Tomporowski, A.; Opielak, M.; Kowalik, K.; Sykut, B. Analiza nierównomierności rozdrabniania biomasy. Część I: Podstawy teoretyczne. *Inż. Apar. Chem.* **2013**, *52*, 42–43.
65. Kruszelnicka, W. Analiza wskaźników stosowanych do oceny rozdrabniania. *Inż. Apar. Chem.* **2017**, *56*, 208–210.
66. Kłos, Z. Classification and measurement of innovations. *J. Mech. Transp. Eng.* **2016**, *68*, 27–42. [CrossRef]
67. Kłos, Z. O kształtowaniu proinnowacyjnych i projakościowych postaw. *J. Mech. Transp. Eng.* **2013**, *65*, 53–63.
68. Flizikowski, J.; Kamyk, W. Efektywność energetyczna specjalnego systemu rozdrabniania. *Teka Kom. Bud. Ekspl. Masz. Elektrotech. Bud.* **2008**, *2*, 17–22.
69. Fuerstenau, D.W.; Kapur, P.C.; Gutsche, O. Comminution of single particles in a rigidly-mounted roll mill part 1: Mill torque model and energy investment. *Powder Technol.* **1993**, *76*, 253–262. [CrossRef]
70. Flizikowski, J. *Rozdrabnianie Tworzyw Sztucznych*; Wydawnictwa Akademii Techniczno-Rolniczej: Bydgoszcz, Poland, 1998.
71. Kruszelnicka, W. Basis of intelligent monitoring system in biomass comminution process. In *Actual problems of Modern Science*; Khmelnytsky National University: Chmielnicki, Ukraina, 2017; pp. 736–747.
72. Opielak, M.; Tomporowski, A. Structural features versus multi-hole grinding efficiency. *Eksploat. Niezawodn. Maint. Reliab.* **2012**, *14*, 222–227.
73. Tomporowski, A. Badanie użyteczności rozdrabniania ziaren ryżu. Część I-model badawczy. *Inż. Apar. Chem.* **2010**, *49*, 119–120.
74. Flizikowski, J.; Bielinski, K. *Technology and Energy Sources Monitoring: Control, Efficiency, and Optimization*; IGI Global: Hershey, PA, USA, 2012; ISBN 978-1-4666-2664-5.
75. Kruszelnicka, W.; Idzikowski, A.; Markowska, K.; Kasner, R. Quality Index of Multi-Disc Grinding Process of Grainy Biomass. *QPI* **2019**, *1*, 503–511. [CrossRef]
76. Tomporowski, A. *Studium efektywności Napędu i Rozwiązań Innowacyjnych Konstrukcji Wielotarczowych Rozdrabniaczy Ziaren Biomasy*; Lubelskie Towarzystwo Naukowe: Lublin, Poland, 2011.
77. Brandt, A.; Erickson, J.K.; Hallett, J.P.; Murphy, R.J.; Potthast, A.; Ray, M.J.; Rosenau, T.; Schrems, M.; Welton, T. Soaking of pine wood chips with ionic liquids for reduced energy input during grinding. *Green Chem.* **2012**, *14*, 1079–1085. [CrossRef]
78. Dukes, C.C.; Baker, S.A.; Greene, W.D. In-wood grinding and screening of forest residues for biomass feedstock applications. *Biomass Bioenergy* **2013**, *54*, 18–26. [CrossRef]
79. Kruszelnicka, W. *Analiza Procesu Wielotarczowego Rozdrabniania Biomasy w Ujęciu Energochłonności i Emisji CO2. Rozprawa Doktorska*; Uniwersytet Technologiczno-Przyrodniczy im. J. J. Śniadeckich w Bydgoszczy: Bydgoszcz, Poland, 2019.
80. Kruszelnicka, W. Study of Physical Properties of Rice and Corn Used for Energy Purposes. In *Springer Proceedings in Energy, Proceedings of the Renewable Energy Sources: Engineering, Technology, Innovation*; Wróbel, M., Jewiarz, M., Szlęk, A., Eds.; Springer International Publishing: Cham, Switzerland, 2020; pp. 149–162.
81. ISO 13322-2:2006. *Particle size Analysis—Image Analysis Methods —Part 2: Dynamic Image Analysis Methods*; International Organization for Standardization: Geneva, Switzerland, 2006.
82. ISO 1446:2001. *Green coffee—Determination of Water Content—Basic Reference Method*; International Organization for Standardization: Geneva, Switzerland, 2001.
83. 35406—How do I Interpret the Shapiro-Wilk Test for Normality in JMP®? Available online: https://www.jmp.com/support/notes/35/406.html (accessed on 9 March 2020).
84. Gutsche, O.; Fuerstenau, D.W. Fracture kinetics of particle bed comminution—Ramifications for fines production and mill optimization. *Powder Technol.* **1999**, *105*, 113–118. [CrossRef]
85. Kruszelnicka, W.; Marczuk, A.; Kasner, R.; Bałdowska-Witos, P.; Piotrowska, K.; Flizikowski, J.; Tomporowski, A. Mechanical and Processing Properties of Rice Grains. *Sustainability* **2020**, *12*, 552. [CrossRef]

12

Sustainable Wind Power Plant Modernization

Robert Kasner, Weronika Kruszelnicka *, Patrycja Bałdowska-Witos, Józef Flizikowski and Andrzej Tomporowski

Department of Manufacturing Technology, Faculty of Mechanical Engineering, University of Science and Technology in Bydgoszcz, 85-796 Bydgoszcz, Poland; robert.kasner@gmail.com (R.K.); patrycja.baldowska-witos@utp.edu.pl (P.B.-W.); fliz@utp.edu.pl (J.F.); a.tomporowski@utp.edu.pl (A.T.)

* Correspondence: weronika.kruszelnicka@utp.edu.pl

Abstract: The production of energy in wind power plants is regarded as ecologically clean because there being no direct emissions of harmful substances during the conversion of wind energy into electricity. The production and operation of wind power plant components make use of the significant potential of materials such as steel, plastics, concrete, oils, and greases. Energy is also used, which is a source of potential negative environmental impacts. Servicing a wind farm power plant during its operational years, which lasts most often 25 years, followed by its disassembly, involves energy expenditures as well as the recovery of post-construction material potential. There is little research in the world literature on models and methodologies addressing analyses of the environmental and energy aspects of wind turbine modernization, whether in reference to turbines within their respective lifecycles or to those which have already completed them. The paper presents an attempt to solve the problems of wind turbine modernization in terms of balancing energy and material potentials. The aim of sustainable modernization is to overhaul: assemblies, components, and elements of wind power plants to extend selected phases as well as the lifecycle thereof while maintaining a high quality of power and energy; high energy, environmental, and economic efficiency; and low harmfulness to operators, operational functions, the environment, and other technical systems. The aim of the study is to develop a methodology to assess the efficiency of energy and environmental costs incurred during the 25-year lifecycle of a 2 MW wind power plant and of the very same power plant undergoing sustainable modernization to extend its lifecycle to 50 years. The analytical and research procedure conducted is a new model and methodological approach, one which is a valuable source of data for the sustainable lifecycle management of wind power plants in an economy focused on process efficiency and the sustainability of energy and material resources.

Keywords: wind power plant; efficiency; sustainable development; modernization; recycling

---

1. Introduction

The production of energy in wind power plants is regarded as ecologically clean because there being no direct emission of harmful substances during the conversion of wind energy into electricity [1,2]. Analyses of the environmental impact of wind power plants have mainly concerned their impact on birds, vibration emissions, noise—both audible and of the infrasound type—as well the impact on the surrounding landscape [3–5]. In a wider context, analyses have been conducted on the impact of wind farm lifecycles with the aid of the Life Cycle Assessment (LCA) method, including areas of potential impact on human health, the quality of the natural environment, and natural resources [6–9]. Analyses have also concerned the impact of the lifecycle of wind power plants on water and soil environment as well as their emissions into the atmosphere [10–12]. Many papers have also dealt with the aspects of the impact of particular elements and structural assemblies of wind power plants during their lifecycle, elements which include rotors, blades, towers, and various types of foundations, both offshore and onshore, with the result being that foundations and towers are the greatest source of potential negative impacts [1,13–17].

Less attention, however, has been given to the relationship between the benefits, costs, and efficiency in a given lifespan of a wind power plant [18]. Kasner [19] introduced the concept of the universal wind power plant integrated efficiency indicator which allows one to assess the efficiency of the use of inputs (costs) on the generation, use, and post-use management of the materials and components of a wind power plant in three areas: potential emissions, energy consumption, and financial costs to obtain benefits in the form of electricity production. Piasecka et al. [20], however, has proposed a method of assessing the destructiveness of costs appearing in the wind power plant lifecycle in four categories: de-ergonomicity, defunctionality, environmental performance breakdown, and biosphere conservation breakdown, showing that the greatest negative impact of both onshore and offshore wind power plants occur in the area of de-ergonomicity. Much research has examined the benefits resulting from the operation of wind power plants, i.e., electricity production, or the ways to both predict or increase wind farm productivity [21–26]. Economic matters relating to the operation of wind power plants have also been addressed as one important aspect thereof [27–30].

The significant potential of construction materials such as steel, plastics, and concrete are used to produce wind power plant components. Energy costs must also be incurred in both the production and operation thereof [1]. In the process of decommissioning wind power plant units, both their material and energy potential must be utilized by means of landfilling or recycling. With regard to the material stages of the lifecycle, production, operation with servicing and power supply, and the post-use utilization of the wind power plant are significant in a complete environmental impact assessment [31–33].

Europe has launched international efforts to protect the environment in the adopted provisions of, inter alia, the Kyoto Protocol on climate change and the focus of European Union (EU) policy on sustainable development and low-carbon economies [34]. The concept of sustainable economic development requires that one take into account the environmental, energy, economic, and social aspects of the lifecycles of technical facilities as well as changes in the management of their lifecycles [35–39], e.g., by modernizing wind power plants and extending their lifecycles [40–42] (Figure 1).

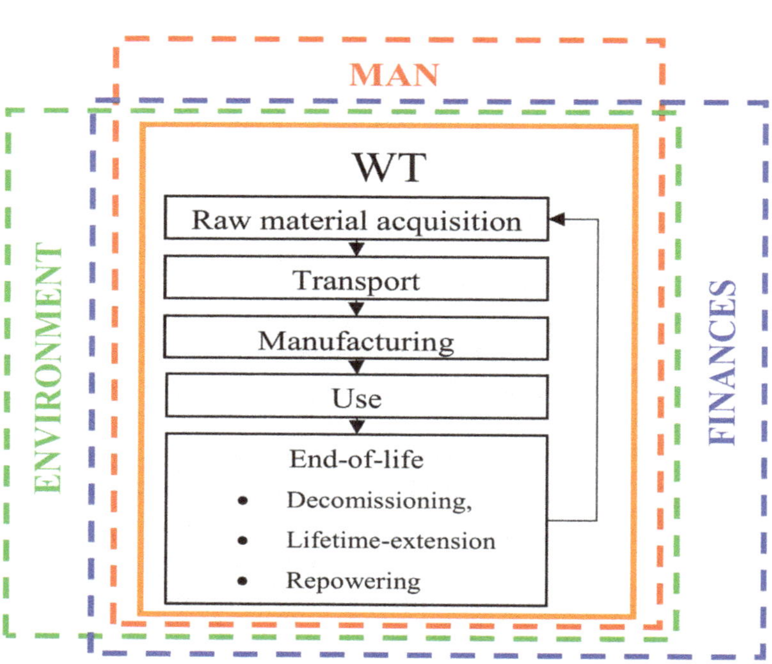

Figure 1. Spaces in sustainable lifecycle management, WT—wind turbine.

The lifespan of wind power plants is accepted to be 25 years and, in most cases, they are disassembled after 25 years [20,43–46]. Many components, including structural elements, can continue to operate for the next 25 years, while the wind power plant itself can operate much longer after the replacement of some components and elements with new ones, which will further reduce the harmful environmental impact of energy production in wind power plants [40,46]. There are basically three distinct strategies for dealing with the wind turbines at the end of their lifecycles: decommissioning, lifetime-extension, and repowering. The aim of decommissioning is to disassemble the wind power plant after it has reached the end of its lifecycle and to then recycle them [41,42,47,48]. Lifetime-extension, in turn, includes measures to extend the wind power plant's lifecycle by replacing worn out elements with new ones. This therefore includes the modernization of mechanical parts and control systems (Table S1 in Supplementary Materials) [40]. The idea behind repowering is to erect a new wind power plant with greater power production capacity at the site of the old one which has ended its lifecycle [40].

The modernization of wind power plants, the replacement of their parts with new ones, and the extension of their lifecycle all fall in line with the aim of sustainable development, and in particular, with the implementation of the slogan: "A resource-efficient Europe." Changes in wind power plant lifecycle management are primarily meant to be harmless, efficient, and of high-quality. Efforts to undertake sustainable modernization make it possible to:

- Reduce both emissions (e.g., carbon dioxide) and natural resource use and energy use per unit of electricity produced;
- Reduce the intensity of resource use when constructing wind power plants;
- Increase energy security by extending the lifecycle and availability of wind power plants.

In the world literature one usually finds analyses concerning mainly the assessment of wind power plant profitability and productivity after lifetime-extension and repowering [41,42,47,48], yet what is missing is a comprehensive assessment of the benefits as well as the ecological, energy, and economic costs arising from modernizations in wind power plant lifecycles. Ziegler et al. [41] performed analysis of the technical, economic, and legal possibilities relating to lifetime-extension showing that the profitability of these processes depend on energy market prices and thus differ depending on the country. Piel et al. [42] proposed a decision-support system for wind power plant operators on how to end the facility's lifecycle, taking into account land topography, wind resources, turbine type, and financial data in the profitability assessment of the processes of decommissioning, repowering, and lifetime extension. A methodology was also developed to assess the possibilities of extending the lifetime of wind turbine towers with special consideration given to wind and stress variables occurring in this structural element [49]. Martinez et al. [50] have shown that the repowering process causes an increase in the wind turbine's production of electricity at the same site with a smaller environmental impact. Studies concerning repowering have shown that such efforts give rise to an increase in both power and productivity [48]. There are no studies on the impact that standard replacements and modernizing replacements have (serving to extend wind power plant lifetime) on integrated efficiency. Therefore, it is worth considering the hypothesis as to whether the benefits and costs of operating a wind power plant after modernization, e.g., in a 50-year operation period, are as effective as in a 25-year lifecycle.

In light of the above, the aim of this paper is to develop a methodology to assess the efficiency of energy and environmental costs incurred in the lifecycle processes of a 2 MW wind power plant in a 25-year use-period and to assess the very same power plant after it has been subject to sustainable modernization to extend its lifecycle to 50 years. The analytical and research procedure conducted is a new model and methodological approach, one which is a valuable source of data for the sustainable lifecycle management of wind power plants in an economy focused on process efficiency and the sustainability of energy and material resources.

2. Materials and Methods

2.1. The Integrated Efficiency of Sustainable Modernization

Efficiency of operation is a goal, a state of energetics that makes it possible to estimate, optimize, modernize, and innovate ideas, constructions, and processes by comparing benefits, costs, and their relation in a given lifecycle. It means effectively using incurred costs to obtain benefits or expected results [19,51,52].

Efficiency in systems engineering is treated as a system feature, one that is measurable, useful for comparing systems of a given class, which expresses different aspects of performance in different time intervals and can be expressed differently depending on the class of the systems, their purpose and conditions [19,51,53].

The assessment of modernization efficiency herein was based on mathematical models which allow for the analyses and assessments consisting of a comparison of benefits and costs in a lifecycle.

To assess the operation of a wind power plant's lifecycle, an integrated efficiency indicator was defined [19], which can describe efficiency in the environmental, economic, and energy spheres depending on the reference point that is established [19]

$$E(t) = \frac{U(t)}{N(t)} \quad (1)$$

where:

E(t)—the integrated efficiency indicator in the lifecycle,
U(t)—the benefits in the lifecycle (environmental, energy, and economic),
N(t)—costs in the lifecycle (environmental, energetic, economic),
t—time of use.

The size of *U(t)* and *N(t)* indicate the values of the benefits obtained (the effects) and the costs incurred up to time t from the beginning of operation (t = 0).

Benefits from the operation of a wind power plant *U(t)* include, inter alia, a product in the form of energy, financial revenue, improvement in quality, reductions of emissions into the environment, the diversification of energy sources, the development and activation of the surroundings, as well as other potential benefits which as of today cannot be defined.

In a given period, the function *U(t)* can take the form of both positive and negative values [19],

$$u(t) = \frac{dU(t)}{dt} \quad (2)$$

while the value of function *U(t)*, on the basis of the elementary values *u(t)*, depending on whether they are continuous functions or discrete ones, can be determined from the following dependencies [19]

$$\begin{gathered} U(\tau) = \int_0^\tau u(t)dt \\ U(\tau) = \sum_i^l u_i \cdot \Delta t_i \end{gathered} \quad (3)$$

Operation costs *N(t)* are to be understood as: negative impacts on the environment, energy use in the entire lifecycle, financial costs incurred, negative societal costs, and potential costs which are currently unknown.

The function *N(t)* in the time interval (0, *t*) is a non-diminishing time function, that is, in each of the elementary intervals (Δt or dt/) the elementary value of costs is not less than zero [19]

$$n(t) = \frac{dN(t)}{dt} \quad (4)$$

while the value of function $N(t)$, depending on whether the function $n(t)$ is continuous or discrete, can be determined from the dependency [19]

$$N(\tau) = \int_0^{\tau} n(t)dt \tag{5}$$

$$N(\tau) = \sum_i^T n_i \cdot \Delta t_i \tag{6}$$

The cost can be expressed with various elements in various units.

For a wind turbine undergoing modernization during its use, the integrated efficiency indicator is defined by a system of equations (Figure 2)

$$E(t) = \begin{cases} \frac{U_r \cdot t}{N_W^1 + N_r \cdot t + N_Z^1} & \text{dla } 0 \le t \le t_{LC1} \\ \frac{U_r \cdot t}{(N_W^1 + N_W^2) + N_r \cdot t + (N_Z^1 + N_Z^2)} & \text{dla } t_{LC1} < t \le t_{LC2} \\ \frac{U_r \cdot t}{(N_W^1 + N_W^2 + N_W^3) + N_r \cdot t + (N_Z^1 + N_Z^2 + N_W^3)} & \text{dla } t_{LC2} < t \le t_{LC3} \\ \vdots & \\ \frac{U_r \cdot t}{\sum_{i=1}^{n+1} N_W^i + N_r \cdot t + \sum_{i=1}^{n+1} N_Z^i} & \text{dla } t_{LC(n)} < t \le t_{LC(n+1)} \end{cases} \tag{7}$$

where: U_r average annual productivity, t – time counted from the beginning of the use-stage, n – number of use-stages throughout the lifecycle, t_{LCn} – the end of the subsequent use-stage, N_W^1, N_W^2, N_W^3, N_W^n – costs at the production-stage of elements in the first, second, third n- nth stages, respectively, N_r – average annual costs at each stage of use, N_Z^1, N_Z^2, N_Z^3, N_Z^n – costs at the post-use management stage of elements utilized in the first, second, third, n-nth use-stages, respectively.

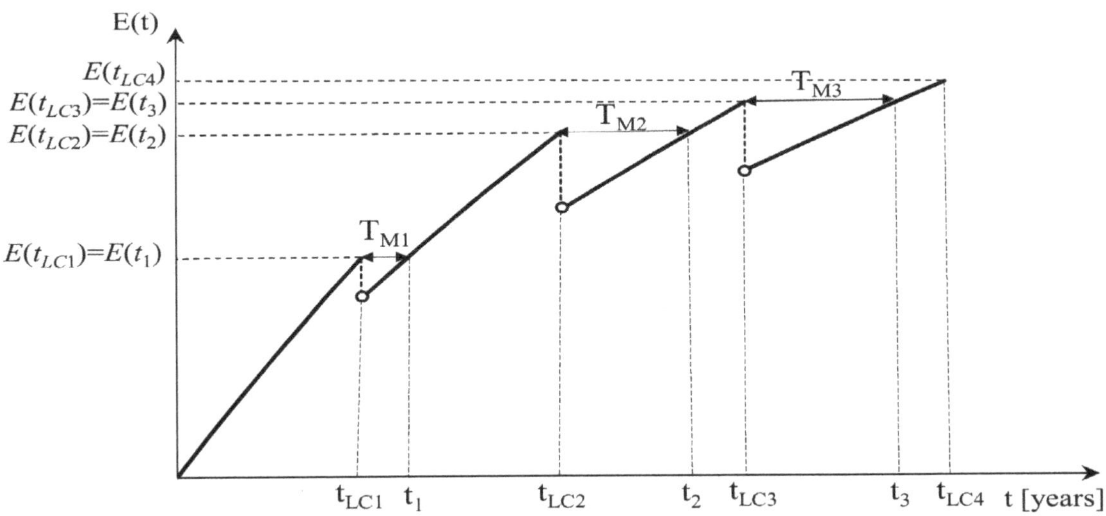

Figure 2. A graphical interpretation of the integrated efficiency indicator in the lifecycle (dependency (7)), including the sustainable modernization indicator (dependence (20)) and the time of the return on costs for modernization (dependence (18)).

The model proposed herein can be applied to assess the efficiency of existing wind power plants as well as those being designed to compare the effects of their operation after the modernization and extension of their lifecycle.

Sustainable modernization, which in essence refers to the process of replacing structura components with new or modernized ones, is a process that encompasses multiple aspects, al of which serve the goal of erecting a wind power plant with the use-properties of a new one. Thi process includes the use of wind power plant materials and resources, and additionally, its executive and energy resources after completion of the use-stage.

2.2. Payback Time for Moderinzation Costs

The first indicator of the assessment of sustainable modernization is time T_{Mn} after which there is a return to the pre-modernization efficiency of the use of costs (Figure 2). The payback time for modernization costs is defined as

$$T_{Mn} = t_n - t_{LCn} \tag{8}$$

where:
t_n – the time after which post-modernization efficiency is equal to pre-modernization efficiency, counted from the beginning of use,
t_{LCn} – the end of a subsequent use-stage
Time T_{Mn}, after which pre-modernization efficiency is achieved, is essentially unknown, but determinable. If one condition is met (Figure 2)

$$E(t_n) = E(t_{LCn}), \quad \text{gdzie } \ t_n \neq t_{LCn} \tag{9}$$

then after modernization at time t_n one achieves the efficiency of the source of energy which is the moment modernization work was begun at time t_{LCn}. The result is the equation

$$\frac{U_r \cdot t_n}{\sum_{i=1}^{n+1} N_W^i + N_r \cdot t_n + \sum_{i=1}^{n+1} N_Z^i} = \frac{U_r \cdot t_{LCn}}{\sum_{i=1}^{n} N_W^i + N_r \cdot t_{LCn} + \sum_{i=1}^{n} N_Z^i} \tag{10}$$

After transformations one obtains

$$t_n\left(\sum_{i=1}^{n} N_W^i + N_r \cdot t_{LCn} + \sum_{i=1}^{n} N_Z^i\right) = t_{LCn}\left(\sum_{i=1}^{n+1} N_W^i + N_r \cdot t_n + \sum_{i=1}^{n+1} N_Z^i\right) \tag{11}$$

$$t_n \cdot \sum_{i=1}^{n} N_W^i + t_n \cdot t_{LCn} \cdot N_r + t_n \cdot \sum_{i=1}^{n} N_Z^i - t_{LCn} \cdot \sum_{i=1}^{n+1} N_W^i - t_n \cdot t_{LCn} \cdot N_r - t_{LCn} \cdot \sum_{i=1}^{n+1} N_Z^i = 0 \tag{12}$$

$$t_n \cdot \sum_{i=1}^{n} N_W^i + t_n \cdot \sum_{i=1}^{n} N_Z^i - t_{LCn} \cdot \left(N_W^{n+1} + \sum_{i=1}^{n} N_W^i\right) - t_{LCn} \cdot \left(N_Z^{n+1} + \sum_{i=1}^{n} N_Z^i\right) = 0 \tag{13}$$

$$t_n \cdot \sum_{i=1}^{n} N_W^i + t_n \cdot \sum_{i=1}^{n} N_Z^i - t_{LCn} \cdot N_W^{n+1} - t_{LCn} \cdot \sum_{i=1}^{n} N_W^i - t_{LCn} \cdot N_Z^{n+1} - t_{LCn} \cdot \sum_{i=1}^{n} N_Z^i = 0 \tag{14}$$

$$(t_n - t_{LCn}) \cdot \sum_{i=1}^{n} N_W^i + (t_n - t_{LCn}) \cdot \sum_{i=1}^{n} N_Z^i = t_{LCn} \cdot N_W^{n+1} + t_{LCn} \cdot N_Z^{n+1} \tag{15}$$

$$(t_n - t_{LCn}) \cdot \left(\sum_{i=1}^{n} N_W^i + \sum_{i=1}^{n} N_Z^i\right) = t_{LCn}\left(N_W^{n+1} + N_Z^{n+1}\right) \tag{16}$$

$$t_n - t_{LCn} = \frac{t_{LCn}\left(N_W^{n+1} + N_Z^{n+1}\right)}{\sum_{i=1}^{n} N_W^i + \sum_{i=1}^{n} N_Z^i} = \frac{t_{LCn}\left(N_W^{n+1} + N_Z^{n+1}\right)}{\sum_{i=1}^{n} (N_W^i + N_Z^i)} \tag{17}$$

Because $T_{Mn} = t_n - t_{LCn}$, the dependency of the payback time on pre-modernization efficiency is expressed as follows:

$$T_{Mn} = \frac{t_{LCn}\left(N_W^{n+1} + N_Z^{n+1}\right)}{\sum_{i=1}^{n} (N_W^i + N_Z^i)} \tag{18}$$

2.3. *The Sustainable Modernization Indicator*

The second indicator of sustainable modernization assessment is the indicator of sustainable modernization E_M, which expresses the relation of the integrated efficiency indicator after the completion of the use-stage after modernization $E(t_{LC(n+1)})$ is performed to the integrated efficiency indicator before the beginning of modernization $E(t_{LCn})$ (Figure 2)

$$E_{Mn} = \frac{E(t_{LC(n+1)})}{E(t_{LCn})} \tag{19}$$

$$E_{Mn} = \frac{\dfrac{t_{LC(n+1)} \cdot U_r}{\sum_{i=1}^{n+1} N_W^i + \sum_{i=1}^{n+1} N_Z^i + t_{LC(n+1)} \cdot N_r}}{\dfrac{t_{LCn} \cdot U_r}{\sum_{i=1}^{n} N_W^i + \sum_{i=1}^{n} N_Z^i + t_{LCn} \cdot N_r}} = \frac{t_{LC(n+1)}\left[\sum_{i=1}^{n} \left(N_W^i + N_Z^i\right) - t_{LCn} \cdot N_r\right]}{t_{LCn}\left[\sum_{i=1}^{n+1} \left(N_W^i + N_Z^i\right) - t_{LC(n+1)} \cdot N_r\right]} \tag{20}$$

where E_{Mn} is the efficiency of subsequent modernizations

Figure 2 presents the dependence of integrated efficiency as a function of the time of use. Integrated efficiency as a function of time is a non-decreasing function. Its increase at the use-stage results from the benefits that grow in time that are generated at this stage. In the wind power plant's lifecycle, an increase in the benefits of its operation is observed (an increase in the electricity produced), with a constant value of both the costs to produce elements and costs relating to post-use management, and a slight increase in costs at the use-stage (maintenance, repairs, energy consumption). The increase in costs at the use-stage is much smaller than the increase in benefits at the same time. Considering the fact that the efficiency indicator is a ratio of benefits to costs, with the assumptions above, said indicator will be an increasing function. At the time of modernization (t_{LCn}), a reduction in integrated efficiency results from the calculation of the costs to manufacture the replaced elements and to undertake their post-use management during modernization, which increases the value of costs throughout the lifecycle.

2.4. *Methodology for Determining the Benefits and Costs in a Wind Power Plant's Lifecycle*

A Vestas V90/105 m wind power plant with a nominal electrical capacity of 2 MW located in central Poland was analyzed in detail. Table 1 presents the basic data of the wind power plant analyzed with the LCA method.

Table 1. Basic data of the Vestas V90 wind turbine.

Description	Unit	Quantity
Lifetime	years	25
Rating per turbine	MW	2
Generator type	-	Three-phase asynchronous generator
Hub height	m	105
Rotor diameter	m	90
Tower type	-	Standard steel
Production	MWh per year	5325
Plan location	-	Poland

The Vestas V90 three-blade rotor with a 90 m diameter and the nacelle housing the main shaft generator, transformer, gearboxes, and brakes are located atop a 105 meter tower. The Vesta V90 wind turbine is an upwind turbine with an adjustable blade pitch system featuring active directionality The turbine is equipped with a 2.0 MW power-rated generator. It uses a OptiTip®microprocessing system for blade pitch control as well as the OptiSpeedTM function (for speed regulation). Owing to these functions, the wind turbine rotor can work at variable rotational speeds, helping to keep power output at or close to the rated power (data from the producer) [54].

The V90 turbine is equipped with a rotor possessing a diameter of 90 m which consists of three blades and a hub. The 44 m blades, laminated with the use of a "prepreg" type material (PP), are constructed from carbon fiber and fiberglass. The blades consist of two panels attached to a support beam (data from the producer) [54].

Wind energy transferred to the turbine is corrected by adjusting the blade pitch, depending on the control strategy that has been adopted. The blades connect to the hub by means of double row four-point contact ball bearings. The hub is used to seat the three blades and transmit the forces of action to the main bearing. The hub design supports the blade bearings and the cylinder that regulates the blade angle. The main gearbox transmits torque from the rotor to the generator. It consists of a planetary gearbox connected to a two-stage parallel gearbox, reaction rods, and vibration dampers The torque is transferred from the high-speed shaft to the generator via a composite clutch behind the disc brake. The generator bearings are lubricated and the grease is fed continuously by an automatic lubricator. The grease flow is approximately 2400 cm^3 per year. The high-speed shaft coupling transmits torque from the high-speed output shaft of the gearbox to the input shaft of the generator It consists of two composite discs, an intermediate sleeve with two aluminum flanges and a fiberglass sleeve. The nacelle housing is made of glass-fiber-reinforced polymers. The nacelle base plate consists of two parts: the front one, made of cast iron, and the back one in the form of a girder construction. A three-phase asynchronous generator with a coiled rotor is connected to the Vestas Converter System (VCS, Vestas Converter System) via a slip ring system (data from the producer) [54].

The Vestas V90 tubular towers have internal flange connections. The tower is erected on a reinforced concrete foundation with an embedded starter ring. The tower is connected to the ring by means of bilateral flanges (internal and external). The list of materials that form the Vestas V90 2 MW wind power plant is presented in Table S2 (in Supplementary Materials).

A comparative analysis of the efficiency of the aforementioned wind power plant in a 25-year operation period [43–46] was performed and the same was done for the same turbine modernized in a 50-year lifecycle. The analysis is primarily intended to describe the existing reality, but also to help model future changes to define recommendations to develop more pro-environmental solutions.

The wind power plant examined herein is a technically mature construction. As part of a service package, the producer guarantees its constant time-availability and productivity throughout the entire 25-year use-period. The periods of reduced reliability at the beginning and end of the use-period of this wind power plant as well as the decreasing operational efficiency throughout its use-period [55,56] are negligible and do not affect the efficiency assessments of the costs incurred.

The energy benefits $U(t)$ are the sum of the average annual values of energy production in the wind power plant U_r over an assumed lifetime t_{LC1} of 25 years [43–46]. The annual environmental and energy benefits U_r are defined as the average annual long-term productivity determined on the basis of data from three years of production (2013–2015) and long-term data that takes into account the 25-year reference period. These benefits came to 5325 MWh/year for the power plant tested [19].

The costs for the 25-year period of use were determined as the sum of costs at individual stages of the wind power plant lifecycle, i.e., costs to generate N_W^1, the average annual costs for operation N_r, and the post-use management N_Z^1 in the form of landfilling and recycling. For costs on post-use management, it was assumed that 90% of materials are recycled and 10% are landfilled, which results from the current possibilities of material processing and available reports on wind power plant recycling [57].

The study took into consideration the management of post-use materials and elements of the wind power plant, which included 90% recycling and 10% landfilling of the waste for materials that could not be reused or processed. As a complex mechanical structure, a wind power plant is built of many different materials, including steel, polymer materials, carbon fibers, fiberglass, copper, iron, and rubber materials, which are subject to various recycling methods to varying degrees, including mechanical recycling and reuse, material recovery, incineration with energy recovery, and pyrolysis [58]. The data presented by the wind power plant manufacturers [59–63] indicate that it is technically feasible to recycle 90% of its materials and to transfer 10% of its materials to landfills. This data correspond to the actual conditions. Some of the materials of the blades, nacelle and rotor housing require a specialized method of energy recycling (incineration with energy recovery, pyrolysis), during which waste intended for the landfill is produced. The analysis in this paper covered the entire lifecycle of the wind power plant, including post-use management; therefore, to calculate the result of the wind power plant's total environmental impact, it is important to take into account the management methods, as—depending on the level of recycling and landfilling—results will vary. This paper took into account data concerning the recyclability of components corresponding to the actual technical possibilities thereof.

2.4.1. Determining Costs with the Aid of the LCA Method

The LCA method was used to determine energy and environmental costs in the lifecycle, a method which is a technique from process management that allows for the assessment of potential environmental threats triggered by proesses arising in the lifecycle of a given technical object. Analysis was conducted in accordance with ISO norm 14044 [64].

Aim and scope of the analysis

The purpose of analyzing environmental impacts is to identify potential negative environmental impacts occurring at particular material stages of the lifecycle, assuming a 25-year wind power plant lifecycle and a 50-year lifecycle for the modernized power plant where the nacelle and rotor have been replaced to determine the values of the integrated efficiency indicators in the context of environmental and energy costs, as well as the time of return on modernization and the sustainable modernization indicator. Environmental impact was determined using the Eco-indicator 99 model using SimaPro software. The Eco-indicator 99 method belongs to a group of methods modelling environmental impact at the level of endpoints of an environmental mechanism. The characterization process takes place for eleven impact categories, falling within three larger groups defined as impact areas or damage categories [65–67]. The following impact areas have been distinguished: human health, ecosystem quality, and raw material resources [68]. The results of the third stage of environmental analysis, i.e., grouping and weighing, were used to determine the environmental costs. Environmental coefficients are the result of this stage, and they are expressed in Pt (environmental points), which are aggregated units that make it possible to compare the results of the ecobalances herein [69,70]. One thousand environmental points are equal to the environmental impact of an average European during a year [65,71]. The cut-off point was equal to 0.05%. The analysis included eleven impact categories, yet in order to determine environmental costs, only the total impact of the wind power plant's lifecycle at particular stages of the lifecycle, i.e., production, use, and post-use development, was taken into account. The analysis also took into account the values of emissions of substances causing acidification SO_2eq (in kg) and emissions of substances causing eutrophication PO_4eq (in kg), which were treated as environmental costs in the determination of sustainable modernization assessment models. The procedure of determining environmental impacts consisted of four stages: defining the aim and scope of the analysis, analysis of the collection of inputs and outputs, LCAI analysis, and interpretation of the results [72,73].

The cumulative energy demand method (CED) makes it possible to define the cumulative energy demand (expressed in GJ) for particular stages of the lifecycle [74,75]. The reults of the CED analysis were the input value for determining the value of the integrated efficiency index from energy costs, from the time of the return on energy costs for modernization, and from the modernization indicato in relation to energy costs.

The Intergovernmental Panel on Climate Change, Global Warming Potential method (IPCC) was used to determine the amount of greenhouse gas emissions in certain stages of the material lifecycle o the wind power plant [76,77]. This gave the value of CO_2eq emissions (in kilograms), which was usec to determine the value of integrated efficiency (Equation 7).

System limit and the functional unit

The borders of Poland were assumed as a territorial limitation. The function of the object was the production of electricity. The productivity of the wind power plant in a 25-year period of use was taken as the functional unit.

The analysis assumed a 25-year wind power plant lifecycle [43–46] and a 50-year lifecycle o the same wind power plant which was subject to modernization. The scope of the modernizatior included the replacement of the nacelle and rotor along with the blades after a 25-year period of use Three material stages of the lifecycle were examined: production, use, and post-use management ir the form of recycling (90%) and landfilling (10%) [57]. Excluded from the system were the stages o transportation, of sale, of technical tests and storage because of a lack of proper data and the impact differences in the means of transport, which to a large degree are dependent on the location of the object. In the Figure 3 the scope of LCA analysis is presented.

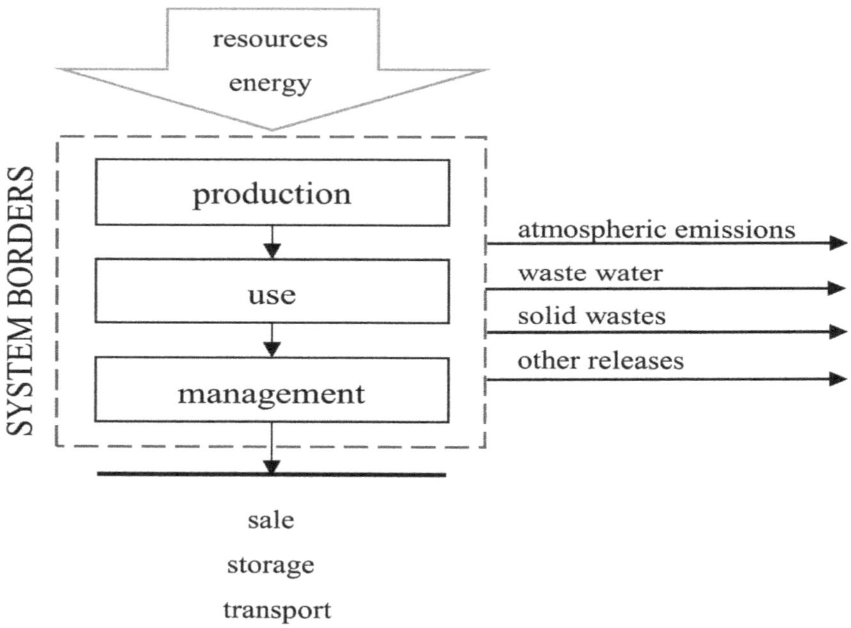

Figure 3. Scope of the LCA analysis of the turbine in a 25-year lifecycle and the same turbine after modernization.

Analysis of the collection of inputs and outputs, data aggregation and validation

Data were collected and divided into particular processes and unit elements with the identification of their inputs and outputs. Process inputs are primarily comprised of materials, natural resources, and energy while outputs are comprised of waste and emissions. The data correspond to one wind power plant. The input data concerned materials and elements of the wind power plant acquired from materials of the producer and from the authors' own research. Data concerning productivity come from a 3 years period (2013 – 2015) of the wind power plant's operation. Less essential data concerning materials and processes were taken from SimaPro databases. In order to create an inventory table, individual environmental impacts of the same type were summed up for all unit processes.

Table S2 (in Supplementary Materials) displays the material and elements that formed the basis of the wind power plant's lifecycle analysis for a 25-year period of use. The total weight of the its materials and elements came to approximately two tons, 76% of which belonged to concrete. Besides concrete, the most important materials used in the wind power plant's lifecycle are steel (approx. 19%), cast iron (approx 3%), polymer materials (approx. 1%) as well as aluminum, copper, and oils (in total approx. 1%). It was assumed that in a 25-year period of use there will be two replacements of the main gearbox [78] and an oil change every 5 years in both the hydraulic system and the main gearbox, with a total weight of 2560 kg.

Table 2 lists the materials and elements of the nacelle and rotor that are subject to replacement after 25 years of use for the scenario in which a 50-year lifecycle was assumed along with the modernization of the test object, while Table 2 lists the weight of the materials replaced. Approximately 48% of the materials and components of the nacelle and over 53% of the materials and components of the rotor undergo replacement. The following elements are replaced with completely ones: the generator and its accompanying cooler, the gearbox, cooler, hydraulic system, and blades (Table 2). This constitutes a small percentage of the total mass of the power plant undergoing modernization. The components in the nacelle that are replaced constitute less than 2% of the total mass, while the rotor components account for slightly over 1% (Table 2). Most of the materials listed are polymeric materials (1.15%), which are replaced in their entirety (Table 3). The material with the second highest share in replacement is cast iron (1.02%), and the third is steel (0.60%).

2.4.2. Determining the Efficiency Indicator from Environmental and Energy Costs

Efficiency of environmental costs is understood as the relation between environmental benefits in the lifecycle and the environmental costs incurred in the form of: the total impact of one thousand Europeans in the course of one year (efficiency of environmental costs), greenhouse gas emissions (environmental efficiency of CO_2 emissions), emissions of substances causing acidification (environmental efficiency of SO_2 emissions), and emissions of substances causing eutrophication (environmental efficiency of PO_4 emissions).

Energy efficiency is understood as the relation between energy benefits in the lifecycle and the energy costs incurred.

To analyze the efficiency of the wind power plant in a 25-year period of use, dependency (7), expanding upon formula (1), was used

$$E(t_{LC1}) = \frac{U(t_{LC1})}{N(t_{LC1})} = \frac{\sum_{i=1}^{t} U_i}{N_W^1 + \sum_{i=1}^{t} N_i + N_Z^1} = \frac{U_r \cdot t_{LC1}}{N_W^1 + N_r \cdot t_{LC1} + N_Z^1} \tag{21}$$

where: $E(t_{LC1})$ – the integrated efficiency indicator during t years of use, U_i – benefits during the i-nth year of use, N_W^1 – costs at the production stage of the 25-year lifecycle, N_i – costs in the i-nth year of use N_Z^1 – costs during post-use management during the 25-year lifecycle, t_{LC1} – time of use, U_r – average annual benefits, N_r – average annual use-related costs.

For a wind power plant that has been modernized during a 50-year period of use, the efficiency indicator is determined by a system of equations

$$\begin{cases} E(t) = \frac{U_r \cdot t}{N_W^1 + N_r \cdot t + N_Z^1} & \text{dla} \quad 0 \le t \le 25 \\ E(t) = \frac{U_r \cdot t}{(N_W^1 + N_W^2) + N_r \cdot t + (N_Z^1 + N_Z^2)} & \text{dla} \quad 25 < t \le 50 \end{cases} \tag{22}$$

where: N_W^2 – costs at the stage of producing new elements used in the modernization process, N_Z^2 – costs at the stage of post-use management of the replaced elements of the wind power plant.

Table 2. List of the materials and elements replaced during wind power plant modernization.

Lp.	Element	Material	Mass [kg]	Share in component, %	Share in unit, %	Share in WPP, %
1	**Nacelle**					
1.1	Generator with the radiator	Copper	1430	19.07	2.10	0.08
		Cast iron	3920	52.27	5.76	0.22
		Steel	2090	27.87	3.07	0.12
		Replacement	7440	99.20	10.94	0.42
1.2	Gearbox	Upgraded steel	2620	15.41	3.85	0.15
		Cast iron	14,060	82.71	20.68	0.80
		Oil	280	1.65	0.41	0.02
		Replacement	16,960	99.76	24.94	0.97
1.3	Radiators	Aluminum	960	100.00	1.41	0.05
1.4	Hydraulic system	Oil	360	9.89	0.53	0.02
		Steel	2660	73.08	3.91	0.15
		Aluminum	420	11.54	0.62	0.02
		Other	200	5.49	0.29	0.01
		Replacement	3640	100.00	5.35	0.21
1.5	Switchgears, inverters, connections	Steel	340	38.64	0.50	0.02
		Copper	240	27.27	0.35	0.01
		Aluminum	180	20.45	0.26	0.01
		Replacement	760	86.36	1.12	0.04
1.6	Nacelle housing	Steel	1100	6.00	1.62	0.06
		Polymer materials	1815	9.90	2.67	0.10
		Replacement	2915	15.89	4.29	0.17
Nacelle total			68,000		100.00	3.87
Replacement			**32,675**		48.05	1.86
2	**ROTOR**					
2.1	Blades	Steel	1750	8.75	4.61	0.10
		Polymer materials reinforced with carbon fiber and glass fiber	18,250	91.25	48.03	1.04
		Replacement	20,000	100.00	52.63	1.14
2.2	Hub	Polymer materials	200	1.11	0.53	0.01
		Replacement	200	1.11	0.53	0.01
Rotor total			38,000		100.00	2.16
Replacement			**20,200**		53.16	1.15

Table 3. Mass of materials replaced during wind power plant (WPP) modernization.

	Mass [kg]	Share in WPP [%]	Aluminum [kg]	Copper [kg]	Steel [kg]	c. Iron [kg]	Polymer Materials [kg]	Concrete [kg]	Oil [kg]	Other [kg]
WPP Total	1,756,810	100	3715	2050	336,300	52,015	20,265	1,339,615	640	2210
Replacement	52,875	3.01	1560	1670	10,560	17,980	20,265	0	640	200
% share of total materials in WPP			0.21	0.12	19.14	2.96	1.15	76.25	0.04	0.13
% share of replacement materials in WPP			0.09	0.10	0.60	1.02	1.15	0	0.04	0.01

In accordance with dependencies (21) and (22), integrated efficiency indicators were determined:

- From environmental costs, where the result of a wind power plant's environmental impact (in environmental points) at certain stages of the lifecycle were assumed as costs;
- From greenhouse gas emissions, where quantitative emissions of CO_2eq (in kg) at certain stages of the wind power plant's lifecycle are assumed as costs, this being obtained as the result of completing LCA analysis;

From the emissions of substances causing acidification, where quantitative emissions of SO_2eq (in kg) at certain stages of the wind power plant's lifecycle are assumed as costs, this being obtained as the result of completing LCA analysis;

- From the emissions of substances causing eutrophication, where quantitative emissions of PO_4eq (in kg) at certain stages of the wind power plant's lifecycle are assumed as costs, this being obtained as the result of completing LCA analysis;
- From energy costs, where potential energy demands (in MJ) at certain stages of the wind power plant's lifecycle are assumed as costs.

2.4.3. Determining the Payback Time for Modernization

For the wind power plant that has undergone modernization analyzed herein, the payback time for modernization (18) is defined by

$$T_{M1} = \frac{t_{LC1}\left(N_W^2 + N_Z^2\right)}{N_W^1 + N_Z^1} \quad (23)$$

Payback times for modernization were determined for ecological costs in the form of greenhouse gas emissions, substances causing acidification, substances causing eutrophication, and energy costs for the wind power plant in a 25-year and 50-year lifecycle.

2.4.4. Determining the Sustainable Modernization Indicator

For the wind power plant that has undergone modernization analyzed herein, the sustainable modernization indicator is defined by:

$$E_{M1} = \frac{E(t_{LC2})}{E(t_{LC1})} = \frac{E(t_{LC2} = 50)}{E(t_{LC1} = 25)} = \frac{\frac{50U_r}{(N_W^1+N_W^2)+50N_r+(N_Z^1+N_Z^2)}}{\frac{25U_r}{N_W^1+25N_r+N_Z^1}} = \frac{2\left(N_W^1 + N_Z^1 + 25N_r\right)}{\left(N_W^1 + N_W^2 + N_Z^1 + N_Z^2 + 50N_r\right)} \quad (24)$$

The value of the of the sustainable modernization indicator was determined for the efficiency of ecological costs, of greenhouse gas emissions, of substances causing acidification, of substances causing eutrophication, and of energy costs for the wind power plant in a 25-year and 50-year lifecycle.

The two dependencies above (23), (24) make it possible to assess the modernization process. They may be applied to assess other machines and devices than the one analyzed in this paper.

3. Results and Discussion

3.1. Costs in the Lifecycle of a Wind Power Plant in a 25-year Lifecycle and in One Subjected to Modernization

Table 4 presents the results of LCA analysis with the use of Eco-indicator 99 modelling (total value of the eco-indicator, acidification, eutrophication) and IPCC modelling (greenhouse gas emissions) at particular stages of the wind power plant's lifecycle, which are the ecological costs in the model of the integrated ecological efficiency indicator.

Table 4. Ecological costs determined with the aid of the LCA method in the material stages of a 25 years and 50-year lifecycle of a wind power plant (the Eco-indicator 99 method).

Lifecycle Stage		Total Eco-Indicator Value [Pt]	Emissions of CO_2eq [kg]	Acidification [kg SO_2eq]	Eutrophication [kg PO_4eq]
Production	N_{W25}	322,460	2,700,559	141,842	728
	N_{W50}	401,663	3,381,947	176,735	907
Use	$N(t=25)$	20,955	589,300	3712	149
	$N(t=50)$	41,910	1,178,600	7424	298
Post-use management	N_{Z25}	−13,423	−528,712	−1203	−67
	N_{Z50}	−30,271	−657,873	−1499	−84

In each of the lifecycle stages considered, higher environmental costs in the form of a negativ impact on the ecosystem and human health were recorded for the 50-year lifecycle of a wind powe plant. However, if the values under consideration were compared not to one 25-year lifecycle of a wind power plant, but to the sum of two lifecycles (with disassembly and recycling after 25 years o landfilling the installation plus the installation of a new one along with 25 years of operation), it i evident that the use of the power plant over a 50-year period with modernization being performed afte 25 years of operation will result in lower values of the eco-indicator and greenhouse gas emissions b approx. 40–50% and lower emissions of substances causing acidification or eutrophication by approx 40% (depending on the stage of the lifecycle) compared to the use of two wind power plants during this period. The highest costs in both scenarios occurred at the stage of production of wind powe plant components, while the lowest were at the stage of post-use management, where the value o costs is generally reduced because of the possibility of recovering materials and energy from element ending their lifecycles. In both cases throughout the entire lifecycle there were costs in the form CO_2 emissions, while the lowest costs were in the form of emissions causing eutrophication.

Table 5 displays the results of energy consumption analysis at different lifecycle stages which constitute energy costs in the energy efficiency indicator model. In each of the lifecycle stages that were assessed, the greatest energy costs were recorded for the 50-year wind power plant lifecycle. However again, if these values were compared not to one 25-year wind power plant lifecycle, but to the sum of two lifecycles, it is evident that the use of the wind power plant over a 50-year period results in a reduction of energy costs from approx. 30 to 50% (depending on the lifecycle stage), where most important thing is to reduce energy consumption at the production stage of wind power plant elements

Table 5. Energy costs determined with the use of the LCA method in the material stages of a 25 years and 50 years wind power plant lifecycle (cumulative energy demand (CED) method).

Lifecycle Stage		Energy Consumption [MJ]
Production	N_{W25}	41,559,527
	N_{W50}	52,740,112
Use	$N(t=25)$	7,661,400
	$N(t=50)$	15,322,800
Post-use management	N_{Z25}	−5,779,943
	N_{Z50}	−8,014,646

3.2. *Efficieny Indicators from Ecological Costs and Energy Costs and Payback Time for Modernization*

The specified benefits and costs, after substituting them into dependencies (21) and (22), made it possible to determine the values of the integrated ecological efficiency indicator from ecological costs, from greenhouse gas emissions, from emissions of substances causing acidification, from emissions of substances causing eutrophication. It was also possible to determine the values of the integrated energy efficiency indicator.

In the case of a modernized wind power plant, for every 1 Pt of environmental impact approximately 2.32 GJ of electricity is produced, for every 1 t of CO_2eq emissions—246 GJ of electricity, for every 1 t of SO_2eq emissions—5247 GJ of electricity, for every 1 t of PO4eq emissions—854,766 GJ of electricity. The energy benefits of a wind power plant after modernization are almost 16 times higher than the costs incurred. What is noticeable is that, in the case of a modernized wind power plant, the values of the efficiency indicator from ecological costs and energy costs increase from 1.4 to 1.6 times in relation to the wind power plant with a 25-year lifecycle (Figures 4–8). These values reflect the better use of costs in the wind power plant's lifecycle to produce benefits when the wind power plant is modernized and its lifecycle is extended by another 25 years.

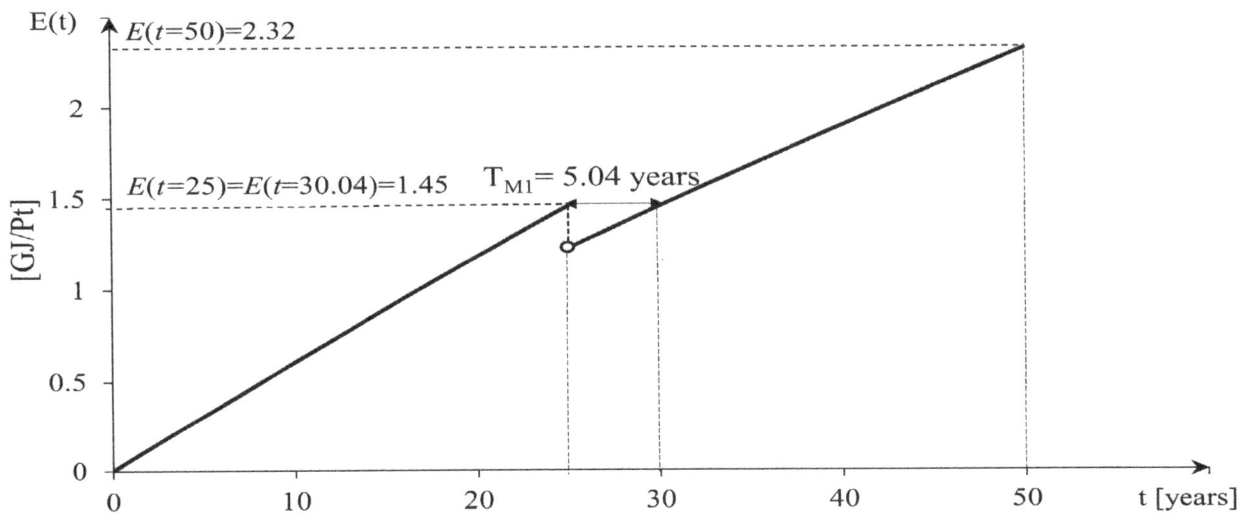

Figure 4. A graphical interpretation of the integrated efficiency indicator from ecological costs for a wind power plant undergoing modernization during a 50-year period of use.

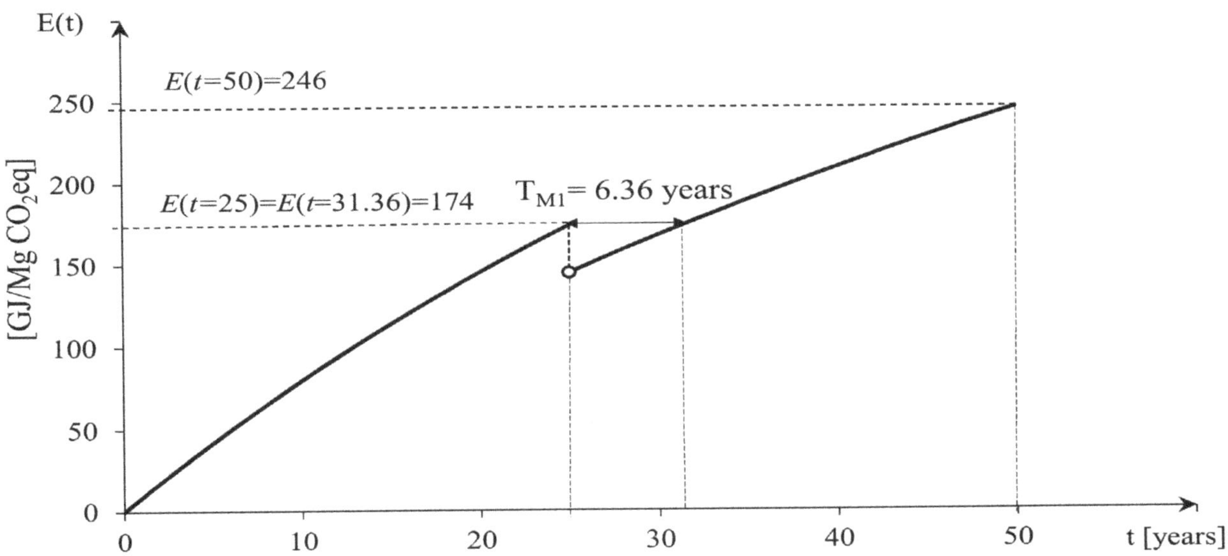

Figure 5. A graphical interpretation of the integrated ecological efficiency indicator from CO_2 emissions for a wind power plant undergoing modernization during a 50-year period of use.

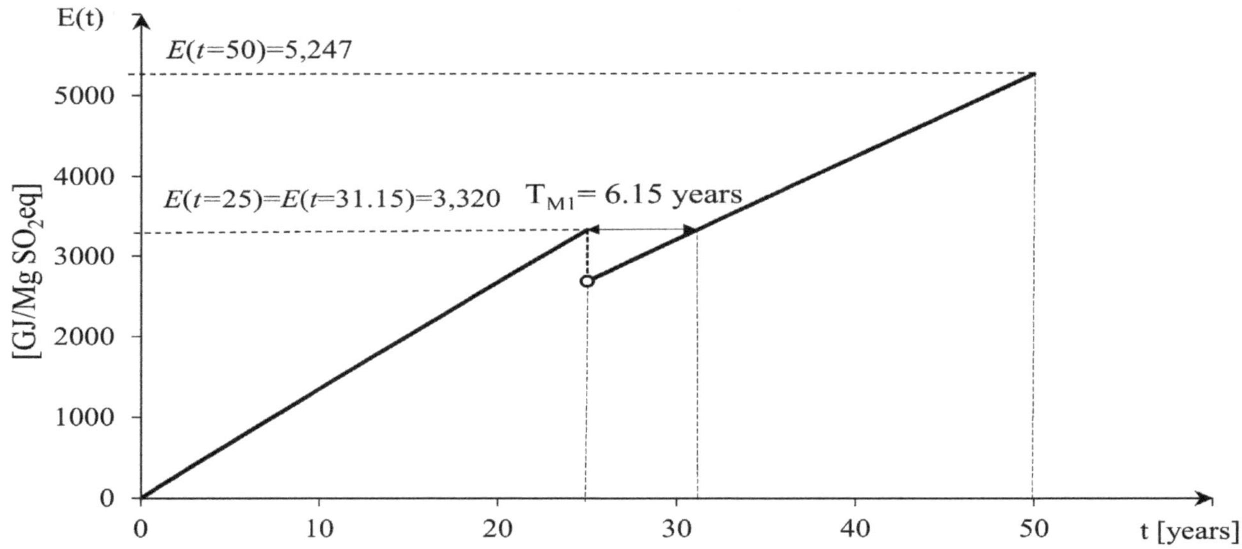

Figure 6. A graphical interpretation of the integrated ecological efficiency indicator from SO_2 emissions for a wind power plant undergoing modernization during a 50-year period of use.

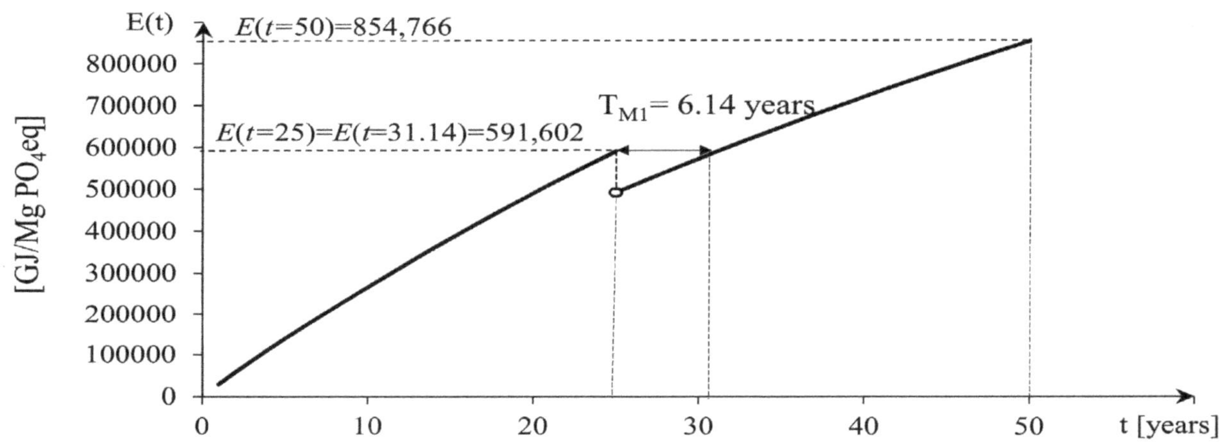

Figure 7. A graphical interpretation of the integrated ecological efficiency indicator from PO_4 emissions for a wind power plant undergoing modernization during a 50-year period of use.

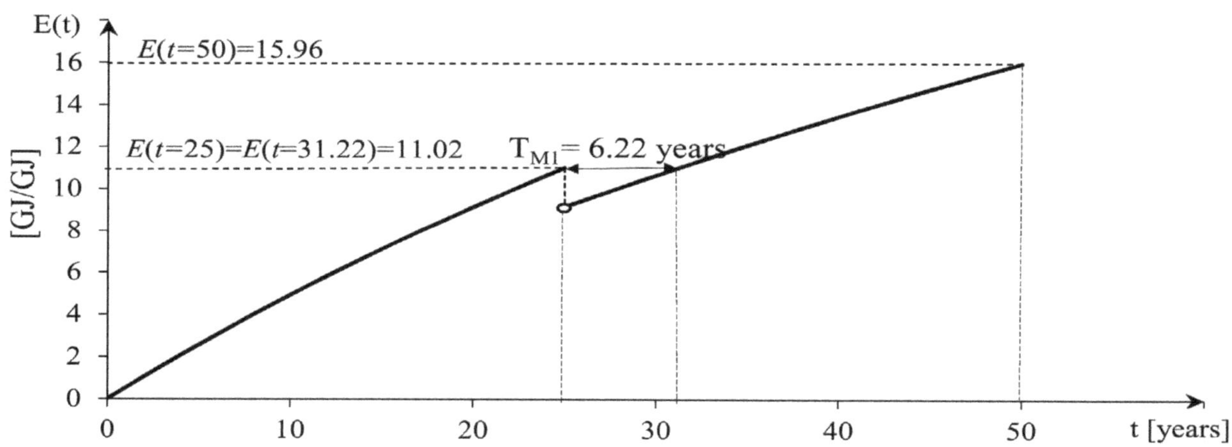

Figure 8. A graphical interpretation of the integrated efficiency indicator from energy costs for a wind power plant undergoing modernization during a 50-year period of use.

Figures 4–8 present a graphical interpretation of the efficiency indicator for a wind power plant undergoing modernization during a 50-year period of use.

Figure 4 presents the dependence of the integrated efficiency from ecological costs for a 50-year period of use with modernization performed after 25 years. For the integrated efficiency from ecological costs, energy production at the use-stage was defined as a benefit, and environmental points resulting from LCA analysis were used as costs. Figure 4 shows that the payback time for environmental costs incurred for wind power plant modernization is 5.04 years, and the integrated ecological efficiency after 50 years of use is higher than after 25 years. Unfortunately, at the time WPP elements are replaced, there is an increase in costs, resulting in a decrease in efficiency below the level for the WPP operating for 25 years.

Figure 5 presents the dependence of the integrated efficiency from CO_2 emissions for a 50-year period of use with modernization performer after 25 years. For the integrated efficiency from the emissions of CO_2 equivalent, energy production at the use-stage was defined as a benefit, and the emissions of CO_2 equivalent resulting from LCA analysis were used as costs. Figure 5 shows that the payback time for costs in the form CO_2 emissions incurred for wind power plant modernization is 6.36 years, and the integrated efficiency from CO_2 emissions after 50 years of use is higher than after 25 years. Unfortunately, at the time WPP elements are replaced, there is an increase in costs, resulting in a decrease in efficiency below the level for the WPP operating for 25 years. In this case, payback with regard to CO_2 emissions occurs more slowly than in the case of total ecological costs. This is primarily the result of a significant increase in costs in the form of CO_2 emissions during the production and use-stage (Table 4).

In the case of integrated efficiency from the emissions of substances causing acidification (SO_2), energy production at the use-stage was defined as a benefit (Figure 6). Costs were constituted by the total emissions of SO_2 equivalent at individual stages of the WPP lifecycle. As in the case of the integrated efficiency indicator from CO_2 emissions, the payback time for modernization was more than 6 years, which may be affected primarily by the increase in the emissions of substances causing acidification (SO_2eq) at the production stage (Table 4). Figure 6 shows that the integrated efficiency from SO_2 emissions after 50 years of use is higher than after 25 years.

In accordance with Figure 7, the integrated efficiency indicator from the emissions of substances causing eutrophication (PO_4) is higher after 50 years of use than after 25. The WPP returns to pre-modernization efficiency after 6.14 years, which is a value close to the payback time from the emissions of substances causing acidification (SO_2). The payback time is mainly affected by the increase in costs in the form of PO_4 emissions at the stage at which the replaced WPP elements are produced and by the emissions generated at the use-stage which are connected to maintenance, to oil changes in particular.

In the case of the integrated efficiency from energy costs, energy production at the use-stage was again defined as a benefit (Figure 8). Costs were constituted by total energy consumption at particular stages of the WPP lifecycle. In this impact area, the payback time for modernization was more than 6 years, which is tied primarily to the increase in energy costs necessary to produce the replaced elements and to the energy consumption at the use-stage relating to the energy consumption for maintenance and for the power plant's own needs.

Values of payback time for modernization were determined on the basis of dependence (23). The wind power plant, after modernization, returns to the efficiency it achieved in the last year before modernization after about 6 years, depending on the efficiency area which is being considered (Figures 4–8). The quickest return to pre-modernization efficiency is in the case of environmental efficiency from ecological costs (approx. 5 years) and the longest such return is in the case of efficiency from greenhouse gas emissions (over 6 years) (Figures 4–8).

By modernizing a wind power plant, one achieves an extension, its time of use, as well as an increase in benefits from its functioning in the form of electricity production. Owing to modernization, there was an increase both in the integrated ecological efficiency indicator and in the integrated energy efficiency indicator throughout the wind power plant's lifecycle, despite the fact that, during the years directly proceeding modernization, there was a drop in these values. Every subsequent modernization and replacement of elements will cause a temporary drop in the integrated efficiency indicator, which is related to an increase in ecological costs and energy costs the moment new elements are added.

The modernization considered in this paper included the replacement of the nacelle, rotor and blades with new ones of the same power. A constant average annual productivity was assumed, one which was the same for the 25-year and 50-year cycles. Replacing the elements with one of greater power would cause a change both in benefits (such as an increase in annual average productivity) and in costs. Costs would potentially be higher than in the case of a rotor of the same power because of an increase in the mass of the elements, and, as previous research has shown [10], the environmental impact of materials used to produce wind power plants is strongly related to their mass (greater mass = higher eco-indicator values). However, considering technological advances and developments in the construction of wind power plants, it is difficult to predict how, over the next 25 years, constructions, production methods, and the materials used to produce such objects will change, which means it is not possible to clearly determine how ecological costs and energy costs will change, the same being true for the values of the integrated efficiency indicators from ecological costs and energy costs.

Of crucial importance is the fact that, as a result of extending the lifecycle of a wind power plant the use of natural resources is thereby limited, as is the energy used to produce its components, and not least of all, the post-use management of a part of its elements is postponed.

3.3. Sustainable Modernization Indicators

Table 6 presents the values of the sustainable modernization indicator in relation to environmental costs and energy costs for wind power plant modernization. Values were determined on the basis of dependence (24). The higher the value of the sustainable modernization indicator, the greater was the increase in the efficiency of the wind power plant because of modernization. The highest value of the sustainable modernization indicator was obtained with respect to the wind power plant's ecological efficiency from ecological costs, and the lowest with respect to the efficiency indicator from the emissions of substances causing eutrophication (Table 6). In the case of all the areas under consideration an increase was recorded in the efficiency of the use of costs by about 1.44–1.6 times, which is primarily the result of an increase in benefits (total electricity production) over a 50-year lifecycle.

Table 6. Values of the sustainable modernization indicator.

The Sustainable Modernization Indicator E_{M1} [-]	
from ecological costs	1.596
from greenhouse gas emissions (CO_2eq)	1.415
from emissions of substances causing acidification (SO_2eq)	1.58
from emissions of substances causing eutrophication (PO_4eq)	1.444
from energy costs	1.448

4. Summary and Conclusions

The aim of the paper was achieved by developing a methodology to assess devices subjected to sustainable modernization, i.e., restoration of an object such that it has the properties of a new device. An assessment of the ecological efficiency and energy efficiency of a modernized wind power plant was carried out with the use of an integrated efficiency indicator. Two indicators were proposed to assess modernization efficiency: the payback time of costs for modernization (23) and the sustainable modernization indicator (24).

In the case of the wind power plant analyzed herein, the payback time for costs incurred for modernization (23) came to 6.36 years from greenhouse gas emissions, 6.22 years from energy costs, and, from the remaining costs analyzed, from 5.04 years to 6.15 years.

In the case of the wind power plant analyzed herein, the sustainable modernization indicator from greenhouse gas emissions came to 1.42, and from energy costs—1.45 (Table 6). This means an over 40% increase in the ecological efficiency of the use of energy costs from greenhouse gases and in the ecological efficiency of the use of energy costs after the completion of the lifecycle of the wind power plant subjected to sustainable modernization.

Analysis of the ecological costs throughout the lifecycle of a modernized wind power plant showed that higher environmental costs in the form of negative impacts on the ecosystem and on human health were recorded in the case of the 50-year lifecycle of a wind power plant. If the values under consideration were compared not to one 25-year lifecycle of a wind power plant, but to the sum of two lifecycles (with disassembly and recycling after 25 years or landfilling the installation plus the installation of a new one along with 25 years of operation), it is evident that the use of the power plant over a 50-year period with modernization being performed after 25 years of operation will result in lower values of the eco-indicator and greenhouse gas emissions by approx. 40–50% and lower emissions of substances causing acidification or eutrophication by approx. 40% (depending on the stage of the lifecycle) compared to the use of two wind power plants during this period.

The indicators of modernization assessment proposed herein fit into a higher assessment of the closed-loop economy, that is, an economic system in which the consumption of raw materials, energy, emissions, and waste volume is minimized by creating a closed process loop. Of crucial importance is the fact that, as a result of extending the lifecycle of a wind power plant, the use of natural resources is thereby limited, as is the energy used to produce its components, and, not least of all, the post-use management of a part of its elements is postponed.

Sustainable modernization also fits into a strategy of sustainable development where an increase in electricity production essential for development results in a reduction of ecological and energy costs per unit of energy produced.

The authors call for the creation of incentive programs via legal changes for wind power plant investors and producers which relate to the implementation of sustainable modernization that results in a reduction of the environmental burdens accompanying energy production.

The assessment methodology proposed herein as well as the models developed to assess the effects of modernization are universal and can be applied to other technical facilities as both the benefits and the costs can be expressed in various units adapter to the specific nature of any technical object's operation. The indicators presented herein constitute one of the elements to assess the effects of modernization and are a form of support for operators and those who manage the lifecycles of technical objects.

Author Contributions: Conceptualization, R.K. and J.F.; methodology, R.K. and W.K.; software, P.B.-W. and R.K.; validation, P.B.-W., R.K., and A.T.; formal analysis, J.F. and A.T.; investigation, R.K., W.K., and P.B.-W.; resources, R.K.; data curation, R.K., P.B.-W., and W.K.; writing—original draft preparation, R.K and W.K.; writing—review and editing, R.K., W.K., J.F., and A.T.; visualization, W.K. and P.B.-W.; supervision, R.K., J.F., A.T.; project administration, R.K. All authors have read and agreed to the published version of the manuscript.

References

1. Haapala, K.R.; Prempreeda, P. Comparative life cycle assessment of 2.0 MW wind turbines. *Int. J. Sustain. Manuf.* **2014**, *3*, 170. [CrossRef]
2. Martínez, E.; Sanz, F.; Pellegrini, S.; Jiménez, E.; Blanco, J. Life cycle assessment of a multi-megawatt wind turbine. *Renew. Energy* **2009**, *34*, 667–673. [CrossRef]

3. Wang, S.; Wang, S. Impacts of wind energy on environment: A review. *Renew. Sustain. Energy Rev.* **2015**, *49*, 437–443. [CrossRef]
4. Saidur, R.; Rahim, N.A.; Islam, M.R.; Solangi, K.H. Environmental impact of wind energy. *Renew. Sustain. Energy Rev.* **2011**, *15*, 2423–2430. [CrossRef]
5. Jeffery, R.D.; Krogh, C.; Horner, B. Adverse health effects of industrial wind turbines. *Can. Fam. Physician* **2013**, *59*, 473–475.
6. Piasecka, I.; Tomporowski, A. Analysis of Environmental and Energetical Possibilitiesof Sustainable Development of Wind and Photovoltaic Power Plants. *Probl. Ekorozw. Probl. Sustain. Dev.* **2018**, *13*, 125–130.
7. Gibon, T.; Hertwich, E.G.; Arvesen, A.; Singh, B.; Verones, F. Health benefits, ecological threats of low-carbon electricity. *Environ. Res. Lett.* **2017**, *12*, 034023. [CrossRef]
8. Basosi, R.; Cellura, M.; Longo, S.; Parisi, M.L. *Life Cycle Assessment of Energy Systems and Sustainable Energy Technologies: The Italian Experience*; Springer: Berlin/Heidelberg, Germany, 2018.
9. Luderer, G.; Pehl, M.; Arvesen, A.; Gibon, T.; Bodirsky, B.L.; de Boer, H.S.; Fricko, O.; Hejazi, M.; Humpenöder, F.; Iyer, G.; et al. Environmental co-benefits and adverse side-effects of alternative power sector decarbonization strategies. *Nat. Commun.* **2019**, *10*, 1–13. [CrossRef]
10. Ozoemena, M.; Cheung, W.M.; Hasan, R. Comparative LCA of technology improvement opportunities for a 1.5-MW wind turbine in the context of an onshore wind farm. *Clean Technol. Environ. Policy* **2018**, *20*, 173–190. [CrossRef]
11. Bonou, A.; Laurent, A.; Olsen, S.I. Life cycle assessment of onshore and offshore wind energy-from theory to application. *Appl. Energy* **2016**, *180*, 327–337. [CrossRef]
12. Mroziński, A.; Piasecka, I. Selected aspects of building, operation and environmental impact of offshore wind power electric plants. *Pol. Marit. Res.* **2015**, *22*, 86–92. [CrossRef]
13. Liu, P.; Barlow, C.Y. The environmental impact of wind turbine blades. *IOP Conf. Ser. Mater. Sci. Eng.* **2016**, *139*, 012032. [CrossRef]
14. Flizikowski, J.; Piasecka, I.; Kruszelnicka, W.; Tomporowski, A.; Mroziński, A. Destruction assessment of wind power plastics blade. *Polimery* **2018**, *63*, 5. [CrossRef]
15. Tomporowski, A.; Piasecka, I.; Flizikowski, J.; Kasner, R.; Kruszelnicka, W.; Mroziński, A.; Bieliński, K. Comparison Analysis of Blade Life Cycles of Land-Based and Offshore Wind Power Plants. *Pol. Marit. Res.* **2018**, *25*, 225–233. [CrossRef]
16. Tsai, L.; Kelly, J.C.; Simon, B.S.; Chalat, R.M.; Keoleian, G.A. Life Cycle Assessment of Offshore Wind Farm Siting: Effects of Locational Factors, Lake Depth, and Distance from Shore. *J. Ind. Ecol.* **2016**, *20*, 1370–1383. [CrossRef]
17. Stavridou, N.; Koltsakis, E.; Baniotopoulos, C.C. A comparative life-cycle analysis of tall onshore steel wind-turbine towers. *Clean Energy* **2019**, *zkz028*, 1–10. [CrossRef]
18. Tomporowski, A.; Flizikowski, J.; Kruszelnicka, W.; Piasecka, I.; Kasner, R.; Mroziński, A.; Kovalyshyn, S. Destructiveness of Profits and Outlays Associated with Operation of Offshore Wind Electric Power Plant. Part 1: Identification of a Model and its Components. *Pol. Marit. Res.* **2018**, *25*, 132–139. [CrossRef]
19. Kasner, R. *Ocena Korzyści i Nakładów Cyklu Życia Elektrowni Wiatrowej*; Rozprawa doktorska, Poznań University of Technology: Poznań, Poland, 2016.
20. Piasecka, I.; Tomporowski, A.; Flizikowski, J.; Kruszelnicka, W.; Kasner, R.; Mroziński, A. Life Cycle Analysis of Ecological Impacts of an Offshore and a Land-Based Wind Power Plant. *Appl. Sci.* **2019**, *9*, 231. [CrossRef]
21. Tomporowski, A.; Flizikowski, J.; Kasner, R.; Kruszelnicka, W. Environmental Control of Wind Power Technology. *Rocz. Ochr. Środowiska* **2017**, *19*, 694–714.
22. Arabian-Hoseynabadi, H.; Oraee, H.; Tavner, P.J. Wind turbine productivity considering electrical subassembly reliability. *Renew. Energy* **2010**, *35*, 190–197. [CrossRef]
23. Mostafaeipour, A. Productivity and development issues of global wind turbine industry. *Renew. Sustain. Energy Rev.* **2010**, *14*, 1048–1058. [CrossRef]
24. Olatayo, K.I.; Wichers, J.H.; Stoker, P.W. Energy and economic performance of small wind energy systems under different climatic conditions of South Africa. *Renew. Sustain. Energy Rev.* **2018**, *98*, 376–392. [CrossRef]
25. Aghbashlo, M.; Tabatabaei, M.; Hosseini, S.S.B.; Dashti, B.; Mojarab Soufiyan, M. Performance assessment of a wind power plant using standard exergy and extended exergy accounting (EEA) approaches. *J. Clean. Prod.* **2018**, *171*, 127–136. [CrossRef]

26. Sağlam, Ü. Assessment of the productive efficiency of large wind farms in the United States: An application of two-stage data envelopment analysis. *Energy Convers. Manag.* **2017**, *153*, 188–214. [CrossRef]
27. Dicorato, M.; Forte, G.; Pisani, M.; Trovato, M. Guidelines for assessment of investment cost for offshore wind generation. *Renew. Energy* **2011**, *36*, 2043–2051. [CrossRef]
28. Laura, C.-S.; Vicente, D.-C. Life-cycle cost analysis of floating offshore wind farms. *Renew. Energy* **2014**, *66*, 41–48. [CrossRef]
29. Myhr, A.; Bjerkseter, C.; Ågotnes, A.; Nygaard, T.A. Levelised cost of energy for offshore floating wind turbines in a life cycle perspective. *Renew. Energy* **2014**, *66*, 714–728. [CrossRef]
30. Snyder, B.; Kaiser, M.J. Ecological and economic cost-benefit analysis of offshore wind energy. *Renew. Energy* **2009**, *34*, 1567–1578. [CrossRef]
31. Alsaleh, A.; Sattler, M. Comprehensive life cycle assessment of large wind turbines in the US. *Clean Technol. Environ. Policy* **2019**, *21*, 887–903. [CrossRef]
32. Garrett, P.; Rønde, K. Life cycle assessment of wind power: Comprehensive results from a state-of-the-art approach. *Int. J. Life Cycle Assess.* **2013**, *18*, 37–48. [CrossRef]
33. Abeliotis, K.; Pactiti, D. Assessment of the environmental impacts of a wind farm in central Greece during its life cycle. *Int. J. Renew. Energy Res.* **2014**, *4*, 580–585.
34. Kruszelnicka, W.; Bałdowska-Witos, P.; Kasner, R.; Flizikowski, J.; Tomporowski, A.; Rudnicki, J. Evaluation of emissivity and environmental safety of biomass grinders drive. *Przemysł Chem.* **2019**, *98*, 1494–1498.
35. Jachimowski, R.; Szczepanski, E.; Klodawski, M.; Markowska, K.; Dabrowski, J. Selection of a Container Storage Strategy at the Rail-road Intermodal Terminal as a Function of Minimization of the Energy Expenditure of Transshipment Devices and CO2 Emissions. *Rocz. Ochr. Sr.* **2018**, *20*, 965–988.
36. Zając, G.; Węgrzyn, A. Analysis of work parameters changes of diesel engine powered with diesel fuel and faee blends. *Eksploat. I Niezawodn. Maint. Reliab.* **2008**, 17–24.
37. Caban, J.; Drozdziel, P.; Vrabel, J.; Sarkan, B.; Marczuk, A.; Krzywonos, L.; Rybicka, I. The Research on Ageing of Glycol-Based Brake Fluids of Vehicles in Operation. *Adv. Sci. Technol. Res. J.* **2016**, *10*, 9–16. [CrossRef]
38. Winiarski, G.; Gontarz, A.; Pater, Z. A new process for the forming of a triangular flange in hollow shafts from Ti6Al4V alloy. *Arch. Civ. Mech. Eng.* **2015**, *15*, 911–916. [CrossRef]
39. Wasiak, A.L. Effect of Biofuel Production on Sustainability of Agriculture. *Procedia Eng.* **2017**, *182*, 739–746. [CrossRef]
40. Flatland, K.; Hove, M.T.; Lavrutich, M.; Nagy, R.L.G. Irreversible Investment in Wind Turbines: Life-Extension Versus Repowering. Available online: http://realoptions.org/openconf2019/data/papers/394.pdf (accessed on 17 January 2020).
41. Ziegler, L.; Gonzalez, E.; Rubert, T.; Smolka, U.; Melero, J.J. Lifetime extension of onshore wind turbines: A review covering Germany, Spain, Denmark, and the UK. *Renew. Sustain. Energy Rev.* **2018**, *82*, 1261–1271. [CrossRef]
42. Piel, J.H.; Stetter, C.; Heumann, M.; Westbomke, M.; Breitner, M.H. Lifetime Extension, Repowering or Decommissioning? Decision Support for Operators of Ageing Wind Turbines. *J. Phys. Conf. Ser.* **2019**, *1222*, 012033. [CrossRef]
43. Siemens, A.G. (Ed.) *Environmental Product Declaration. A Clean Energy Solution-from Cradle to Grave. Offshore Wind Power Plant Employing SWT-6.0-154*; Siemens A.G. Wind Power: Hamburg, Germany, 2015.
44. Shafiee, M.; Brennan, F.; Espinosa, I.A. A parametric whole life cost model for offshore wind farms. *Int. J. Life Cycle Assess.* **2016**, *21*, 961–975. [CrossRef]
45. McCulloch, M.; Raynolds, M.; Laurie, M. *Life-Cycle Value Assessment of a Wind Turbine*; The Pembina Institute: Calgary, AB, Canada, 2000.
46. Jensen, J.P. Routes for Extending the Lifetime of Wind Turbines. Available online: https://www.plateconference.org/routes-extending-lifetime-wind-turbines/ (accessed on 19 December 2018).
47. Sun, H.; Gao, X.; Yang, H. Investigation into offshore wind farm repowering optimization in Hong Kong. *Int. J. Low Carbon Technol.* **2019**, *14*, 302–311. [CrossRef]
48. Lacal-Arántegui, R.; Uihlein, A.; Yusta, J.M. Technology effects in repowering wind turbines. *Wind Energy* **2020**, *23*, 660–675. [CrossRef]
49. Amiri, A.K.; Kazacoks, R.; McMillan, D.; Feuchtwang, J.; Leithead, W. Farm-wide assessment of wind turbine lifetime extension using detailed tower model and actual operational history. *J. Phys. Conf. Ser.* **2019**, *1222*, 012034. [CrossRef]

50. Martínez, E.; Latorre-Biel, J.I.; Jiménez, E.; Sanz, F.; Blanco, J. Life cycle assessment of a wind farm repowerin process. *Renew. Sustain. Energy Rev.* **2018**, *93*, 260–271. [CrossRef]
51. Flizikowski, J.; Tomporowski, A.; Kasner, R.; Mroziński, A.; Kruszelnicka, W. Machinery Life Cycle Efficienc Models for their Sustainable Development. *Syst. Saf. Hum. Technol. Facil. Environ.* **2019**, *1*, 363–37 [CrossRef]
52. Wasiak, A. Technologies of Biofuel Production. In *Modeling Energetic Efficiency of Biofuels Production*; Springe Berlin/Heidelberg, Germany, 2019.
53. Wasiak, A.; Orynycz, O. The Effects of Energy Contributions into Subsidiary Processes on Energetic Efficiency of Biomass Plantation Supplying Biofuel Production System. *Agric. Agric. Sci. Procedia* **2015**, *7*, 292–30 [CrossRef]
54. Garrett, P.; Rønde, K. *Life Cycle Assessment of Elektricity Production from V90-2.0MW Gridstreamer Wind Plan* Vestas Wind Systems A/S: Randers, Danmark, 2011.
55. Ding, Y. *Data Science for Wind Energy*; CRC Press: Boca Raton, FL, USA, 2019.
56. Byon, E.; Ntaimo, L.; Singh, C.; Ding, Y. Wind energy facility reliability and maintenance. In *Handbook o Wind Power Systems*; Springer: Berlin/Heidelberg, Germany, 2013.
57. Guezuraga, B.; Zauner, R.; Pölz, W. Life cycle assessment of two different 2 MW class wind turbines. *Renew Energy* **2012**, *37*, 37–44. [CrossRef]
58. Razdan, P.; Garrett, P. Life Cycle Assessment of Electricity Production from an Onshore V136-4.2 MW Wind Plant. Available online: http://communityrenewables.org.au/wp-content/uploads/2013/02/Vestas-Lifecycle Analysis_2006.pdf (accessed on 4 March 2020).
59. Vestas Wind Systems A/S. *Life Cycle Assessment of Offshore and Onshore Sited Wind Power Plants Based on Vesta V90–3.0 MW Turbines*; Vestas Wind Systems A/S: Randers, Denmark, 2006.
60. Fox, T.R. Recycling Wind Turbine Blade Composite Material as Aggregate in Concrete. Available online: https://www.semanticscholar.org/paper/Recycling-wind-turbine-blade-composite-material-as-Fox_ 75a4ed7264c6680b3c0cf80201019d3b0e202c82 (accessed on 10 March 2020).
61. Karavida, S.; Nõmmik, R. *Waste Management of End-of-Service Wind Turbines*; Aalborg University: Aalborg, Denmark, 2015.
62. Goodship, V. *Management, Recycling and Reuse of Waste Composites*; Woodhead Publishing Limited: Cambridge UK, 2010.
63. Rasul, M.G.; Azad, A.K.; Sharma, S.C. *Clean Energy for Sustainable Development. Comparisons and Contrasts of New Approaches*; Elsevier: Amsterdam, The Netherlands, 2017.
64. International Organization for Standardization ISO. ISO 14044:2006—Environmental Management—Life Cycle Assessment—Requirements and Guidelines 2006. Available online: https://www.iso.org/obp/ui/#iso std:iso:14044:en (accessed on 12 December 2019).
65. Dreyer, L.C.; Niemann, A.L.; Hauschild, M.Z. Comparison of Three Different LCIA Methods: EDIP97, CML2001 and Eco-indicator 99. *Int. J. Life Cycle Assess.* **2003**, *8*, 191–200. [CrossRef]
66. Alemam, A.; Cheng, X.; Li, S. Treating design uncertainty in the application of Eco-indicator 99 with Monte Carlo simulation and fuzzy intervals. *Int. J. Sustain. Eng.* **2018**, *11*, 110–121. [CrossRef]
67. Piasecka, I.; Tomporowski, A.; Piotrowska, K. Environmental analysis of post-use management of car tires. *Przem. Chem.* **2018**, *97*, 1649–1653.
68. Mannheim, V.; Fehér, Z.S.; Siménfalvi, Z. Innovative solutions for the building industry to improve sustainability performance with Life Cycle Assessment modelling. In *Solutions for Sustainable Development*; Taylor & Francis Group: Milton Park, UK, 2019.
69. Mannheim, V.; Siménfalvi, Z. Determining a Priority Order between Thermic Utilization Processes for Organic Industrial Waste with LCA. *Waste Manag. Environ. VI* **2012**, *163*, 153–166.
70. Klos, Z. Ecobalancial assessment of chosen packaging processes in food industry. *Int. J. Life Cycle Assess.* **2002**, *7*, 309. [CrossRef]
71. Piotrowska, K.; Kruszelnicka, W.; Baldowska-Witos, P.; Kasner, R.; Rudnicki, J.; Tomporowski, A.; Flizikowski, J.; Opielak, M. Assessment of the Environmental Impact of a Car Tire throughout Its Lifecycle Using the LCA Method. *Materials* **2019**, *12*, 4177. [CrossRef] [PubMed]
72. Finnveden, G.; Hauschild, M.Z.; Ekvall, T.; Guinée, J.; Heijungs, R.; Hellweg, S.; Koehler, A.; Pennington, D.; Suh, S. Recent developments in Life Cycle Assessment. *J. Environ. Manag.* **2009**, *91*, 1–21. [CrossRef] [PubMed]

73. Kirchain Jr, R.E.; Gregory, J.R.; Olivetti, E.A. Environmental life-cycle assessment. *Nat. Mater.* **2017**, *16*, 693–697. [CrossRef]
74. Puig, R.; Fullana-I-Palmer, P.; Baquero, G.; Riba, J.-R.; Bala, A. A cumulative energy demand indicator (CED), life cycle based, for industrial waste management decision making. *Waste Manag.* **2013**, *33*, 2789–2797. [CrossRef]
75. Röhrlich, M.; Mistry, M.; Martens, P.N.; Buntenbach, S.; Ruhrberg, M.; Dienhart, M.; Briem, S.; Quinkertz, R.; Alkan, Z.; Kugeler, K. A method to calculate the cumulative energy demand (CED) of lignite extraction. *Int. J. Life Cycle Assess.* **2000**, *5*, 369–373.
76. O'Brien, D.; Shalloo, L.; Patton, J.; Buckley, F.; Grainger, C.; Wallace, M. Evaluation of the effect of accounting method, IPCC v. LCA, on grass-based and confinement dairy systems' greenhouse gas emissions. *Animal* **2012**, *6*, 1512–1527. [CrossRef]
77. Roscoe, P. Method, Measurement, and Management in IPCC Climate Modeling. *Hum. Ecol.* **2016**, *44*, 655–664. [CrossRef]
78. Andersen, N.; Eriksson, O.; Hillman, K.; Wallhagen, M. Wind Turbines' End-of-Life: Quantification and Characterisation of Future Waste Materials on a National Level. *Energies* **2016**, *9*, 999. [CrossRef]

13

Classifying the Level of Energy-Environmental Efficiency Rating of Brazilian Ethanol

Nilsa Duarte da Silva Lima, Irenilza de Alencar Nääs, João Gilberto Mendes dos Reis * and Raquel Baracat Tosi Rodrigues da Silva

Postgraduate Program in Production Engineering, Universidade Paulista - UNIP, Dr. Bacelar Street 1212, 04026002 São Paulo, Brazil; nilsa.lima@stricto.unip.br (N.D.d.S.L.); irenilza.naas@docente.unip.br (I.d.A.N.); raquel.silva@stricto.unip.br (R.B.T.R.d.S.)
* Correspondence: joao.reis@docente.unip.br.

Abstract: The present study aimed to assess and classify energy-environmental efficiency levels to reduce greenhouse gas emissions in the production, commercialization, and use of biofuels certified by the Brazilian National Biofuel Policy (RenovaBio). The parameters of the level of energy-environmental efficiency were standardized and categorized according to the Energy-Environmental Efficiency Rating (E-EER). The rating scale varied between lower efficiency (D) and high efficiency + (highest efficiency A+). The classification method with the J48 decision tree and naive Bayes algorithms was used to predict the models. The classification of the E-EER scores using a decision tree using the J48 algorithm and Bayesian classifiers using the naive Bayes algorithm produced decision tree models efficient at estimating the efficiency level of Brazilian ethanol producers and importers certified by the RenovaBio. The rules generated by the models can assess the level classes (efficiency scores) according to the scale discretized into high efficiency (Classification A), average efficiency (Classification B), and standard efficiency (Classification C). These results might generate an ethanol energy-environmental efficiency label for the end consumers and resellers of the product, to assist in making a purchase decision concerning its performance. The best classification model was naive Bayes, compared to the J48 decision tree. The classification of the Energy Efficiency Note levels using the naive Bayes algorithm produced a model capable of estimating the efficiency level of Brazilian ethanol to create labels.

Keywords: biofuel policy; efficiency rating; ethanol; data mining

---

1. Introduction

Developing renewable energy is one of the leading global interests in promoting sustainability and environmental quality, including modern electricity grids worldwide, which have begun to rely more heavily on renewable energy sources [1–3]. Ethanol is a renewable fuel produced by the fermentation of sugarcane extract and molasses. The product has a lower carbon footprint, is biodegradable, and has greater energy-environmental efficiency (renewable energy) compared to oil due to its sustainability in the production chain with better use of natural resources [4–10]. Ethanol is one of the main biofuels consumed in Brazil.

Biofuel partially (or entirely) replaces fossil fuels in engines (flex vehicles) [11]. The addition of 27% ethanol in gasoline (Cgasoline, with the addition of anhydrous ethanol fuel) has been mandatory in Brazil since 2015 [8,12–14]. Such an initiative expanded the country's consumption of biofuels and increased the energy matrix's sustainability. It also supported the goal of reducing 37% of GHG emissions by 2025, compared to 2005 levels [2,12,15,16]. Differently, some European Union countries also include electromobility (electric cars) in their GHG emission reduction forecasts [17].

The production of biofuel from sugarcane is supported by the National Agency of Petroleum, Natural Gas and Biofuels' (ANP) sustainable development programs to achieve goals to reduce greenhouse gas emissions. Therefore, annual national targets were developed by the Brazilian National Biofuel Policy (RenovaBio) [12,15]. These goals are established in units of Decarbonization Credits (CBios) calculated from annual mandatory targets defined individually for each producer and distributor according to their participation in the fossil fuel market [12,16,18–21].

The certification of biofuel production assigns different marks to each biofuel producer and importer, in a value inversely related to the carbon amount of biofuel produced. The note reflects the individual contribution of each producing agent to mitigate a specific amount of GHG concerning its fossil substitute (in tons of CO_2 equivalent). The total emission is compared with that of the equivalent fossil fuel (gasoline, for ethanol), resulting in a final score (Energy-Environmental Efficiency Rating), characterizing the mitigation of emissions. This note generates CBios for biofuel producers and importers; with the decarbonization of the Brazilian energy matrix, there is a mechanism for the commercialization of these CBios linked to the carbon intensity of biofuels [2,12,16,18].

The incentive to reduce pollutant emissions in the biofuel chain goes far beyond the use of flex vehicles by consumers [9,11,22], but it is directly linked to decarbonization credits given to biofuel producers and distributors, because although renewable, it depends on how sugarcane is produced [5,8,23–26]. Alkimim and Clarke [27] showed that the carbon debt of deforestation in Brazilian biomes for ethanol production was equivalent to 608 Mg CO_2 ha^{-1} in the Amazon, 142 Mg CO_2 ha^{-1} in the Cerrado, and 212 Mg CO_2 ha^{-1} to the Atlantic Forest with the respective return time of 62, 15, and 22 years. In this context, it is essential to integrate dissemination strategies, clear and comprehensive, on what is the level of energy-environmental efficiency of biofuel producers and distributors [22,28,29], in order to raise awareness and increase the number of certificates at RenovaBio, the membership of which is not mandatory.

Predictive models are essential for the biofuel chain for both the consumer and the distributor as this classification of the level of efficiency can generate labels of their energy-environmental performance, increasing transparency and environmental responsibility concerning a product such as ethanol. The use of data mining to assess the performance of classifiers in the sugarcane sector is quite broad. It ranges from classifiers for mapping sugarcane planting [30] to deep learning techniques for the detection of sugarcane diseases [31] and classification of crop yield characteristics with neural networks used both in the recognition and in the grading of satellite images of sugarcane plantations [32].

The present study aimed to assess and classify energy-environmental efficiency levels to reduce greenhouse gas emissions in the production, commercialization, and use of biofuels certified by the Brazilian National Biofuel Policy (RenovaBio).

2. Material and Methods

2.1. *Characterization of the Study*

The study considered certification data for the production or efficient import of biofuels from the Brazilian National Biofuel Policy (RenovaBio) regulated by the National Agency of Petroleum, Natural Gas and Biofuels [12,33,34]. The Certificate of Efficient Production of Biofuels includes an Energy-Environmental Efficiency Note, resulting from the technical profile informed by a calculation validation spreadsheet (RenovaCalc) [2,12,33,34], and is linked to the volume of biofuel produced and marketed, generating the Decarbonization Credits (CBio) of the RenovaBio to reduce carbon emissions and improve the performance of biofuels. In addition to offering CBio, it is generated by the difference between the CO_2 equivalent emission of fossil fuels (baseline) and its biofuel substitute [12,18]. The biofuels evaluated in the study were anhydrous ethanol and hydrous ethanol. Only the database of the certificates of the efficient production or import of biofuels approved in 2019 was evaluated [12].

2.2. *Classifying the Biofuel Energetic-Environmental Efficiency*

The classification levels of the biofuel environmental efficiency were performed based on the Energy-Environmental Efficiency Rating. Such a rating is the result of the certificates of the production or efficient import of biofuels informed through the calculation (RenovaCalc), linked to the volume of biofuel produced and commercialized, generating Decarbonization Credits (CBio) in the RenovaBio's program.

The energy-environmental efficiency level parameters were standardized according to Table 1. The dataset was categorized according to the Energy-Environmental Efficiency Rating (RenovaBio) [12], the rating scale varying between lower efficiency and high efficiency + (highest efficiency), and an example of the label's design of the energy-environmental efficiency for the performance of a certificate is given (Figure 1).

Table 1. Parameters of the level of the Energy-Environmental Efficiency Rating.

Energy-Environmental Efficiency Rating [1]	EC [2]	Description
81–100	A+	High efficiency + (highest efficiency)
61–80	A	High efficiency
51–60	B	Average efficiency
31–50	C	Standard efficiency
1–30	D	Lower efficiency

[1] Based on the RenovaBio; [2] Efficiency Classes.

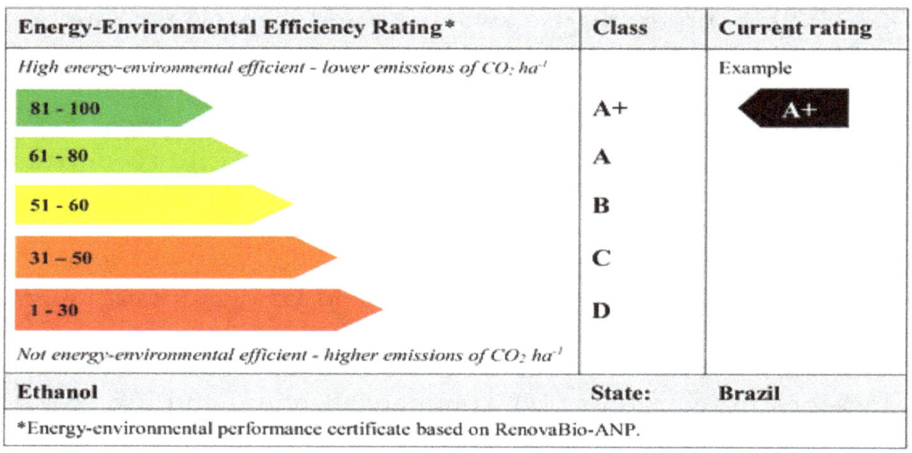

Figure 1. Energy-environmental efficiency label design for performance certificates.

The calculation and indication of the environmental energy efficiency score (E-EER) of the certification for efficient production of biofuel, made available by the ANP, was discretized into categories of levels (classes) (pre-processing of the dataset) for the classification (data mining) and to create one of the labels of the energy-environmental performance (post-processing) of Brazilian ethanol. This labeling system allowed it to be classified into five classes (A+, A, B, C, and D) to provide consumers with a differentiation of the ethanol consumed from different producers, regions, or states.

The RenovaBio Program allows producers and importers to be able to declare the energy-environmental efficiency of their product, which is economically attractive for decarbonization and the competitiveness of biofuels in the oil market, with a complex and solid structure (Figure 2) [2,12,18,33,34]. The label may be shown at fuel pumps to consumers with a validity of one to three years, a validity that is applied to the Certification of Efficient Production of Biofuels when approved by the ANP. It can also endorse the information and increase transparency in the biofuel market at the consumer level, helping to make a purchase decision. The objective of energy-environmental labeling is to encourage Brazilian sugar and alcohol industries to develop innovations and improvements beyond the minimum levels of efficiency. However, it is expected that more ethanol producers will be able to adhere to the ANP certifications of RenovaBio [2,12,19,33,34], and consequently, the labeling system can be improved with the inclusion of more data on the platform.

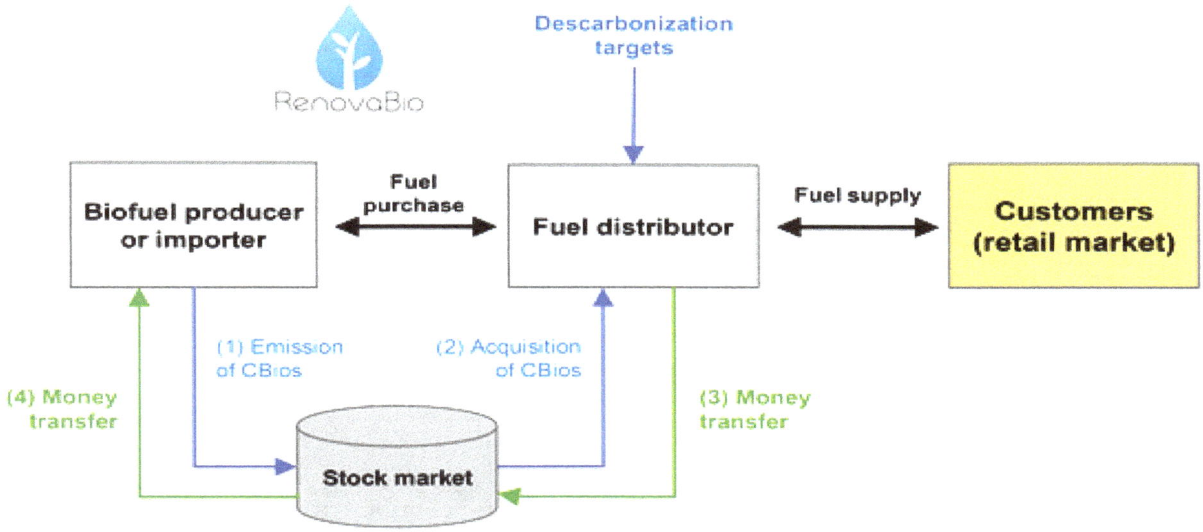

Figure 2. Plan of the emission and trade of decarbonization credits (CBios) created by the RenovaBioProgram, adapted by Klein et al. [19].

2.3. *Classification of Model Prediction*

Data mining applies to this study, through techniques (algorithms), for the classification of the levels (classes) of energy-environmental efficiency in the search of strategic information that allows the extraction of implicit information existing in the databases, contributing to the process of identifying and classifying new patterns [35–37]. The steps of the data mining method were selection, pre-processing, data mining, and post-processing (knowledge filtering, interpretation and explanation, evaluation, and knowledge integration) for knowledge discovery from the classifiers [36,37]. The results obtained could be used in information management, information processing, decision making, and process control.

The data contained in the databases could be used to learn a specific target concept [35–38]. The tasks performed by data mining techniques and machine learning, the classification, build models that can be applied to unclassified data to categorize them into classes, to relate the meta attribute (whose value will be predicted) and a set of forecasting attributes [35–38].

The data were assessed in the ANP database for the registration of certificates of the production or efficient import of biofuels approved and included in the RenovaBio program in 2019 [12,33,34]. We considered only anhydrous ethanol and hydrated ethanol products, generating two products for the same biofuel producer and importer. The data pre-processing was performed in Excel spreadsheets for further processing in the data mining software Weka © (Waikato Environment for Knowledge Analysis) Version 3.8.4 [39–42]. The attributes used to build the predictive model were: "biofuel-type", "state", "eligible-volume (%)", "emission factor", "Energy-Environmental Efficiency Rating", and "LER" (Level of Efficiency Rating). Figure 3 presents the modeling process used to classify the Energy-Environmental Efficiency Rating.

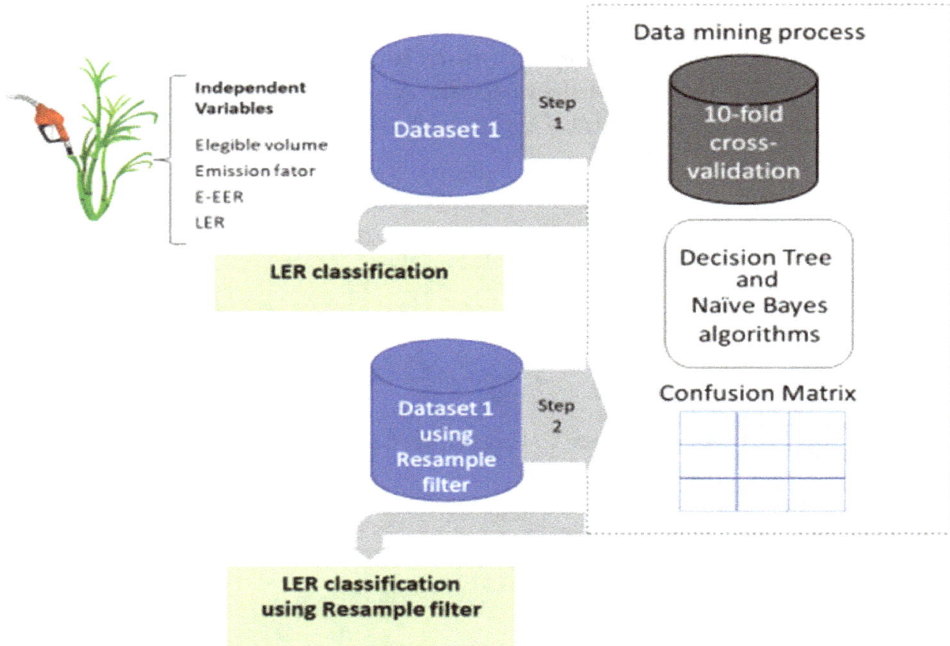

Figure 3. Schematic of the modeling process used to classify the Energy-Environmental Efficiency Rating of Brazilian ethanol. E-EER, Energy-Environmental Efficiency Rating; LER, Level of Efficiency Rating.

A classifier is a mapping from unlabeled instances to (discrete) classes. Classifiers have a form (classification tree) plus an interpretation procedure (including how to handle unknown values). Most classifiers also can provide probability estimates (or other likelihood scores), which can be thresholded to yield a discrete class decision, thereby taking into account a cost/benefit or utility function [43,44].

During the pre-processing of the data, the dataset was extracted from the RenovaBio Program (ANP) platform, selecting only the data on the ethanol product and organized in a spreadsheet. The implementation of the supervised filter Resample was applied to maintain the distribution of classes in the subsample and to reach a uniform distribution for comparing the data not submitted to the filter (noResample). The filter Resample produces a random subsample of a dataset using either sampling with replacement or without replacement. The filter was made to maintain the class distribution in the subsample or to bias the class distribution toward a uniform distribution.

The J48 decision tree classification algorithms ($''weka.classifiers.trees.J48 - C0.25 - M2''$ without the filter and in relation to $''data - biofuel - weka.filters.supervised.instance.Resample - B0.0 - S1 - Z100.0''$ using the resample filter) and the naive Bayes algorithm ($''weka.classifiers.bayes.NaiveBayes''$ without the filter and in relation to $''data - biofuel - weka.filters.supervised.instance.Resample - B0.0 - S1 - Z100.0''$ using the resample filter) were used. The model validation was done using cross-validation (10 folds) applied to the dataset of 27 instances and six attributes ("biofuel-type", "state", "eligible-volume (%)", "emission-factor", "energy-environmental-efficiency-rating", and "LER" (Level of Efficiency Rating)). For each sample, the known class label was compared with the prediction of the learned class model.

In supervised learning, each data input object is preassigned a class label. The main task of supervised algorithms (J48 and naive Bayes) is to learn a model that produces the same labeling for the provided data [43–45]. The decision tree algorithm is a widely used algorithm for classification, which uses attribute values to partition the decision space into smaller subspaces in an iterative manner, and the decision processes can be represented graphically as a tree. Each possible decision is covered and represented as a branch, and a complete decision process is essentially a path or branch from the root node to a leaf [43,44]. Naive Bayes is a classification algorithm widely used in problems due to its simplicity, effectiveness, and robustness, being a probabilistic approach based on assumptions that resources are independent of each other and that their weights are equally important [46]. They can better represent the complex relationships between input variables found in real problems [46]. Probabilistic inference can be studied as an approach based on the assumption that decision variables follow probable distributions. The essence of a Bayesian classifier is to estimate the probabilities of all alternative models or hypotheses, given data as evidence, and then to find the most probable classification to be assigned to each new input [44,46].

In the step of the post-processing for filtering, interpretation and explanation, evaluation, and knowledge integration generated by the algorithms for knowledge discovery from the classifiers, the metrics of the performance of the algorithms were used. Post-processing procedures usually include various pruning routines, rule quality processing, rule filtering, rule combination, model combination, or even knowledge integration [47].

The last step is to evaluate the prediction, and such an analysis was made based on the performance values obtained through the test of the prediction model [48]. The performance evaluation measures of the prediction models used were the confusion matrix, accuracy, precision, and recall and the correlation coefficient between classes (Matthews Correlation Coefficient (MCC)) for testing with the resample filter. The Kappa statistic measured the learning capacity of the algorithm.

The confusion matrix presented a matrix with results obtained during the test phase of the model, and it was used in models that used classification algorithms. Considering a confusion matrix of a hypothesis, it offered an adequate measure of the classification model, by showing the number of correct classifications versus the predicted rankings for each class, over a set of instances. The number of correct answers, for each class, was located on the main diagonal of the matrix, and the other elements represented errors in the classification.

The precision represents what has been classified correctly. The values obtained in correctly classified instances and incorrectly classified instances are determinant for predicted accuracy, since they display the values of correct classification and incorrect classification obtained by the algorithm (Equation (1)).

$$Precision = \frac{TP}{(TP + FP)} \quad (1)$$

where TP = True Positive; FP = False Positive . The sensitivity (recall) signifies the proportion of wrong classifications or the occurrence of defects. In addition to accuracy and precision, its value varies from 0 to 1, with values closer to 1 being indicators of a good performance prediction model obtained by Equation (2).

$$Sensitivity(Recall) = \frac{TP}{(TP + FN)} \quad (2)$$

where TP = True Positive; FN = False Negative.

The Kappa statistic is a metric that compares an observed precision with an expected precision (random chance). It is a measure used to deal with multi-class and unbalanced class problems. The Kappa statistic can be defined as a measure of the degree of agreement between two categorized datasets. The Kappa result varies between 0 and 1. The higher the Kappa value, the stronger the bond [49] (Equation (3)).

$$Kappastatistic = \frac{P_O - P_E}{1 - P_E} \quad (3)$$

where P_O = proportion of observed agreements; P_E = proportion of agreements expected by chance.

The Matthews Correlation Coefficient (MCC) is a correlation coefficient between the dependent classes and represents a measure of quality. Unlike accuracy, precision, and sensitivity, its value ranges from −1 to 1, where values closer to −1 are indicators of a poor prediction model, values equal to 0 indicate that the prediction model is entirely random, and values closer to 1 are indicators of a prediction model with good performance (Equation (4)).

$$MCC = \frac{TP * TN - FP * FN}{(TP + FP)(TP + FN)(TN + FP)(TN + FN)} \quad (4$$

where TP = True Positive; TN = True Negative; P = False Positive; FN = False Negative.

3. Results

3.1. Classification of the Energy-Environmental Efficiency Level for Biofuels

The **If-Then** classification rules are presented related to the energy-environmental efficiency leve for biofuels.

The decision tree generated by the J48 algorithm presented the following classification rules (Figure 4): **If** the energy-environmental efficiency deficiency (E-EER) was higher than 60.3, **then** the classification was A (high efficiency). **If** the Energy-Environmental Efficiency Rating (E-EER) was less than or equal to 60.3, **then** the rating depended on the state where the ethanol was produced. **If** the state was Goiás (GO), **then** the classification was C (standard efficiency). **If** the state was São Paulo (SP), **then** the classification was B (average efficiency). **If** the state was Mato Grosso do Sul (MS), **then** the rating was B (average efficiency). The results indicated that the model classified with precision above 60.

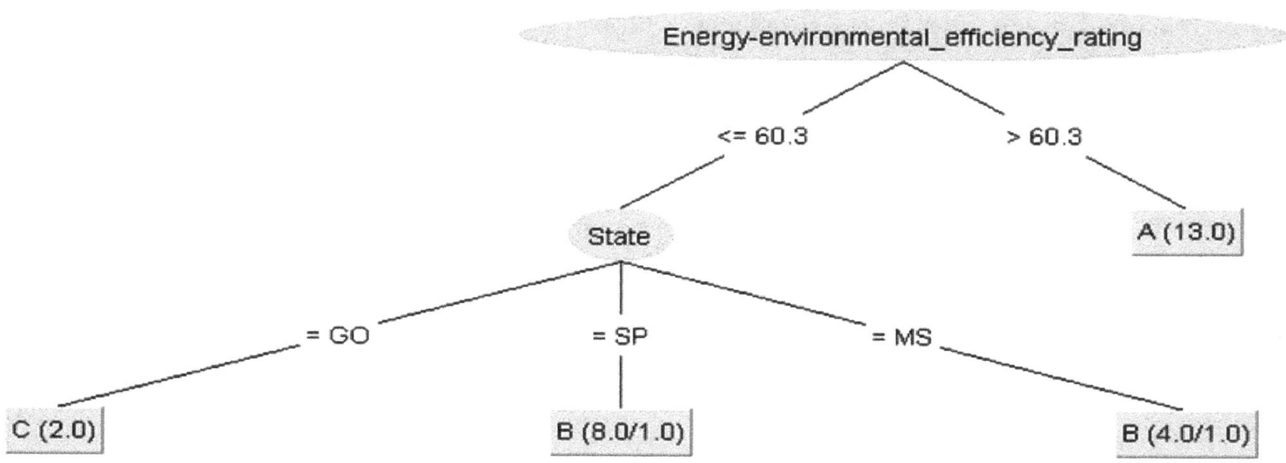

Figure 4. Decision tree for classifying the level of the energy-environmental efficiency of biofuels.

The performance of the classifiers in predicting the classes of the E-EER level using the J48 decision tree algorithm (Table 2) showed 74.07% of instances classified correctly and 25.93% for those classified incorrectly, and the learning capacity of the algorithm was 0.56 for the Kappa statistic. Class A showed the best 100% True Positive (TP) rate, with a 0.07 False Positive (FP) rate, a precision of 0.93, and high recall. However, for Class C, the precision was zero, not classifying any.

Table 2. Classifier performance (J48 decision tree) for scale class prediction models of the level of the energy-environmental efficiency of biofuel.

	Metrics				
Classifier Model	**Correctly Classified Instances (%)**	**Incorrectly Classified Instances (%)**	**Kappa Statistic**	**Test Mode**	
J48 Decision Tree	74.07	25.93	0.56	10-Fold Cross-Validation	
	Detailed Accuracy by Class-Classifier Performance				
Class	**TP Rate**	**FP Rate**	**Precision**	**Recall**	**MCC**
A	1.00	0.07	0.93	1.00	0.93
B	0.70	0.23	0.64	0.70	0.46
C	0	0.09	0	0	0.12
WA	0.740	0.130	0.680	0.740	0.600

WA = Weighted Average. TP rate = True Positive rate. FP rate = False Positive rate. MCC = Matthews Correlation Coefficient.

The performance of the naive Bayes algorithm showed 81.48 of instances correctly classified and 18.52% for those classified incorrectly, with a learning capacity of 0.70 for the Kappa statistic. It classified Classes A and B with high performance, with an accuracy of 1.00 and 0.73, respectively (Table 3). Comparing the two algorithms, naive Bayes presented better performance indexes concerning the J48 decision tree.

Table 3. Naive Bayes classifier performance for scale class prediction models of the level of the energy-environmental efficiency of biofuel.

	Metrics				
Classifier Model	**Correctly Classified Instances (%)**	**Incorrectly Classified Instances (%)**	**Kappa Statistic**	**Test Mode**	
Naive Bayes	81.48	18.52	0.7	10-Fold Cross-Validation	
	Detailed Accuracy by Class-Classifier Performance				
Class	**TP Rate**	**FP Rate**	**Precision**	**Recall**	**MCC**
A	0.92	0	1.00	0.92	0.93
B	0.80	0.18	0.73	0.80	0.61
C	0.50	0.09	0.50	0.50	0.41
WA	0.81	0.08	0.82	0.81	0.73

WA = Weighted Average. TP rate = True Positive rate. FP rate = False Positive rate. MCC = Matthews Correlation Coefficient.

The decision tree generated by the J48 algorithm for the dataset with the application of the resample filter (Figure 5) presented the following classification rules (Figure 4). **If** the Energy-Environmental Efficiency Rating (E-EER) is higher than 60.1, **then** the classification is A (high efficiency). **If** the Energy-Environmental Efficiency Rating (E-EER) is less than or equal to 60.1, **then** the classification depends on the eligible volume (%) of the biomass. **If** the eligible volume is higher than 97.43, **then** the rating is B (average efficiency). **If** the eligible volume is less than or equal to 97.43, **then** the rating is C (standard efficiency) (Figure 5). The results indicated that the use of the resample filter had higher weight for the appropriate attribute volume during the evaluation of the E-EER and that ethanol production depended directly on this volume.

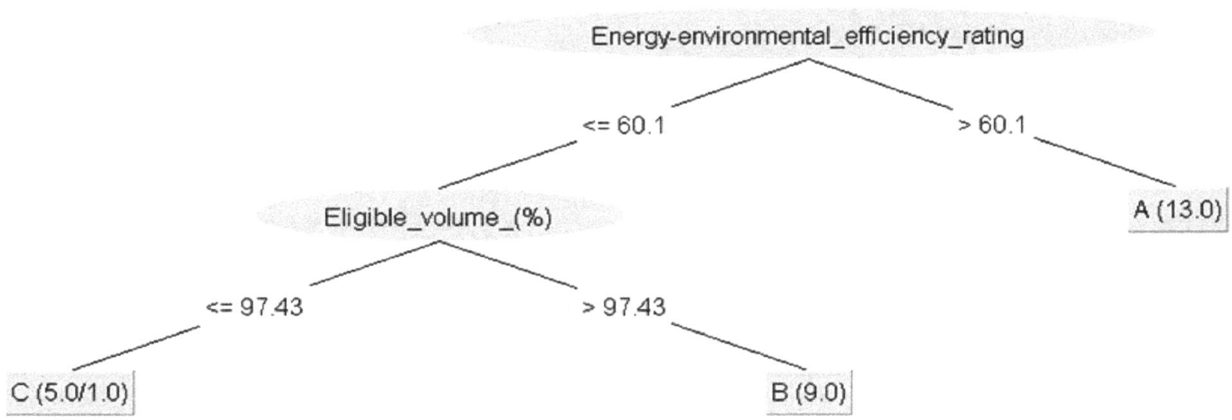

Figure 5. Decision tree for the energy-environmental level classification of biofuels using the Resample filter. GO, Goiás; SP, São Paulo; MS, Mato Grosso do Sul.

The performance of the classifiers in the prediction of the classes of the E-EER level using the J48 decision tree algorithm and with the application of the resample filter (Table 4) presented 81.48% of instances correctly classified and 18.52% for those classified incorrectly, and the algorithm learning ability was 0.70 for the Kappa statistic. Classes A and B showed high prediction with values of 1.00 and 0.78, respectively. However, the forecast for the C class was 0.40.

Table 4. Classifier performance for scale class prediction models of the level of the environmental efficiency of biofuel with the resample filter.

	Metrics-Classifier Performance				
Classifier Model	**Correctly Classified Instances (%)**	**Incorrectly Classified Instances (%)**	**Kappa Statistic**	**Test Mode**	**Filters**
J48 decision tree	81.48	18.52	0.70	10-fold cross-validation	supervised instance resample
	Detailed Accuracy by Class-Classifier Performance				
Class	**TP Rate**	**FP Rate**	**Precision**	**Recall**	**MCC**
A	1.00	0	1.00	1.00	1.00
B	0.70	0.12	0.78	0.70	0.85
C	0.50	0.13	0.40	0.50	0.79
WA	0.81	0.06	0.83	0.81	0.91

WA = Weighted Average. TP rate = True Positive rate. FP rate = False Positive rate. MCC = Matthews Correlation Coefficient.

The performance of the naive Bayes algorithm showed 77.78% of instances classified correctly and 22.22% for those classified incorrectly, with a learning capacity of 0.64 for the Kappa statistic (Table 5). It classified Classes A and B with high performance, with an accuracy of 0.77 and 0.90, respectively (Table 5). The J48 decision tree showed better performance indexes when compared to naive Bayes.

Table 5. Classifier performance (naive Bayes) for scale class prediction models of the level of the environmental efficiency of biofuel using the resample filter.

	Metrics-Classifier Performance				
Classifier Model	**Correctly Classified Instances (%)**	**Incorrectly Classified Instances (%)**	**Kappa Statistic**	**Test Mode**	**Filters**
Naive Bayes	77.78	22.22	0.64	10-fold cross-validation	supervised instance resample
	Detailed Accuracy by Class-Classifier Performance				
Class	**TP Rate**	**FP Rate**	**Precision**	**Recall**	**MCC**
A	0.77	0	1	0.77	0.80
B	0.90	0.23	0.69	0.90	0.64
C	0.50	0.09	0.50	0.50	0.41
WA	0.78	0.10	0.81	0.78	0.68

WA = Weighted Average. TP rate = True Positive rate. FP rate = False Positive rate. MCC = Matthews Correlation Coefficient.

The confusion matrix for the J48 decision tree algorithm presented results with positive gains for he classes when the resample filter was applied, going from 20 to 22 strikes, specifically for Class C Table 6). This gain in the resample increased the normal balance or distribution of the dataset, shown in Table 4. The results of the Matthews Correlation Coefficient (MCC) proved this gain with values of 1.00 for A, 0.85 for B, and 0.79 for C.

Table 6. Confusion matrix of the classification of the level of energy-environmental efficiency with the J48 decision tree.

Classifier Model No Instance Filter-J48 Decision Tree				
A	**B**	**C**	**Total**	**Classified as**
13	0	0	13	A
1	7	2	10	B
0	4	0	4	C
14	11	2	27	
Classifier Model with the Supervised-Filter Instance: Resample				**Classified as**
13	0	0	13	A
0	7	3	10	B
0	2	2	4	C
13	9	5	27	

The confusion matrix for the naive Bayes algorithm also showed results with positive gains for the classes when the resample filter was applied, going from 20 to 21 hits, specifically for Class C, with two hits (Table 7). However, this gain did not have good accuracy when we observed the results of Matthews Correlation Coefficient (MCC) (Table 5), which presented values of 0.80 for A, 0.64 for B, and 0.41 for C, when compared to the J48 decision tree algorithm in the same analysis condition, which performed better.

Table 7. Confusion matrix of the classification of the level of energy-environmental efficiency with naive Bayes.

Classifier Model No Instance Filter-Naive Bayes				
A	**B**	**C**	**Total**	**Classified as**
13	0	0	13	A
1	7	2	10	B
0	4	0	4	C
14	11	2	27	
Classifier Model with the Supervised-Filter Instance: Resample				**Classified as**
10	13	A		
0	9	1	10	B
0	2	2	4	C
10	13	4	27	

The results showed that the classification using the naive Bayes algorithm was better than J48 in the approach without the resample filter. It classified the minority class (C) better and presented a higher degree of agreement (Kappa statistic). It also indicated high performance by the Matthews correlation coefficient concerning the J48 decision tree algorithm. However, the use of the resample filter in both algorithms improved the distribution of classes in the confusion matrix (Tables 6 and 7), especially Class C.

4. Discussion

The decision tree signalized the production state importance. The E-EER value was higher for the most productive states (SP, MS). However, Brazilian ethanol producers' certification was still small concerning the high number of states and the volume of ethanol produced in Brazil. The classification of the energy-environmental efficiency of Goiás state did not show its status as one of the states that produces the most ethanol in Brazil [50]. The number of certified producers was still less than it should be. Such a public policy (RenovaBio) was implemented in 2018, so far with little adhesion by the producers and importers in producing states.

The comparison of the J48 decision tree and naive Bayes algorithms, applied to the dataset without using filters, showed differences in the performance indicators. The naive Bayes algorithm, when compared to the J48 decision tree, presented better performance results, predicted Class C better (50% higher in the TP rate), and had better learning capacity (higher Kappa statistic value), indicating that the prediction model had a good performance with this algorithm. However, the sensitivity of the model decreased. The application of the Resample filter in the pre-processing of data for classification with the J48 decision tree algorithm was the one that showed the best performance compared to naive Bayes, as the Kappa statistic was high and the MMC for all classes was higher, indicating that the prediction model performed well.

The data mining technique is also applied in the area of environmental impacts of sugarcane production, for predicting the energy produced and the environmental impacts. Artificial intelligence methods, artificial neural networks, and adaptive neural fuzzy inference system models are also used to predict the environmental impacts of the life cycle and energy output of sugarcane production on planted farms [51].

Integrated approaches based on complex systems for forecasting the growth of sugarcane based on meteorological parameters using extreme machine learning and neural networks were able to show a more generalized model of forecasting for the growth of sugarcane, bringing benefits to industry and the community [52].

Brazil is a major producer of sugarcane and a major consumer of ethanol, with intensive production in order to meet the demands of biofuel (clean energy) and to reduce the use of fossil fuel (oil). However, several efforts in public policy must come into synergy regarding the growing consumption of biofuel and the sustainable development of the sugarcane chain, mainly regarding its energy-environmental efficiency, use of inputs, and agricultural processes [5]. The ethanol energy-environmental performance labels are applicable in this context, as they would help consumers understand how the chain is evaluated mainly in terms of environmental impact and also collaborating with the transparency of public sector policies.

The most evaluated and used energy efficiency labels are for buildings, appliances, equipment, and lamps. These labeling systems are part of public energy-saving policies; since their implementation, there have been improvements in the evaluation standard, and they have also impacted consumption behavior [53–60]. The implementation of energy efficiency measures can guarantee a sustainable economy. In this context, the energy efficiency labeling program for buildings is generally designed with performance processes and standards, with a rigorous database. Eco-labeling is another system with an approach based on the environmental performance of products that also influences consumer choice [61]. Batista et al. [57] investigated the contribution of labeling to reducing the electricity consumption of buildings and noted that conventional buildings that adopt measures such as painting walls and ceilings white, in addition to using smoked glass, were sufficient to raise the rating to an A level.

The evaluation of buildings generally includes energy classification schemes and shows the difference between the Brazilian scheme and those applied by other countries to improve the labeling method [62]. Other studies have included a review of international building energy efficiency codes and labeling schemes to establish standards for the assessment and classification of buildings in terms of energy performance [63]. Lopes et al. [60] reviewed studies on energy efficiency policies and regulations for buildings, highlighting how the Brazilian program can be improved compared to the American and Portuguese programs; however, this labeling system does not inform about the reduction of GHG emissions.

Another study evaluated two new proposals for an energy efficiency label and a new method for assessing the energy efficiency of public lighting systems. The main difference between the proposals was the number of parameters evaluated. However, the current evaluation system evaluates only one parameter (energy efficiency index), and the study's proposals recommend five parameters: lamps, energy efficiency index, light pollution, renewable energy contribution, and light dimming [64].

All of these studies demonstrated that the labeling system can be implemented and improved for consolidation and active contribution to consumer behavior. However, it could also increase the contribution of the ethanol production sector with greater participation with the goal of reducing greenhouse gases and reducing energy use. The biofuel sector could benefit from the implementation of an energy-environmental labeling system. With a successful approach, it could increase adherence to the RenovaBio program and consequently increase the sector's decarbonization credits.

5. Conclusions

The Brazilian National Biofuel Policy (RenovaBio) is one of the main strategies to encourage the reduction of pollutants in the renewable energy sector from sugarcane. For this reason, the implementation of simple labeling can impact consumer behavior and increase the transparency of the incentive program to reduce environmental impacts.

After testing two classifiers, the best model evaluated was naive Bayes without the use of the resample filter compared to the J48 decision tree, also with the use of the resample filter. The classification of the Energy Efficiency Note (RenovaBio) levels using a Bayesian classifier, the naive Bayes algorithm, produced a model capable of predicting the efficiency level of Brazilian ethanol producers and importers certified to create labeling. The rules generated by the models were capabl of estimating the classes according to the scale discretized into high efficiency (Classification A) average efficiency (Classification B), and standard efficiency (Classification C), with more accurat forecasts for the observed classes. These results could generate an ethanol energy-environmenta efficiency label for the end consumers and resellers of the product.

However, RenovaBio's database of ethanol was small, concerning the records of efficiency scores registered in the program, as adherence to the program is still voluntary, and the implementation is recent, hindering deeper learning in the classification of labels. Besides, a more in-depth analysis could improve the model's forecast in the generation of energy-environmental labels for biofuels.

Author Contributions: Conceptualization, N.D.d.S.L. and J.G.M.d.R.; methodology, N.D.d.S.L. and I.d.A.N. software, N.D.d.S.L. and I.d.A.N.; validation, N.D.d.S.L., I.d.A.N., J.G.M.d.R., and R.B.T.R.d.S.; formal analysis N.D.d.S.L. and I.d.A.N.; investigation, N.D.d.S.L., J.G.M.d.R., and R.B.T.R.d.S.; resources, I.d.A.N. and R.B.T.R.d.S. data curation, N.D.d.S.L. and J.G.M.d.R.; writing, original draft preparation, N.D.d.S.L. and I.d.A.N.; writing review and editing, I.d.A.N., J.G.M.d.R., and R.B.T.R.d.S.; visualization, I.d.A.N., J.G.M.d.R., and R.B.T.R.d.S. supervision, I.d.A.N. and J.G.M.d.R.; project administration, N.D.d.S.L. and R.B.T.R.d.S.; funding acquisition I.d.A.N. All authors read and agreed to the published version of the manuscript.

References

1. Cabral, O.M.; Freitas, H.C.; Cuadra, S.V.; de Andrade, C.A.; Ramos, N.P.; Grutzmacher, P.; Galdos, M. Packer, A.P.C.; da Rocha, H.R.; Rossi, P. The sustainability of a sugarcane plantation in Brazil assessed by the eddy covariance fluxes of greenhouse gases. *Agric. For. Meteorol.* **2020**, *282–283*, 107864. [CrossRef]
2. Grassi, M.; Pereira, G. Energy-cane and RenovaBio: Brazilian vectors to boost the development of Biofuels *Ind. Crop. Prod.* **2019**, *129*, 201–205. [CrossRef]
3. Tucki, K.; Orynycz, O.; Wasiak, A.; Świć, A.; Dybaś, W. Capacity Market Implementation in Poland Analysis of a Survey on Consequences for the Electricity Market and for Energy Management. *Energies* **2019** *12*, 839. [CrossRef]
4. Filoso, S.; Carmo, J.B.d.; Mardegan, S.F.; Lins, S.R.M.; Gomes, T.F.; Martinelli, L.A. Reassessing the environmental impacts of sugarcane ethanol production in Brazil to help meet sustainability goals *Renew. Sustain. Energy Rev.* **2015**, *52*, 1847–1856. [CrossRef]
5. de Andrade Junior, M.A.U.; Valin, H.; Soterroni, A.C.; Ramos, F.M.; Halog, A. Exploring future scenarios of ethanol demand in Brazil and their land-use implications. *Energy Policy* **2019**, *134*, 110958. [CrossRef]
6. Cardozo, N.P.; de Oliveira Bordonal, R.; La Scala, N. Sustainable intensification of sugarcane production under irrigation systems, considering climate interactions and agricultural efficiency. *J. Clean. Prod.* **2018**, *204*, 861–871. [CrossRef]
7. Kumar, A.; Ogita, S.; Yau, Y.Y. (Eds.) *Biofuels: Greenhouse Gas Mitigation and Global Warming*; Springer: New Delhi, India, 2018. [CrossRef]
8. Sarangi, P.K.; Nanda, S.; Mohanty, P. (Eds.) *Recent Advancements in Biofuels and Bioenergy Utilization*; Springer: Singapore, 2018. [CrossRef]
9. Isabella, G.; Yu, A.S.O.; Silva, A.M.d.; Pegetti, A.L. Another driver of the Brazilian fuel ethanol supply chain: The consumers' preferences. *Rev. De Adm.* **2017**, *52*, 304–316. [CrossRef]
10. Souza, S.P.; Seabra, J.E. Environmental benefits of the integrated production of ethanol and biodiesel. *Appl. Energy* **2013**, *102*, 5–12. [CrossRef]
11. Santos, A.S.; Gilio, L.; Halmenschlager, V.; Diniz, T.B.; Almeida, A.N. Flexible-fuel automobiles and CO_2 emissions in Brazil: Parametric and semiparametric analysis using panel data. *Habitat Int.* **2018**, *71*, 147–155. [CrossRef]
12. ANP. Política Nacional de Biocombustíveis. Available online: http://www.anp.gov.br/producao-de-biocombustiveis/renovabio (accessed on 10 December 2019).
13. Brito, T.L.F.; Islam, T.; Stettler, M.; Mouette, D.; Meade, N.; Moutinho dos Santos, E. Transitions between technological generations of alternative fuel vehicles in Brazil. *Energy Policy* **2019**, *134*, 110915. doi:10.1016/j.enpol.2019.110915. [CrossRef]

4. Brazilian Ministério do Meio Ambiente Plano Setorial de Mitigação e de Adaptação às Mudanças Climáticas para a Consolidação de uma Economia de Baixa Emissão de Carbono na Agricultura. Available online: http://www.mma.gov.br/images/arquivo/80076/Plano_ABC_VERSAO_FINAL_13jan2012.pdf. (accessed on 10 December 2019).

5. UNICA. Política Nacional de Biocombustíveis (RenovaBio). Available online: http://unicadata.com.br/index.php (accessed on 10 December 2019).

6. Pischke, E.C.; Solomon, B.; Wellstead, A.; Acevedo, A.; Eastmond, A.; De Oliveira, F.; Coelho, S.; Lucon, O. From Kyoto to Paris: Measuring renewable energy policy regimes in Argentina, Brazil, Canada, Mexico and the United States. *Energy Res. Soc. Sci.* **2019**, *50*, 82–91. doi:10.1016/j.erss.2018.11.010. [CrossRef]

7. Tucki, K.; Orynycz, O.; Świć, A.; Mitoraj-Wojtanek, M. The Development of Electromobility in Poland and EU States as a Tool for Management of CO2 Emissions. *Energies* **2019**, *12*, 2942. doi:10.3390/en12152942. [CrossRef]

8. Salina, F.H.; de Almeida, I.A.; Bittencourt, F.R. RenovaBio Opportunities and Biofuels Outlook in Brazil. In *Renewable Energy and Sustainable Buildings*; Sayigh, A., Ed.; Springer International Publishing: Cham, Switzerland, 2020; pp. 391–399. doi:10.1007/978-3-030-18488-9_30. [CrossRef]

9. Klein, B.C.; Chagas, M.F.; Watanabe, M.D.B.; Bonomi, A.; Maciel Filho, R. Low carbon biofuels and the New Brazilian National Biofuel Policy (RenovaBio): A case study for sugarcane mills and integrated sugarcane-microalgae biorefineries. *Renew. Sustain. Energy Rev.* **2019**, *115*, 109365. doi:10.1016/j.rser.2019.109365. [CrossRef]

20. Novato, M.; Lacerda, M.I. RenovaBio-Towards a New National Biofuel Policy and a Truly Sustainable World. *Innov. Energy Res.* **2017**, *6*. doi:10.4172/2576-1463.1000164. [CrossRef]

21. Santini, G.A.; Pinto, L.d.B.; Queiroz, T.R. Cana-de-açúcar como base da matriz energética nacional. *Rev. PolíTica AgríCola* **2011**, *20*, 89–99.

22. Benites-Lazaro, L.; Giatti, L.; Sousa Junior, W.; Giarolla, A. Land-water-food nexus of biofuels: Discourse and policy debates in Brazil. *Environ. Dev.* **2020**, 100491. doi:10.1016/j.envdev.2019.100491. [CrossRef]

23. Moreno, A.D.; Susmozas, A.; Oliva, J.M.; Negro, M.J. Overview of bio-based industries. In *Biobased Products and Industries*; Elsevier: Amsterdam, The Netherlands, 2020; pp. 1–40. doi:10.1016/B978-0-12-818493-6.00001-4. [CrossRef]

24. Benites-Lazaro, L.L.; Giatti, L.; Giarolla, A. Sustainability and governance of sugarcane ethanol companies in Brazil: Topic modeling analysis of CSR reporting. *J. Clean. Prod.* **2018**, *197*, 583–591. doi:10.1016/j.jclepro.2018.06.212. [CrossRef]

25. Caldarelli, C.E.; Gilio, L. Expansion of the sugarcane industry and its effects on land use in São Paulo: Analysis from 2000 through 2015. *Land Use Policy* **2018**, *76*, 264–274. doi:10.1016/j.landusepol.2018.05.008. [CrossRef]

26. Taheripour, F.; Hertel, T.W.; Liu, J. The role of irrigation in determining the global land use impacts of biofuels. *Energy Sustain. Soc.* **2013**, *3*. doi:10.1186/2192-0567-3-4. [CrossRef]

27. Alkimim, A.; Clarke, K.C. Land use change and the carbon debt for sugarcane ethanol production in Brazil. *Land Use Policy* **2018**, *72*, 65–73. doi:10.1016/j.landusepol.2017.12.039. [CrossRef]

28. Nogueira, L.A.H.; Souza, G.M.; Cortez, L.A.B.; Brito Cruz, C.H.d. Biofuels for Transport. In *Future Energy*; Elsevier: Amsterdam, The Netherlands, 2020; pp. 173–197. doi:10.1016/B978-0-08-102886-5.00009-8. [CrossRef]

29. Pereira, A.O.; Pereira, A.S.; La Rovere, E.L.; Barata, M.M.d.L.; Villar, S.d.C.; Pires, S.H. Strategies to promote renewable energy in Brazil. *Renew. Sustain. Energy Rev.* **2011**, *15*, 681–688. doi:10.1016/j.rser.2010.09.027. [CrossRef]

30. Wang, M.; Liu, Z.; Ali Baig, M.H.; Wang, Y.; Li, Y.; Chen, Y. Mapping sugarcane in complex landscapes by integrating multi-temporal Sentinel-2 images and machine learning algorithms. *Land Use Policy* **2019**, *88*, 104190. doi:10.1016/j.landusepol.2019.104190. [CrossRef]

31. Militante, S.V.; Gerardo, B.D.; Medina, R.P. Sugarcane Disease Recognition using Deep Learning. In Proceedings of the 2019 IEEE Eurasia Conference on IOT, Communication and Engineering (ECICE), Yunlin, Taiwan, 3–6 October 2019; pp. 575–578. doi:10.1109/ECICE47484.2019.8942690. [CrossRef]

32. Cevallos, J.C.; Villagomez, J.A.; Andryshchenko, I. Convolutional Neural Network in the Recognition of Spatial Images of Sugarcane Crops in the Troncal Region of the Coast of Ecuador. *Procedia Comput. Sci.* **2019**, *150*, 757–763. doi:10.1016/j.procs.2019.02.001. [CrossRef]

33. ANP. Resolução ANP Nº 758 de 2018—Regulamenta a CertificaçãO da ProduçãO ou ImportaçãO Eficient de BiocombustíVeis de que trata o art. 18 da lei Nº 13.576, de 26 de Dezembro de 2017, e o Credenciament de Firmas Inspetoras. Available online: http://legislacao.anp.gov.br/?path=legislacao-anp/resol-anp/201 novembro&item=ranp-758-2018 (accessed on 10 December 2019).
34. Brazil. Decreto Nº 9.308, 15 de Março de 2018. Metas CompulsóRias Anuais de Redução de Emissões d Gases Causadores do Efeito Estufa para a Comercialização de Combustíveis de que trata a Lei nº 13.57 de 26 de Dezembro de 2017. Available online: http://www.planalto.gov.br/ccivil_03/_ato2015-2018/201 Decreto/D9308.htm (accessed on 18 March 2019).
35. Larose, D.T.; Larose, C.D. *Data Mining and Predictive Analytics*, 2nd ed.; Wiley Series on Method and Applications in Data Mining; John Wiley & Sons Inc: Hoboken, NJ, USA, 2015.
36. Frank, E.; Hall, M.; Holmes, G.; Kirkby, R.; Pfahringer, B.; Witten, I.H.; Trigg, L. Weka-A Machine Learnin Workbench for Data Mining. In *Data Mining and Knowledge Discovery Handbook*; Maimon, O., Rokach, L., Eds Springer: Boston, MA, USA, 2010; pp. 1269–1277. doi:10.1007/978-0-387-09823-4_66. [CrossRef]
37. Ngai, E.W.T.; Xiu, L.; Chau, D.C.K. Application of data mining techniques in customer relationshi management: A literature review and classification. *Expert Syst. Appl.* **2009**, *36*, 2592–260 doi:10.1016/j.eswa.2008.02.021. [CrossRef]
38. Naik, A.; Samant, L. Correlation Review of Classification Algorithm Using Data Mining Too WEKA, Rapidminer, Tanagra, Orange and Knime. *Procedia Comput. Sci.* **2016**, *85*, 662–66 doi:10.1016/j.procs.2016.05.251. [CrossRef]
39. Holmes, G.; Donkin, A.; Witten, I.H. WEKA: A machine learning workbench. In Proceedings o the ANZIIS '94—Australian New Zealnd Intelligent Information Systems Conference, Brisbane, Australi 29 November–2 December 1994; pp. 357–361. doi:10.1109/ANZIIS.1994.396988. [CrossRef]
40. Witten, I.H.; Frank, E.; Trigg, L.E.; Hall, M.A.; Holmes, G.; Cunningham, S.J. Weka: Practical machine learnin tools and techniques with Java implementations. In *Working Paper*; University of Waikato, Department o Computer Science: Hamilton, New Zealand, 1999.
41. Markov, Z.; Russell, I. An introduction to the WEKA data mining system. *ACM SIGCSE Bull.* **2006** *38*, 367–368. doi:10.1145/1140123.1140127. [CrossRef]
42. Hall, M.; Frank, E.; Holmes, G.; Pfahringer, B.; Reutemann, P.; Witten, I.H. The WEKA data mining software An update. *ACM SIGKDD Explor. Newsl.* **2009**, *11*, 10. doi:10.1145/1656274.1656278. [CrossRef]
43. Provost, F.; Fawcett, T. *Data Science for Business: What You Need to Know about Data Mining and Data-Analyti Thinking*, 1st ed.; O'Reilly: Beijing, China, 2013.
44. Yang, X.S. *Introduction to Algorithms for Data Mining and Machine Learning*; Academic Press: London, UK, 2019
45. Pérez-Ortiz, M.; Jiménez-Fernández, S.; Gutiérrez, P.; Alexandre, E.; Hervás-Martínez, C.; Salcedo-Sanz, S A Review of Classification Problems and Algorithms in Renewable Energy Applications. *Energies* **2016** *9*, 607. doi:10.3390/en9080607. [CrossRef]
46. Arar, M.F.; Ayan, K. A feature dependent Naive Bayes approach and its application to the software defect prediction problem. *Appl. Soft Comput.* **2017**, *59*, 197–209. doi:10.1016/j.asoc.2017.05.043. [CrossRef]
47. Maimon, O.; Last, M. Post-Processing of Data Mining Results. In *Knowledge Discovery and Data Mining*; Series Title: Massive Computing; Springer: Boston, MA, USA, 2001; Volume 1, pp. 53–59 doi:10.1007/978-1-4757-3296-2_4. [CrossRef]
48. Bowes, D.; Hall, T.; Gray, D. Comparing the performance of fault prediction models which report multiple performance measures: recomputing the confusion matrix. In Proceedings of the 8th International Conference on Predictive Models in Software Engineering—PROMISE '12, Lund, Sweden, 21–22 September 2012 pp. 109–118. doi:10.1145/2365324.2365338. [CrossRef]
49. Sim, J.; Wright, C.C. The Kappa Statistic in Reliability Studies: Use, Interpretation, and Sample Size Requirements. *Phys. Ther.* **2005**, *85*, 257–268. doi:10.1093/ptj/85.3.257. [CrossRef]
50. ANP. Dados Estatísticos. Available online: http://www.anp.gov.br/dados-estatisticos (accessed on 20 March 2020).
51. Kaab, A.; Sharifi, M.; Mobli, H.; Nabavi-Pelesaraei, A.; Chau, K.W. Combined life cycle assessment and artificial intelligence for prediction of output energy and environmental impacts of sugarcane production. *Sci. Total. Environ.* **2019**, *664*, 1005–1019. doi:10.1016/j.scitotenv.2019.02.004. [CrossRef] [PubMed]

52. Taherei Ghazvinei, P.; Hassanpour Darvishi, H.; Mosavi, A.; Yusof, K.B.W.; Alizamir, M.; Shamshirband, S.; Chau, K.W. Sugarcane growth prediction based on meteorological parameters using extreme learning machine and artificial neural network. *Eng. Appl. Comput. Fluid Mech.* **2018**, *12*, 738–749. doi:10.1080/19942060.2018.1526119. [CrossRef]
53. Vine, E.; du Pont, P.; Waide, P. Evaluating the impact of appliance efficiency labeling programs and standards: process, impact, and market transformation evaluations. *Energy* **2001**, *26*, 1041–1059. doi:10.1016/S0360-5442(01)00053-6. [CrossRef]
54. Wiel, S.; McMahon, J.E. *Energy-Efficiency Labels and Standards: A Guidebook for Appliances, Equipment, and Lighting*, 2nd ed.; Technical Report LBNL–45387—2nd-Edition, LBNL-45387, 877316; Collaborative Labeling and Appliance Standards Program (CLASP): Washington, DC, USA, 2005. doi:10.2172/877316. [CrossRef]
55. Lee, S.E.; Rajagopalan, P. Building energy efficiency labeling programme in Singapore. *Energy Policy* **2008**, *36*, 3982–3992. doi:10.1016/j.enpol.2008.07.014. [CrossRef]
56. Mills, B.; Schleich, J. What's driving energy efficient appliance label awareness and purchase propensity? *Energy Policy* **2010**, *38*, 814–825. doi:10.1016/j.enpol.2009.10.028. [CrossRef]
57. Batista, N.d.N.; Rovere, E.L.L.; Aguiar, J.C.R. Energy efficiency labeling of buildings: An assessment of the Brazilian case. *Energy Build.* **2011**, *43*, 1179–1188. doi:10.1016/j.enbuild.2010.11.010. [CrossRef]
58. Lopes, M.N.; Iwamoto, G.; Riella, H.L.; Lamberts, R.; Campos, L.M.S. The building simulation as an alternative method in brazilian energy efficiency labeling of commercial buildings. In Proceedings of the Building Simulation 2011 Conference, Sydney, Australia, 14–16 November 2011.
59. Newell, R.G.; Siikamäki, J. Nudging Energy Efficiency Behavior: The Role of Information Labels. *J. Assoc. Environ. Resour. Econ.* **2014**, *1*, 555–598. doi:10.1086/679281. [CrossRef]
60. Lopes, A.d.C.P.; Oliveira Filho, D.; Altoe, L.; Carlo, J.C.; Lima, B.B. Energy efficiency labeling program for buildings in Brazil compared to the United States' and Portugal's. *Renew. Sustain. Energy Rev.* **2016**, *66*, 207–219. doi:10.1016/j.rser.2016.07.033. [CrossRef]
61. Banerjee, A.; Solomon, B.D. Eco-labeling for energy efficiency and sustainability: A meta-evaluation of US programs. *Energy Policy* **2003**, *31*, 109–123. doi:10.1016/S0301-4215(02)00012-5. [CrossRef]
62. Scalco, V.A.; Fossati, M.; de Souza Versage, R.; Sorgato, M.J.; Lamberts, R.; Morishita, C. Innovations in the Brazilian regulations for energy efficiency of residential buildings. *Archit. Sci. Rev.* **2012**, *55*, 71–81. doi:10.1080/00038628.2011.641731. [CrossRef]
63. Fossati, M.; Scalco, V.A.; Linczuk, V.C.C.; Lamberts, R. Building energy efficiency: An overview of the Brazilian residential labeling scheme. *Renew. Sustain. Energy Rev.* **2016**, *65*, 1216–1231. doi:10.1016/j.rser.2016.06.048. [CrossRef]
64. Gutierrez-Escolar, A.; Castillo-Martinez, A.; Gomez-Pulido, J.M.; Gutierrez-Martinez, J.M.; González-Seco, E.P.D.; Stapic, Z. A review of energy efficiency label of street lighting systems. *Energy Effic.* **2017**, *10*, 265–282. doi:10.1007/s12053-016-9454-7. [CrossRef]

Permissions

All chapters in this book were first published in MDPI; hereby published with permission under the Creative Commons Attribution License or equivalent. Every chapter published in this book has been scrutinized by our experts. Their significance has been extensively debated. The topics covered herein carry significant findings which will fuel the growth of the discipline. They may even be implemented as practical applications or may be referred to as a beginning point for another development.

The contributors of this book come from diverse backgrounds, making this book a truly international effort. This book will bring forth new frontiers with its revolutionizing research information and detailed analysis of the nascent developments around the world.

We would like to thank all the contributing authors for lending their expertise to make the book truly unique. They have played a crucial role in the development of this book. Without their invaluable contributions this book wouldn't have been possible. They have made vital efforts to compile up to date information on the varied aspects of this subject to make this book a valuable addition to the collection of many professionals and students.

This book was conceptualized with the vision of imparting up-to-date information and advanced data in this field. To ensure the same, a matchless editorial board was set up. Every individual on the board went through rigorous rounds of assessment to prove their worth. After which they invested a large part of their time researching and compiling the most relevant data for our readers.

The editorial board has been involved in producing this book since its inception. They have spent rigorous hours researching and exploring the diverse topics which have resulted in the successful publishing of this book. They have passed on their knowledge of decades through this book. To expedite this challenging task, the publisher supported the team at every step. A small team of assistant editors was also appointed to further simplify the editing procedure and attain best results for the readers.

Apart from the editorial board, the designing team has also invested a significant amount of their time in understanding the subject and creating the most relevant covers. They scrutinized every image to scout for the most suitable representation of the subject and create an appropriate cover for the book.

The publishing team has been an ardent support to the editorial, designing and production team. Their endless efforts to recruit the best for this project, has resulted in the accomplishment of this book. They are a veteran in the field of academics and their pool of knowledge is as vast as their experience in printing. Their expertise and guidance has proved useful at every step. Their uncompromising quality standards have made this book an exceptional effort. Their encouragement from time to time has been an inspiration for everyone.

The publisher and the editorial board hope that this book will prove to be a valuable piece of knowledge for researchers, students, practitioners and scholars across the globe.

List of Contributors

Rial A. Rajagukguk, Raden A. A. Ramadhan and Hyun-Jin Lee
Department of Mechanical Engineering, Kookmin University, 77 Jeongneung-ro, Seongbuk-gu, Seoul 02727, Korea

Adam Kupczyk, Joanna Mączyńska-Sęczek, Ewa Golisz and Piotr F. Borowski
Institute of Mechanical Engineering, Warsaw University of Life Sciences, 166 Nowoursynowska Str., 02-787 Warsaw, Poland

Marian Piwowarski and Krzysztof Kosowski
Faculty of Mechanical Engineering, Gdansk University of Technology, Gabriela Narutowicza Street 11/12, 80-233 Gdansk, Poland

Prasert Nonthakarn, Mongkol Ekpanyapong and Erik Bohez
School of Engineering and Technology, Asian Institute of Technology, Pathum Thani 12120, Thailand

Udomkiat Nontakaew
Faculty of Engineering, King Mongkut's University of Technology North Bangkok, Bangkok 10800, Thailand

Chalita Kaewbuddee, Ekarong Sukjit, Jiraphon Srisertpol, Somkiat Maithomklang, Khatha Wathakit, Niti Klinkaew, Pansa Liplap and Weerachai Arjharn
Suranaree University of Technology, 111 University Avenue, Suranaree Sub-District, Muang District, Nakhon Ratchasima 30000, Thailand

Farid Antonio Barrozo Budes and Guillermo Valencia Ochoa
Mechanical Engineering Department, Universidad del Atlántico, Carrera 30 Número 8-49, Puerto Colombia, Barranquilla 080007, Colombia

Luis Guillermo Obregon
Research Group on Sustainable Chemical and Biochemical Processes, Chemical Engineering Department, Universidad del Atlántico, Carrera 30 Número 8-49, Puerto Colombia, Barranquilla 080007, Colombia

Adriana Arango-Manrique
Electrical and Electronic Engineering Department, Universidad del Norte, Km 5 Vía Puerto Colombia, Barranquilla 080007, Colombia

José Ricardo Núñez Álvarez
Energy Department, Universidad de la Costa, Calle 58 Número 55- 66, Barranquilla 080002, Colombia

Leizhen Wang and Guorong Wang
School of Mechanical Engineering, Southwest Petroleum University, Chengdu 610500, Sichuan, China

Ya-Long Ding
Guangzhou Institute of Energy Conversion, Chinese Academy of Sciences, Guangzhou 510641, China
College of Chemistry and Pharmaceutical Engineering, Huanghuai University, Zhumadian 463000, China

Hua-Qin Wang
College of Chemistry and Pharmaceutical Engineering, Huanghuai University, Zhumadian 463000, China

Chun-Gang Xu and Xiao-Sen Li
Guangzhou Institute of Energy Conversion, Chinese Academy of Sciences, Guangzhou 510641, China
CAS Key Laboratory of Gas Hydrate, Guangzhou 510640, China
Guangdong Provincial Key Laboratory of New and Renewable Energy Research and Development, Guangzhou 510640, China
Guangzhou Center for Gas Hydrate Research, Chinese Academy of Sciences, Guangzhou 510640, China

Tomasz Jaworski and Małgorzata Kajda-Szcześniak
Department of Technologies and Installations for Waste Management, Silesian University of Technology, 44-100 Gliwice, Poland

Naga Kavitha Kommuri, Andrew McGordon, Antony Allen and Dinh Quang Truong
WMG, University of Warwick, Coventry CV4 7AL, UK

Weronika Kruszelnicka and Robert Kasner
Faculty of Mechanical Engineering, University of Science and Technology in Bydgoszcz, Al. Prof. S. Kaliskiego 7, 85-796 Bydgoszcz, Poland

Robert Kasner, Weronika Kruszelnicka, Patrycja Bałdowska-Witos, Józef Flizikowski and Andrzej Tomporowski
Department of Manufacturing Technology, Faculty of Mechanical Engineering, University of Science and Technology in Bydgoszcz, 85-796 Bydgoszcz, Poland

Nilsa Duarte da Silva Lima, Irenilza de Alencar Nääs, João Gilberto Mendes dos Reis and Raquel Baracat Tosi Rodrigues da Silva
Postgraduate Program in Production Engineering, Universidade Paulista - UNIP, Dr. Bacelar Street 1212, 04026002 São Paulo, Brazil

Index

www.ingramcontent.com/pod-product-compliance
Lightning Source LLC
LaVergne TN
LVHW060531210926
848485LV00005B/1182
* 9 7 8 1 6 4 1 1 6 8 2 9 8 *